PREFACE TO THE THIRD EDITION

This book is an introduction to real-time systems. It is intended not as a cookbook, but, rather, as a stimulus for thinking about hardware and software in a different way. It is necessarily broader than deep. It is a survey book, designed to heighten the reader's awareness of real-time issues.

This book is the culmination of more than 20 years of building, studying, and teaching real-time systems. The author's travels have taken him to NASA, UPS, Lockheed Martin, the Canadian and Australian Defense Forces, MIT's Charles Stark Draper Labs, and many other places. These visits and interactions with literally hundreds of students from such places as Boeing, Motorola, and Siemens have resulted in a wider understanding of real-time systems and particularly their real application. This book is, in essence, a compendium of these experiences. The author's intent is to provide a practical framework for software engineers to design and implement real-time systems. This approach is somewhat different from that of other texts on the subject.

Because of the pragmatic approach, a few of the results and viewpoints presented book's may be controversial. The author has adapted many of the formal definitions from their traditional rigid form into words that are more compatible with practical design. In many places theoretical treatments have been omitted where they would have obscured applied results. In these cases, the reader is referred to additional reading. This author is a great believer in research in this area, and in many places has indicated where research needs to be done or is being done.

Although the book may appear simplistic, it is subtly complex. Consider the semaphore operators. They can be written with a minimum amount of code, yet they are fraught with danger for the real-time designer. In the same way, this book has a kind of Zen-like simplicity and complexity: a yin and a yang.

INTENDED AUDIENCE

This text is an introductory-level book intended for junior–senior level and graduate computer science and electrical engineering students, and practicing software engineers. It can be used as a graduate-level text if it is supplemented with an advanced reader, such as one by the author [Laplante00]. This book is especially useful in an industrial setting for new real-time systems designers who need to get "up to speed" very quickly. This author has used earlier editions of this book in this way to teach short courses for several clients.

The reader is assumed to have some experience in programming in one of the more popular languages, but other than this, the prerequisites for this text are minimal. Some familiarity with discrete mathematics is helpful in understanding some of the formalizations, but it is not essential. A background in basic calculus and probability theory will assist in the reading of Chapter 7.

PROGRAMMING LANGUAGES

Although there are certain "preferred" languages for real-time system design, such as C, C++, Ada 95, and increasingly Java, many real-time systems are still written in Fortran, assembly language, and even Visual BASIC. It would be unjust to focus this book on one language, say C, when the theory should be language independent. However, for uniformity of discussion, points are illustrated, as appropriate, in generic assembly language and C. While the C code is not intended to be ready-to-use, it can be easily adapted with a little tweaking for use in a real system.

ORGANIZATION OF THE BOOK

Real-time software designers must be familiar with computer architecture and organization, operating systems, software engineering, programming languages, and compiler theory. The text provides an overview of these subjects from the perspective of the real-time systems designer. Because this is a staggering task, depth is occasionally sacrificed for breadth. Again, suggestions for additional readings are provided where depth has been sacrificed.

The book is organized into chapters that are essentially self-contained. Thus, the material can be rearranged or omitted, depending on the background and interests of the audience or instructor. Each chapter contains both easy and challenging exercises that stimulate the reader to confront actual problems. The exercises, however, cannot serve as a substitute for practical experience.

The first chapter provides an overview of the nature of real-time systems. Much of the basic vocabulary relating to real-time systems is developed along with a discussion of the challenges facing the real-time system designer. Finally, a brief historical review is given. The purpose of this chapter is to foreshadow the rest of the book as well as quickly acquaint the reader with pertinent terminology.

The second chapter presents a more detailed review of basic computer architecture concepts from the perspective of the real-time systems designer and some basic concepts of electronics. Specifically, the impact of different architectural features on real-time performance is discussed. The remainder of the chapter discusses different memory technologies, input/output techniques, and peripheral support for real-time systems. The intent here is to increase the reader's awareness of the impact of the computer architecture on design considerations.

Chapter 3 provides the core elements of the text for those who are building practical real-time systems. This chapter describes the three critical real-time kernel services: scheduling/dispatching, intertask communication, and memory management. It also covers special problems inherent in these designs, such as deadlock and the priority inheritance problem. This chapter also highlights issues in POSIX compliance of real-time kernels.

In Chapter 4, the nature of requirements engineering is discussed. Next, structured analysis and object-oriented analysis are discussed as paradigms for requirements writing. An extensive design case study is provided.

Chapter 5 surveys several commonly used design specification techniques used in both structural and object-oriented design. Their applicability to real-time systems is emphasized throughout. No one technique is a silver bullet, and the reader is encouraged to adopt his or her own formulation of specification techniques for the given application. A design case study is also provided.

Chapter 6 begins with a discussion of the language features desirable in good software engineering practice in general and real-time systems design in particular. A review of several of the most widely used languages in real-time systems design, with respect to these features, follows. The intent is to provide criteria for rating a language's ability to support real-time systems and to alert the user to the possible drawbacks of using each language in real-time applications.

Chapter 7 discusses several techniques for improving the response time of real-time systems. Many of the ideas discussed in this chapter are well-known but unwritten laws of programming. Some are compiler optimization techniques that can be used to improve our code. Others are tricks that have been passed down by word of mouth. This chapter can help wring out that extra bit of performance from a critical system.

The final chapter discusses general software engineering considerations, including the use of metrics and techniques for improving the fault-tolerance and reliability of real-time systems. Later in the chapter, techniques for improving reliability through rigorous testing are discussed. Systems integration is also discussed. The chapter also reviews some special techniques that are needed in real-time systems.

While the difference between the first and second editions of this book is incremental, the third edition is essentially a new book. During the intervening eight years since the second edition, so many changes have taken place that more than a face-lift was needed. Approximately 50% of the material from the previous editions has been discarded and the remainder entirely rewritten. Hence, about 50% of the book is new material.

When this course is taught in a university setting, typically students are asked to build a real-time multitasking system of their choice. Usually, it is a game on a PC, but some students can be expected to build embedded hardware controllers of surprising complexity. The author's "assignment" to the reader would be to build such a game or simulation, using at least the coroutine model. The application should be useful or at least pleasing, so some sort of a game is a good choice. The project should take no more than 15 hours and cover all phases of the software life-cycle model discussed in the text. Hence, those readers who have never built a real-time system will have the benefit of the experience.

A NOTE ON REFERENCES

Real-Time Systems Engineering is based on more than 50 years of experience and work by many individuals. Rather than clutter the text with endless citations for the origin of each idea, the author chose to cite only the most key ideas where the reader would want to seek out the source for further reading. Some of the text is adapted from two other books written by the author on software engineering and computer architecture [Laplante03c] [Gilreath03]. Where this has been done, it is so noted. *Note:* In all cases where some sections of this text, particularly the author's own, appear as "adapted" or "paraphrased," it means that the work is being reprinted with both major and minor differences. However, rather than confuse the issue with intermittent quotation marks for verbatim text, the reader should attribute all ideas to cited authors from the point where the usage is noted to the ending reference. This author, however, retains responsibility for any errors. In all cases, permission to reprint this material has been obtained.

Many good theoretical treatments of real-time systems exist, and they are noted where applicable. However, these books are sometimes too theoretical for practicing software engineers and students who are often too impatient to wade through the derivations for the resultant payoff. These readers want results that they can use now in the trenches, and they want to see how they can be used, not just know that they exist. In this text, an attempt is made to distill the best of the theoretical results, combined with practical experience to provide a toolkit for the real-time designer.

This book contains an extensive bibliography. Where verbatim phrases were used or where a figure came from another source, the author tried to cite it appropriately. However, if any sources were inadvertently overlooked, the author wishes to correct the error. In addition, in a book of this magnitude and complexity, errors are bound to occur. Please notify the author if you find any errors of omission, commission, citation, and so on by email, at plaplante@psu.edu and they will be corrected at the next possible opportunity.

ACKNOWLEDGMENTS

The author wishes to acknowledge and thank the many individuals who assisted in the preparation of this book. Dr. Purnendu Sinha of Concordia University, wrote much of Chapter 3 and various parts of other chapters relating to scheduling theory, contributed many exercises, and classroom tested the material. Dr. Colin

Neill, a Penn State colleague, co-wrote Chapters 4 and 5, reviewed and contributed to other chapters, and single-handedly changed the author's mind concerning the value of object-oriented methods. Research collaborator William Gilreath reviewed parts of the manuscript, contributed to parts on computer architecture, and provided some of the more interesting exercises. Dr. David Russell of Penn State reviewed the manuscript and provided a most supportive environment at the Great Valley School of Graduate Professional Studies where the author works. Valuable reviews were also provided by Dr. Mike Hinchey, Dr. Dave Sinha, and Patricia Feingold. The acquisition and editorial team at IEEE Press/John Wiley, in particular Tony VenGratis and John Griffin, provided terrific support and encouragement. The author's wife, Nancy, typed and edited much of the material.

The author also wishes to thank the many students who, over the last 20 years, have contributed ideas to this book through discussions, projects, and classes. While an exhaustive list is impossible, for the third edition the author must single out Michael Barnes, David Cloutier, Jim Goldman, Dana Gryger, Michael Lutz, Dr. Jeff Nash, Mike Rapa, and Fred Woolsey. In particular, the author thanks Fred, Dana, and Mike for contributing the excellent case study on the traffic controller system found in Chapters 4 and 5, and David for contributing to the discussion on the use of object-oriented languages in Chapter 6.

The author is grateful for the success of the first editions of this book, with more than 15,000 copies sold to the college text and professional markets. The only thing more gratifying than its adoption at such prestigious universities as Carnegie Mellon University, University of Illinois at Urbana-Champaign, Princeton University, the United States Air Force Academy, Polytechnic University, and many others, has been the feedback received from individuals thankful for the influence that the book has had on them.

Finally, the author wishes to thank his wife Nancy, and his children, Christopher and Charlotte, for putting up with the seemingly endless work on this manuscript and too many other projects to mention. This book is dedicated to them with love.

PHILLIP A. LAPLANTE

Chester County, Pennsylvania
September, 2003

1

BASIC REAL-TIME
CONCEPTS

Consider a system in which data need to be processed at a regular and timely rate. For example, an aircraft uses a sequence of accelerometer pulses to determine its position. In addition, systems other than aeronautic ones require a rapid response to events that occur at nonregular rates, such as an overtemperature failure in a nuclear plant. In some sense it is understood that these events require real-time processing.

Now consider a situation in which a passenger approaches an airline reservation counter to pick up his ticket for a certain flight from New York to Boston, which is leaving in 5 minutes. The reservation clerk enters the appropriate information into the computer and a few seconds later a boarding pass is generated. Is this a real-time system?

Indeed, all three systems – aircraft, nuclear plant, and airline reservations – are real-time because they must process information within a specified interval or risk system failure. Although these examples provide an intuitive definition of a real-time system, it is necessary to clearly define when a system is real-time and when it is not. This chapter answers the preceding questions, defines a number of terms, and introduces issues that are examined further later.

1.1 TERMINOLOGY

The fundamental definitions of real-time systems engineering can vary depending on the resource consulted. The following definitions have been collected and refined to the smallest common subset of agreement to form the vocabulary of

Real-Time Systems Design and Analysis, By Phillip A. Laplante
ISBN 0-471-22855-9 © 2004 Institute of Electrical and Electronics Engineers

this text. Moreover, these definitions are presented in a form that is intended to be most useful to the practicing engineer, as opposed to the theorist.

1.1.1 Systems Concepts

The hardware of the general-purpose computer solves problems by repeated execution of macroinstructions, collectively known as software. Software is traditionally divided into system programs and application programs.

System programs consist of software that interfaces with the underlying computer hardware, such as schedulers, device drivers, dispatchers, and programs that act as tools for the development of application programs. These tools include compilers, which translate high-order language programs into assembly code; assemblers, which translate the assembly language into a special binary format called object or machine code; and linkers, which prepare the object code for execution. An operating system is a specialized collection of system programs that manage the physical resources of the computer. As such, a real-time operating system is a systems program.

Application programs are programs written to solve specific problems, such as payroll preparation, inventory, and navigation. Certain design considerations play a role in the design of certain systems programs and application software intended to run in real-time environments.

The notion of a "system" is central to software engineering, and indeed to all engineering, and warrants formalization.

Definition: A system is a mapping of a set of inputs into a set of outputs.

When the internal details of the system are not of interest, the mapping function can be considered as a black box with one or more inputs entering and one or more outputs exiting the system (see Figure 1.1).

Every real-world entity, whether synthetic or occurring naturally, can be modeled as a system. In computing systems, the inputs represent digital data from

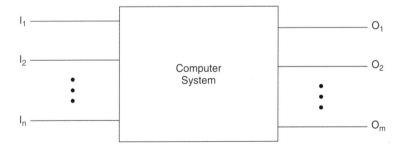

Figure 1.1 A system with n inputs and m outputs.

hardware devices and other software systems. The inputs are often associated with sensors, cameras, and other devices that provide analog inputs, which are converted to digital data, or provide direct digital input. The digital output of the computer system can be converted to analog outputs to control external hardware devices such as actuators and displays (Figure 1.2).

Modeling a real-time system, as in Figure 1.2, is somewhat different from the more traditional model of the real-time system as a sequence of jobs to be scheduled and performance to be predicted, which is very similar to that shown in Figure 1.3. The latter view is simplistic in that it ignores the fact that the input sources and hardware under control are complex. Moreover, there are other, sweeping software engineering considerations that are hidden by the model shown in Figure 1.3.

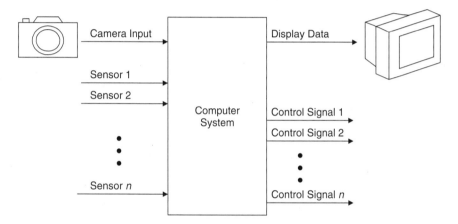

Figure 1.2 Typical real-time control system including inputs from sensors and imaging devices and producing control signals and display information [Laplante03b].

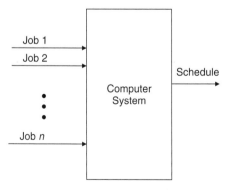

Figure 1.3 A classic representation of a real-time system as a sequence of jobs to be scheduled.

Look again at to the model of a real-time system shown in Figure 1.2. Note that in its realization there is some delay between presentation of the inputs (stimulus) and appearance of the outputs (response). This fact can be formalized as follows:

Definition: The time between the presentation of a set of inputs to a system (stimulus) and the realization of the required behavior (response), including the availability of all associated outputs, is called the response time of the system.

How fast the response time needs to be depends on the purpose of the system.

1.1.2 Real-Time Definitions

The previous definitions set the stage for a formal definition of a real-time system.

Definition: A real-time system is a system that must satisfy explicit (bounded) response-time constraints or risk severe consequences, including failure.

What is a "failed" system? In the case of the space shuttle or a nuclear plant, it is painfully obvious when a failure has occurred. For other systems, such as an automatic bank teller machine, the notion of failure is less clear. For now, failure will be defined as the "inability of the system to perform according to system specification," or, more formally:

Definition: A failed system is a system that cannot satisfy one or more of the requirements stipulated in the formal system specification.

Because of this definition of failure, precise specification of the system operating criteria, including timing constraints, is important. This matter is discussed later.

Various other definitions exist for real-time, depending on which source is consulted. Nonetheless, the common theme among all definitions is that the system must satisfy deadline constraints in order to be correct. For example, an alternative definition might be:

Definition: A real-time system is one whose logical correctness is based on both the correctness of the outputs and their timeliness.

In any case, note that by making unnecessary the notion of timeliness, every system becomes a real-time system.

Real-time systems are often reactive or embedded systems. Reactive systems are those in which scheduling is driven by ongoing interaction with their environment; for example, a fire-control system reacts to buttons pressed by a pilot. Embedded systems are those that are found in a system that is not itself a computer. For example, a modern automobile contains many embedded computers that control fuel injection, airbag deployment, braking, climate control, and so forth. Today, many household items such as televisions, stereos, washing machines, even toys contain embedded computers. It is clear that sophisticated systems such as aircraft, spacecraft, and industrial machines must contain many embedded, reactive computer systems.

The three systems mentioned earlier satisfy the criteria for a real-time system precisely. An aircraft must process accelerometer data within a certain period that depends on the specifications of the aircraft; for example, every 10 milliseconds. Failure to do so could result in a false position or velocity indication and cause the aircraft to go off-course at best or crash at worst. For a nuclear reactor thermal problem, failure to respond swiftly could result in a meltdown. Finally, an airline reservation system must be able to handle a surge of passenger requests within the passenger's perception of a reasonable time (or before the flights leave the gate). In short, a system does not have to process data in microseconds to be considered real-time; it must simply have response times that are constrained.

1.1.2.1 When Is a System Real-Time? It can be argued that all practical systems are real-time systems. Even a batch-oriented system – for example, grade processing at the end of a semester or a bimonthly payroll run – is real-time. Although the system may have response times of days or weeks (e.g., the time that elapses between submitting the grade or payroll information and issuance of the report card or check), it must respond within a certain time or there could be an academic or financial disaster. Even a word-processing program should respond to commands within a reasonable amount of time (e.g., 1 second), or it will become torturous to use. Most of the literature refers to such systems as soft real-time systems.

> **Definition:** A soft real-time system is one in which performance is degraded but not destroyed by failure to meet response-time constraints.

Conversely, systems where failure to meet response-time constraints leads to complete and catastrophic system failure are called hard real-time systems.

> **Definition:** A hard real-time system is one in which failure to meet a single deadline may lead to complete and catastrophic system failure.

Table 1.1   A sampling of hard, soft, and firm real-time systems

	Real-Time Classification	Explanation
Automated teller machine	Soft	Missing even many deadlines will not lead to catastrophic failure, only degraded performance.
Embedded navigation controller for autonomous robot weed killer	Firm	Missing critical navigation deadlines causes the robot to veer hopelessly out of control and damage crops.
Avionics weapons delivery system in which pressing a button launches an air-to-air missile	Hard	Missing the deadline to launch the missile within a specified time after pressing the button can cause the target to be missed, which will result in catastrophe.

Firm real-time systems are those systems with hard deadlines where some arbitrarily small number of missed deadlines can be tolerated.

> **Definition:** A firm real-time system is one in which a few missed deadlines will not lead to total failure, but missing more than a few may lead to complete and catastrophic system failure.

As noted, all practical systems minimally represent soft real-time systems. Table 1.1 gives a sampling of hard, firm, and soft real-time systems.

Note that there is a great deal of latitude for interpretation of hard, firm, and soft real-time systems. For example, in the automated teller machine, missing too many deadlines will lead to significant customer dissatisfaction and potentially even enough loss of business to threaten the existence of the bank. This extreme scenario represents the fact that every system can probably be characterized any way – soft, firm, or hard – real-time by the construction of a supporting scenario. The careful construction of systems requirements (and, hence, expectations) is the key to setting and meeting realistic deadline expectations. In any case, it is a principal goal of real-time systems engineering to find ways to transform hard deadlines into firm ones, and firm ones into soft ones.

Since this text is mostly concerned with hard real-time systems, it will use the term real-time system to mean embedded, hard real-time system, unless otherwise noted.

1.1.2.2 The Nature of Time   It is typical, in studying real-time systems, to consider the nature of time, because deadlines are instants in time. But the question arises, "Where do the deadlines come from?" Generally speaking, deadlines are based on the underlying physical phenomena of the system under control. For example, in animated displays, images must be updated at approximately 30 frames per second to provide continuous motion, because the human eye can resolve updating at a slower rate. In navigation systems, accelerations must be read at a rate that is based on the maximum velocity of the vehicle, and so on. In some cases, systems have deadlines that are imposed on them that are based on nothing less than guessing or on some forgotten and since eliminated requirement. The problem in these cases is that the undue constraints may be placed on the systems. This is a primary maxim of real-time systems design – to understand the basis and nature of the timing constraints, so that they can be relaxed if necessary.

Many real-time systems utilize time-stamping and global clocks for synchronization, task initiation, and data marking. It must be noted, however, that clocks keep inaccurate time; even the official U.S. atomic clock must be adjusted. Moreover, there is an associated digitization error with clocks, which may need to be considered when using them for data time-stamping.

1.1.3 Events and Determinism

In software systems, a change in state results in a change in the flow-of-control of the computer program. Consider the flowchart in Figure 1.4. The decision block represented by the diamond suggests that the stream of program instructions, can take one of two paths, depending on the response in question. if-then, goto, and case statements in any language represent a possible change in flow-of-control. Invocation of procedures in C and Ada represent changes in

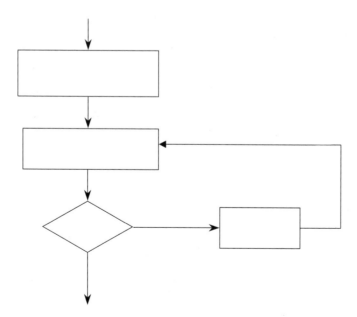

Figure 1.4 A simple program flowchart showing a branch as a change in flow-of-control, represented by the diamond icon.

flow-of-control. In object-oriented languages, instantiation of an object or the invocation of a method causes the change in sequential flow-of-control. In general, consider the following definition.

> **Definition:** Any occurrence that causes the program counter to change nonsequentially is considered a change of flow-of-control, and thus an event.

In scheduling theory, the "release" time of a "job" is similar to an event.

> **Definition:** The release time is the time at which an instance of a scheduled task is ready to run, and is generally associated with an interrupt.

Events are slightly different from jobs in that events can be caused by interrupts as well as conditional and unconditional branches.

1.1.3.1 Synchronous and Asynchronous Events
An event can be either synchronous or asynchronous. Synchronous events are those that occur at predictable times in the flow-of-control, such as that represented by the decision box in the flowchart of Figure 1.4. The change in flow-of-control, represented by a conditional branch instruction, or by the occurrence of an internal trap interrupt, can be anticipated (although it may not always occur).

Asynchronous events occur at unpredictable points in the flow-of-control and are usually caused by external sources. A clock that pulses "regularly" at 5 milliseconds is not a synchronous event. While it represents a periodic event, even if the clock were able to tick at a perfect 5 milliseconds without drift (which it cannot for physical reasons), the point where the tick occurs with the flow-of-control is subject to many factors. These factors include the time at which the clock starts relative to the program and propagation delays in the computer system itself. An engineer can never count on a clock ticking exactly at the rate specified, and so a clock-driven event must be treated as asynchronous.

Events that do not occur at regular intervals (or periods) are called aperiodic. Aperiodic events that tend to occur very infrequently are called sporadic.[1] Table 1.2 characterizes a sampling of events.

For example, an interrupt generated by a periodic external clock represents a periodic but asynchronous event. A periodic but synchronous event is one

---

[1] Scheduling theorists define aperiodic events as those nonperiodic events with soft deadlines, and sporadic events as nonperiodic events with hard deadlines. At the same time, they treat periodic tasks as having hard deadlines only. These restrictions are usually made because they promote theoretical formulations. No such distinction is made in this text.

Table 1.2 Taxonomy of events and some examples

	Periodic	Aperiodic	Sporadic
Synchronous	Cyclic code	Typical branch instruction	Branch instruction, e.g., error recovery
	Processes scheduled by internal clock	Garbage collection	Traps
Asynchronous	Clock-generated interrupt	Regular, but not fixed-period interrupt	Externally generated exception
			"Random events"

Note: Many of these items will be discussed later, or can be found in the glossary.

represented by a sequence of invocation of tasks in a repeated, circular fashion, otherwise known as cyclic code. A typical conditional or unconditional branching instruction[2] that is not part of a code block and that runs repeatedly at a regular rate represents a synchronous but aperiodic event. A branch instruction that happens infrequently, say, on the detection of some exceptional condition, is both sporadic and synchronous. Finally, interrupts that are generated irregularly (randomly) by an external device are classified as either asynchronous aperiodic or sporadic, depending on whether the interrupt is generated frequently or not with respect to the system clock.

1.1.3.2 Determinism
In every system, and particularly in an embedded real-time system, maintaining control is extremely important. For any physical system certain states exist under which the system is considered to be out of control; the software controlling such a system must therefore avoid these states. For example, in certain aircraft guidance systems, rapid rotation through a 180° pitch angle can cause loss of gyroscopic control. The software must be able to anticipate and avert all such scenarios.

Another characteristic of a software-controlled system is that the CPU continues to fetch and execute instructions from the program area of memory, rather than from data or other unwanted memory regions. The latter scenario can occur in poorly tested systems and is a catastrophe from which there is almost no hope of recovery.

Software control of any real-time system and associated hardware is maintained when the next state of the system, given the current state and a set of inputs, is predictable. In other words, the goal is to anticipate how a system will behave in all possible circumstances.

---

[2] "Branching" means both a single macroinstruction that causes a conditional or unconditional jump, or the sequence of such instructions that is generated by a compiler due to a procedure call, object instantiation, or method invocation.

> **Definition:** A system is deterministic if, for each possible state and each set of inputs, a unique set of outputs and next state of the system can be determined.

Event determinism means the next states and outputs of a system are known for each set of inputs that trigger events. Thus, a system that is deterministic is event deterministic. Although it would be difficult for a system to be deterministic only for those inputs that trigger events, this is plausible, and so event determinism may not imply determinism.[3]

It is interesting to note that while it is a significant challenge to design systems that are completely event deterministic, and as mentioned it is possible to inadvertently to end up with a system that is nondeterministic, it is also hard to design systems that are deliberately nondeterministic. This situation arises from the difficulties in designing completely random number generators. Deliberately nondeterministic systems would be desirable, for example, as casino gambling machines.

Finally, if in a deterministic system the response time for each set of outputs is known, then, the system also exhibits temporal determinism.

A side benefit of designing deterministic systems is that guarantees can be given that the system will be able to respond at any time, and in the case of temporally deterministic systems, when they will respond. This reinforces the association of control with real-time systems.

1.1.4 CPU Utilization

The final and most important term to be defined is a critical measure of real-time system performance. Because in the von Neumann paradigm, the CPU continues to fetch, decode, and execute instructions as long as power is applied, the CPU will execute either no-ops or instructions or instructions that are not related to the satisfaction of a deadline (for example, noncritical "housekeeping"). The measure of the time spent doing idle processing, in a sense, indicates how much real-time processing is occurring.

> **Definition:** The (CPU) utilization or time-loading factor, U, is a measure of the percentage of nonidle processing.

A system is said to be time-overloaded if $U > 100\%$. Systems that are too highly utilized are undesirable because changes or additions cannot be made to the system without risk of time-overloading. Systems that are not sufficiently utilized

---

[3] This definition implies that the system must have a finite number of states. It is reasonable to make this assumption in a digital computer system where all inputs are digitized to within a finite range.

Table 1.3   CPU utilization zones and typical applications and recommendations

Utilization (%)	Zone Type	Typical Application
0–25	Significant excess processing power – CPU may be more powerful than necessary	Various
26–50	Very safe	Various
51–68	Safe	Various
69	Theoretical limit	Embedded systems
70–82	Questionable	Embedded systems
83–99	Dangerous	Embedded systems
100+	Overload	Stressed systems

are not necessarily good, because this implies that the system was overengineered and that costs can be reduced with less expensive hardware. While a utilization of 50% is common for new products, 80% might be acceptable for systems that do not expect growth. However, 70% as a target for U is one of the most celebrated and potentially useful results in the theory of real-time systems where tasks are periodic and independent – a result that will be examined later. Table 1.3 gives a summary of certain CPU utilizations and typical situations in which they are associated.

U is calculated by summing the contribution of utilization factors for each (periodic or aperiodic) task. Suppose a system has $n \geq 1$ periodic tasks, each with an execution period of p_i, and hence, execution frequency, $f_i = 1/p_i$. If task i is known to have (or has been estimated to have) a maximum (worst case) execution time of e_i, then the utilization factor, u_i, for task e_i is

$$u_i = e_i/p_i \tag{1.1}$$

Then the overall system utilization is

$$U = \sum_{i=1}^{n} u_i = \sum_{i=1}^{n} e_i/p_i \tag{1.2}$$

Note that the deadline for periodic task i, d_i, is typically the next cycle or time period, and is a critical design factor that is constrained by e_i. The determination of the e_i either prior to or after the code has been written can be extremely difficult, and in many cases, impossible, in which case estimation must be used. For aperiodic and sporadic tasks u_i is calculated by assuming some worst-case execution period, usually the worst-case delay between event occurrences. Such approximations can inflate the utilization factor unnecessarily or can lead to overconfidence because of the tendency to "not worry" about its excessive contribution. The danger is to discover later that a higher frequency of occurrence than budgeted has led to a time-overload and system failure.

The utilization factor differs from CPU throughput, which is a measure of the number of macroinstructions per second that can be processed based on some predetermined instruction mix. This type of measurement is typically used to compare CPU horsepower for a particular application.

The choice of task deadlines, calculation and reduction of execution times, and other factors that influence CPU utilization will be discussed at great length in Chapter 7.

1.2 REAL-TIME SYSTEM DESIGN ISSUES

Real-time systems are a complex subdiscipline of computer systems engineering that is strongly influenced by control theory, software engineering, and operations research (via scheduling theory). Figure 1.5 depicts just some of the disciplines of computer science and electrical engineering that affect the design and analysis of real-time systems. Thus, because real-time systems engineering is so multidisciplinary, it stands out as a highly specialized area.

The design and implementation of real-time systems requires attention to numerous problems. These include:

- The selection of hardware and software, and evaluation of the trade-off needed for a cost-effective solution, including dealing with distributed computing systems and the issues of parallelism and synchronization.

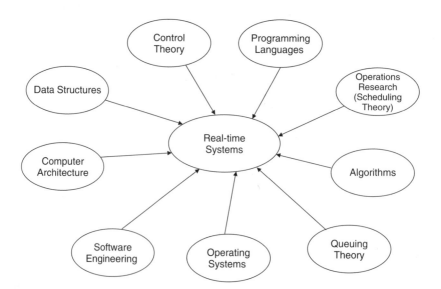

Figure 1.5 Disciplines that impact on real-time systems engineering.

- Specification and design of real-time systems and correct representation of temporal behavior.
- Understanding the nuances of the programming language(s) and the real-time implications resulting from their translation into machine code.
- Maximizing of system fault tolerance and reliability through careful design.
- The design and administration of tests, and the selection of test and development equipment.
- Taking advantage of open systems technology and interoperability. An open system is an extensible collection of independently written applications that cooperate to function as an integrated system. For example, a number of versions of the open operating system, Linux, have emerged for use in real-time applications. Interoperability can be measured in terms of compliance with open system standards, such as the CORBA real-time standard.
- Finally, measuring and predicting response time and reducing it. Performing a schedulability analysis, that is, determining and guaranteeing deadline satisfaction, a priori, is the focus of most of scheduling theory.

Of course, the engineering techniques used for hard real-time systems can be used in the engineering of all other types of systems, with an accompanying improvement of performance and robustness. Perhaps this alone is reason enough to study the engineering of real-time systems.

1.3 EXAMPLE REAL-TIME SYSTEMS

Embedded real-time systems are so pervasive that they are even found in household appliances and toys. A small sampling of real-time domains and their applications is given in Table 1.4.

In the introduction some real-time systems were mentioned. The following descriptions provide more details for each system and others provide examples and exercises. Clearly, the descriptions are not intended as formal specifications. The process of specifying systems clearly and concisely is discussed later.

Consider the inertial measurement system for an aircraft. The software specification states that the software will receive x, y, and z accelerometer pulses at a 10-millisecond rate from special hardware. The software will determine the accelerations in each direction and the roll, pitch, and yaw of the aircraft. Figure 1.6 illustrates these movements.

The software will also receive information such as temperature at a 1-second rate. The task of the software is to compute the actual velocity vector based on the orientation, accelerometer readings, and various compensation factors (such as for temperature effects) at a 40-millisecond rate. The system is to output true acceleration, velocity, and position vectors to a pilot's display every 40 milliseconds, but using a different clock.

These tasks execute at four different rates in the inertial measurement system and need to communicate and synchronize. The accelerometer readings must be

Table 1.4 Some typical real-time domains and applications

Domain	Applications
Avionics	Navigation Displays
Multimedia	Games Simulators
Medicine	Robot surgery Remote surgery Medical imaging
Industrial Systems	Robotic assembly lines Automated inspection
Civilian	Elevator control Automotive systems

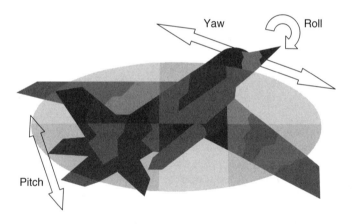

Figure 1.6 Movements of an aircraft: roll, pitch, and yaw movements.

time-relative or correlated; that is, it is undesirable to mix an x accelerometer pulse from t with z and y pulses from time $t + 1$. These are critical design issues for this system.

Next, consider a monitoring system for a nuclear plant that will be handling three events signaled by interrupts. The first event is triggered by any of several signals at various security points, which will indicate a security breach. The system must respond to this signal within 1 second. The second (and most important) event indicates that the nuclear core has reached an overtemperature. This signal must be dealt with within 1 millisecond. Finally, an operator's display is to be updated at approximately 30 times per second. The nuclear plant system requires a mechanism to ensure that the "meltdown imminent" indicator can interrupt any other processing. How is this accomplished?

As another example, recall the airline reservation system mentioned earlier. Management has decided that to prevent long lines and customer dissatisfaction, turnaround time for any transaction must be less than 15 seconds, and no overbooking will be permitted (how lovely this would be). At any time, several agents may try to access the database and perhaps book the same flight simultaneously. Here record-locking and communications mechanisms are needed to protect against the alteration of the database containing the reservation information by more than one clerk simultaneously. How is this done?

Now consider a computer system that controls all aspects of the bottling of jars of pasta sauce[4] as they travel along a conveyor belt. The empty jars are microwaved to disinfect them. A mechanism fills each jar with a precise serving of sauce as it passes beneath. Another station caps the bottles. Of course, there is an operator's display that provides an animated rendering of the production line activities. There are numerous events triggered by exceptional conditions such as the conveyor belt jamming, a bottle overflowing or breaking. If the conveyor belt travels too fast, the bottle will move past its designated station prematurely. Therefore there is a wide range of events both synchronous and asynchronous to be dealt with.

As a final example consider a system used to control a set of traffic lights at a four-way traffic intersection (north-, south-, east-, and west-bound traffic). This system controls the lights for auto and foot traffic at a four-way intersection in a busy city like Philadelphia. Input may be taken from sensors under the ground, push buttons, cameras, and so on. The traffic lights need to operate in a synchronized fashion, and yet react to asynchronous events (such as a pedestrian pressing a button at a crosswalk). Failure to operate in a proper fashion can result in auto accidents and even fatalities.

The challenge presented by each of these systems is to determine the appropriate design approach with respect to the issues discussed in Section 1.2.

1.4 COMMON MISCONCEPTIONS

As a part of truly understanding the nature of real-time systems, it is important to address a number of frequently cited misconceptions. These are summarized as follows:

1. Real-time systems are synonymous with "fast" systems.
2. Rate-monotonic analysis has solved "the real-time problem."

---

[4] The author's mother, who is Italian, calls sautéed tomatoes "sauce," while his wife, who is also Italian, calls it "gravy." Definitions can vary.

3. There are universal, widely accepted methodologies for real-time systems specification and design.

4. There is never a need to build a real-time operating system, because many commercial products exist.

5. The study of real-time systems is mostly about scheduling theory.

The first misconception, that real-time systems must be fast, arises from the fact that many hard real-time systems indeed deal with deadlines in the tens of milliseconds, such as the aircraft navigation system. But the pasta sauce jars can move along the conveyor belt past a given point at a rate of one every 2 seconds. The airline reservation system has a deadline of 15 seconds. These deadlines are not particularly fast, but satisfying them determines the success or failure of the system.

The second misconception is that rate-monotonic systems provide a simple recipe for building real-time systems. Rate-monotonic systems – a periodic system in which interrupt priorities are assigned such that the faster the frequency of execution, the higher the interrupt priority – have received a lot of attention since the 1970s. While they provide much guidance in the design of real-time systems, and while there is abundant theory surrounding them, they are not a panacea. Rate-monotonic systems will be discussed in great detail.

What about the third misconception? Unfortunately, there are no universally accepted and foolproof methods for the design and specification of real-time systems. This is not a failure of researchers or the software industry, but is because of the difficulty of discovering universal solutions. Even after more than 30 years of research there is no methodology available that answers all of the challenges of real-time specification and design all the time and for all applications.

The fourth misconception is that there is never a need to build a real-time operating system from scratch. While there are a number of viable, popular, and cost-effective commercial real-time operating systems, these, too, are not a panacea. Commercial solutions have their place, but choosing when to use an off-the-shelf solution and choosing the right one are challenges that will be discussed later.

Last, while it may be challenging and scholarly to study scheduling theory, from an engineering standpoint, most published results require impractical simplifications and clairvoyance in order to make the theory work. Because this is an engineering text, it avoids any theoretical results that resort to these measures.

1.5 BRIEF HISTORY

The history of real-time systems, as characterized by developments in the United States, is loosely tied to the evolution of the computer. Modern real-time systems, such as those that control nuclear power stations, military aircraft weapons systems, or medical monitoring equipment, are complex, yet many still exhibit characteristics of systems developed in the 1940s through the 1960s.

1.5.1 Theoretical Advances

Much of the theory of real-time systems is derived from the many underlying disciplines shown in Figure 1.5. In particular, aspects of operations research, which emerged in the late 1940s, and queuing systems, which emerged in the early 1950s, have influenced most of the more theoretical results.

Martin published one of the earliest and certainly the most influential early book on real-time systems [Martin67]. Martin's book was soon followed by several others (e.g., [Stimler69[5]]), and the influence of operations research (scheduling) and queuing systems can be seen in these works. It is also interesting to study these texts in the context of the limitations of the hardware of the time.

In 1973 Liu and Layland published their work on rate-monotonic theory [Liu73]. Over the last 30 years significant refinement of this theory has made it a more practical theory for use in designing real systems.

The 1980s and 1990s saw a proliferation of theoretical work on improving predictability and reliability of real-time systems and on solving problems related to multiprocessing systems. Today, a rather limited group of experts continue to study issues of scheduling and performance analysis, even as a wider group of generalist systems engineers tackle broader issues relating to the implementation of real, practical systems. An important paper by Stankovic et al. [Stankovic95] described some of the difficulties in conducting research on real-time systems – even with significant restriction of the system, most problems relating to scheduling are too difficult to solve by analytic techniques.[6]

1.5.2 Early Systems

The origin of the term real-time computing is unclear. It was probably first used either with project Whirlwind, a flight simulator developed by IBM for the U.S. Navy in 1947, or with SAGE, the Semiautomatic Ground Environment air defense system developed for the U.S. Air Force in the early 1950s. Both projects qualify as real-time systems by today's standards. In addition to its real-time contributions, the Whirlwind project included the first use of ferrite core memory and a form of high-order language compiler that predated Fortran.

Other early real-time systems were used for airline reservations, such as SABRE (developed for American Airlines in 1959), as well as for process control, but the advent of the national space program provided even greater opportunities for the

---

[5] By coincidence, the author met Saul Stimler in 1995. He was still vibrant and actively thinking about real-time systems.

[6] At a 1992 NATO Advanced Study Institute that the author attended, Professor C. L. Liu (co-discoverer of the rate-monotonic theory) stood up at a keynote talk and began by stating, "There are no useful results in optimal scheduling for real-time systems." The crowd was stunned (except the author). There is no reason to believe that this situation has changed since then.

development of real-time systems for spacecraft control and telemetry. It was not until the 1960s that rapid development of such systems took place, and then only as significant nonmilitary interest in real-time problem solutions become coupled with the availability of equipment well adapted to real-time processing.

1.5.3 Hardware Developments

Weak processors and exceptionally slow and sparse memories handicapped many of the earliest systems. Whirlwind introduced the ferrite core memory, a vast improvement over its predecessor, the vacuum tube.

In the early 1950s the asynchronous interrupt was introduced and incorporated as a standard feature in the Univac Scientific 1103A. The middle 1950s saw a distinct increase in the speed and complexity of large-scale digital computers designed for scientific computation, without an increase in size. These developments made it possible to apply "real-time" computation in the field of control systems. Such improvements were particularly noticeable in IBM's development of SAGE.

In the 1960s and 1970s, advances in integration levels and processing speeds enhanced the spectrum of real-time problems that could be solved. In 1965 alone it was estimated that more than 350 real-time process control systems existed.

The 1980s and 1990s have seen multiprocessing systems and other non–von Neumann architectures utilized in real-time applications.

Finally, the late 1990s and early 2000s have seen new trends in real-time embedded systems in consumer goods and Web-enabled devices. The availability of small processors with limited memory and functionality has rejuvenated some of the challenges faced by early real-time systems designers. Fortunately, 50 years of experience is available to draw upon.

1.5.4 Early Software

Early real-time systems were written directly in microcode, assembly language, and later in higher-level languages. As previously noted, Whirlwind used an early form of high-order language called an algebraic compiler to simplify coding. Later systems employed Fortran, CMS-2, and JOVIAL, the preferred languages in the U.S. Army, Navy, and Air Force, respectively.

In the 1970s, the Department of Defense mandated the development of a single language that all services could use, and that provided high-level language constructs for real-time programming. After a selection and refinement process, the Ada language appeared as a standard in 1983. Shortfalls in the language were identified and a new, improved version of the language, Ada 95, appeared in 1995.

Today, however, only a small number of systems are developed in Ada 95. Most systems are written in C and C++, even assembly language and Fortran. In the last 5 years, there has been an increase in the use of object-oriented methodologies and languages like Java in embedded real-time systems. Of course, other

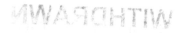

languages are used in various settings. The real-time aspects of programming languages are discussed later in the text.

1.5.5 Commercial Operating System Support

The first commercial operating systems were designed for mainframe computers. IBM developed the first real-time executive, its Basic Executive in 1962, which provided diverse real-time scheduling. By 1963, its Basic Executive II had disk resident user/systems programs.

By the mid 1970s more affordable minicomputer systems could be found in many engineering environments. In response, a number of important real-time operating systems were developed by the minicomputer manufacturers. Notable among these were the Digital Equipment Corporation (DEC) family of real-time multitasking executives (RSX) for the PDP-11 and Hewlett-Packard's Real-Time Executive (RTE) series of operating systems for its HP 2000 series.

By the late 1970s and early 1980s, the first operating systems for microprocessor-based systems appeared. These included RMX-80, MROS 68K, VRTX, and several others. Over the last 20 years many real-time operating systems have appeared and many have disappeared.

A summary of some of the landmark events in the field of real-time systems in the United States is given in Table 1.5.

Table 1.5   Landmarks in real-time systems history in the United States

Year	Landmark	Developer	Development	Innovations
1947	Whirlwind	MIT/US Navy	Flight simulator	Ferrite core memory, "real response times"
1957	SAGE	IBM	Air defense	Specifically designed for real-time
1958	Scientific 1103A	Univac	General purpose	Hardware interrupt
1959	SABRE	IBM	Airline reservation	Hub-go-ahead policy
1962	Basic Executive	IBM	General purpose	First real-time executive
1963	Basic Executive II	IBM	General purpose	Diverse real-time scheduling, Disk resident user/systems programs

(continued)

Table 1.5   Landmarks in real-time systems history (*continued*)

Year	Landmark	Developer	Development	Innovations
1970s	RSX, RTE	DEC, HP	Real-time operating systems	Hosted by minicomputers
1973	Rate-monotonic system	Liu and Layland	Theory	Stated upper bound on utilization for schedulable systems
1980s	RMX-80, MROS 68K, VRTX, etc.	Various	Real-time operating system	Hosted by microprocessors
1983	Ada 83	U.S. Department of Defense	Programming language	Intended for mission-critical, embedded, real-time systems
1995	Ada 95	Community	Programming Language	Refinement to Ada 83

1.6   EXERCISES

1.1   Consider a payroll processing system for a small manufacturing firm. Describe three different scenarios in which the system can be justified as hard, firm, or soft real-time.

1.2   Discuss whether the following are hard, soft, or firm real-time systems:

(a)   The Library of Congress print manuscript database system.

(b)   A police database that provides information on stolen automobiles.

(c)   An automatic teller machine.

(d)   A coin-operated video game.

(e)   A university grade-processing system.

(f)   A computer-controlled routing switch used at a local telephone company branch exchange.

1.3   Consider a real-time weapons control system aboard a fighter aircraft. Discuss which of the following events would be considered synchronous and which would be considered asynchronous to the computing system.

(a)   A 10-millisecond, externally generated clock interrupt.

(b)   A divide-by-zero trap.

(c)   A built-in-test software failure.

    (d)   A discrete signal generated by the pilot pushing a button to fire a rocket.

    (e)   A discrete signal indicating "low on fuel."

1.4   Describe a system that is completely non-real-time, that is, there are no bounds whatsoever for any response time.

1.5   For the following systems concepts, fill in the cells of Table 1.2 with descriptors for possible events. Estimate event periods for the periodic events.

    (a)   Elevator control: this system controls all aspects of a bank of elevators that service a 50-story building in a busy city like Philadelphia.

    (b)   Automotive control: this on-board crash avoidance system uses data from a variety of sensors and makes decisions and affects behavior to avoid collision, or protect the occupants in the event of an imminent collision. The system might need to take control of the automobile from the driver.

1.6   For the systems in Exercise 1.2, what are reasonable response times for all the events?

1.7   For the example systems introduced (inertial measurement, nuclear plant monitoring, airline reservation, pasta bottling, and traffic-light control) enumerate some possible events and note whether they are periodic, episodic, or sporadic. Discuss reasonable response times for the events.

2

HARDWARE
CONSIDERATIONS

Understanding the underlying hardware of the real-time system allows for efficient hardware and software resource utilization. Although it is desirable for the programming language to abstract away hardware details from the designers, this is usually impossible to achieve – if not at the design state, certainly at the hardware/software integration stages. Therefore, an understanding of computer architecture is essential to the real-time systems engineer. While it is not the intent here to provide a complete review of computer architecture, a brief survey of the most important issues is appropriate. For a more thorough treatment, see, for example, [Gilreath03]. Some of the following discussion is adapted from that resource.

2.1 BASIC ARCHITECTURE

In its simplest form, a computer system consists of a CPU and memory interconnected by a bus (Figure 2.1).

There are three systemwide buses: power, address, and data. The power bus refers to the distribution of power to the various components of the computer system; the address bus is the medium for exchanging individual memory addresses, and therein the data bus is used to move data between the various components in the system. When referring to the system bus, the address and data buses collectively are generally what are meant.

For the most part, this book deals with single-processor (uniprocessing) systems. Some real-time systems are multiprocessing in the sense that there are

Real-Time Systems Design and Analysis, By Phillip A. Laplante
ISBN 0-471-22855-9 © 2004 Institute of Electrical and Electronics Engineers

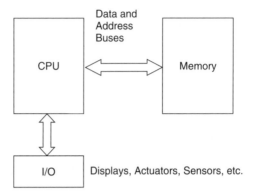

Figure 2.1   A "10,000" foot view of a von Neumann architecture.

many processors distributed in the system, and that these are loosely coupled through messaging. Other real-time systems use multiprocessing in a way that allows the processing system to schedule tasks across the different processors. These types of systems are far more complex, and the theory and tools available are generally impractical. Therefore, this text will concentrate on uniprocessing real-time systems. Even here the challenges are significant.

2.2   HARDWARE INTERFACING

The following sections contain discussions that are slanted toward the software or systems engineer rather than the electrical engineer. That is, the intent is not to be able to design these hardware structures, but rather to understand their behavior in the context of embedded real-time systems.

2.2.1   Latching

In signaling between devices is it is important to have a mechanism for "record-ing" the appearance of that signal for later processing. This process is called latching. In essence, latching involves setting a flip-flop corresponding to some event. For example, interrupt signals are latched into the programmable interrupt controller so that they can be serviced at an appropriate time.

Once the latch is read, it needs to be reset so that a new signal can be received. Thus, for example, in the case of an interrupt, if a second interrupt is signaled on the same input but the previous interrupt has not been serviced (and the latch reset), the second interrupt will be lost, leading to a missed deadline. Therefore, in servicing latched inputs of any kind it is important to read and clear the latch as soon as possible.

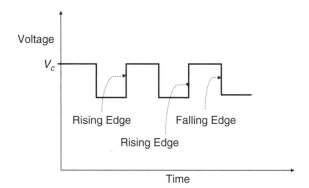

Figure 2.2 A fictitious time-varying signal (typically, a clock) showing two rising edges, each of which represents a single event, and a falling edge. V_c represents a critical or threshold voltage.

2.2.2 Edge versus Level Triggered

Logic devices can either be edge or level triggered, and the difference matters in the sense that it can affect the way the system recognizes true events or false ones. A transition from low to high is called a rising edge, and from high to low it is called a falling edge (Figure 2.2).

In other cases, the signal is represented by the voltage exceeding a certain threshold. When the signal reaches that level, an event is triggered and latched so that another event cannot be triggered until the latch is reset. For example, in Figure 2.2 if such level-triggered logic is used, only a single event will be recorded unless the latch is reset during the time period shown.

The differences between edge-based and level-based logic are important, because in the course of manipulating certain signals it is possible to create "false" events by prematurely resetting logic, which will be shown shortly.

2.2.3 Tristate Logic

When multiple devices are connected to the same bus structure it is important that those devices that are not currently involved in data interchange remain, essentially, unconnected. To achieve this effect, those devices that are not involved are placed into a high-impedance state at their bus interconnections, that is, they are "tristated." Hence a particular electrical signal can be in one of three levels, high, low, or tristated. Tristate logic is essential in the design of computer systems.

Signals that are improperly tristated will be in an unknown state in which the signal is "floating," that is, arbitrarily high or low. Floating signals can be the source of many insidious problems, such as falsely indicated interrupts and improper setting of switches.

2.2.4 Wait States

When a microprocessor must interface with a slower peripheral or memory device, the normal timing of the microprocessor may need to be altered. Specifically, in some cases a wait state may need to be added to the bus cycles that access that peripheral or memory. Wait states extend the microprocessor read or write cycle by a certain number of processor clock cycles to allow the device or memory to "catch up."

For example, in a certain system EEPROM, RAM, and ROM all have different memory access times. These memory technologies will be discussed shortly. Since RAM memory is typically faster than ROM, wait states would need to be inserted when accessing RAM if the timing is to be made uniform across all regions of memory. Of course, wait states degrade overall systems performance, but do preserve determinism because it can be assumed that each memory access takes the same amount of time.

2.2.5 Systems Interfaces and Buses

A typical microprocessor-based system will have a common group of 8, 16, 32, or 64 or more signals for reading and writing data. These signals are collectively referred to as the system bus. Within the CPU there is another common bus for intraprocessor communication between components.

The system bus is used for communication between the CPU, memory, and device. Transmit/receive hybrid devices, or transceivers, provide communication services to other devices joined by a common bus.

When accessing devices that require serial interfacing, or if the number of lines on the bus is less than those internal to the device, then a multiplexer (or MUX) is needed to enable communication to the serial device over the parallel bus. The MUX circuitry is responsible for ensuring that all transmitted and received data conform to an appropriate protocol. This process includes parallel to serial conversion for transmittal and vice versa with receipt and other circuitry (Figure 2.3). The standard universal asynchronous relay terminal (UART) is typically used for parallel-to-serial-bus interfaces and is seen in many commercial applications. Such a scheme is compatible with 8-, 16-, 32-bit parallel buses and any device-internal representation.

Of course, there are numerous standardized and customized systems interfaces and bus types. Three particularly common ones for embedded real-time systems are introduced.

2.2.5.1 MIL-STD-1553B
A widely used standard in both military and commercial avionics applications is the MIL-STD-1553B bus standard that specifies a hardware configuration and transmission and receipt protocols. The 1553B bus protocol is a master–slave protocol. "Master–slave" indicates that one device orchestrates all activities on the bus through directives and the other devices simply follow those directives, that is, they cannot initiate activity on their own.

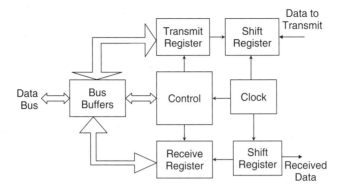

Figure 2.3 A transmitter/receiver device used to multiplex parallel data to serial. To receive, the parallel data are captured from the bus into a receive register and then shifted into a serial stream of bits. To transmit, the data are loaded into a shift register, then shifted into a parallel transmit–receive buffer for transmission.

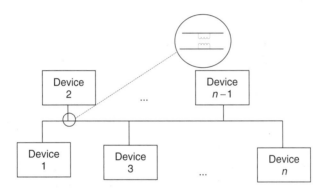

Figure 2.4 MIL-STD 1553B configuration. The inset shows the inductive coupling connection specified. Such a connection helps the system withstand electrical failure of one of the devices.

The 1553B standard is arranged so that one module on the bus acts as a bus controller (the master) and the others respond to its commands (the slaves). This type of configuration is common in many commercial networks and computer subsystems, and is illustrated in Figure 2.4.

The 1553B protocol is based on a list of activities to be performed by each of the devices connected to the bus. The master device maintains this list. Each message essentially consists of a device ID number, a directive and, possibly, a set of data. All devices are listening to the bus at all times, awaiting broadcast of a message directive with their unique bus ID (or a broadcast message intended for all devices on the bus).

The following example is a slightly modified version of how the 1553B works. Suppose the master wants device number 5 to send 10 packets of data to device number 6. The master puts a message out on the bus that essentially says, "Device number 5 put 10 packets of data on the bus for device number 6." Device number 5 then places the data on the bus. When device number 6 sees the data on the bus with their ID in the header, it captures them. All other devices on the bus can "see" the data, but only device number 6 takes the data.

Besides electrical fault-tolerance provisions, the 1553B protocol provides for fault tolerance in that if the master device is disabled, provisions can be made for another device in the system to take over as master.

2.2.5.2 *Small Computer Systems Interface*

The small computer systems interface (SCSI) or "scuzzy" is a widely used, PC-based, parallel interface that supports many kinds of devices. There have been three generations of SCSI (1, 2, and 3) and other variants such as Narrow, Wide, Fast, Ultra, Ultra-2, and Ultra160 SCSI.

This synopsis focuses on SCSI 3, which is made up of at least 14 separate standards documents. These standards resolve many of the conflicts in previous versions and add significant functionality and performance improvements. SCSI 3 also supports Fiber Channel, and FireWire® instead of the familiar ribbon cable connection. The new interfaces are backward compatible with SCSI-2 as well as SCSI-1 via the single-ended interface. SCSI supports devices connected in a daisy-chained fashion (Figure 2.5).

Although the devices are daisy chained and appear to be dependent, they are independent and each can directly communicate with the others as well as with the host. Each device is uniquely configured by connecting one end to the host adapter and then setting the device ID with a plug-in terminator, jumpers, or switches. Id number 0 (zero) is set for the boot device, and the higher the ID number, the higher the priority of the device in bus access arbitration.

2.2.5.3 *IEEE 1394 Firewire*

The IEEE 1394 bus standard describes a very fast external bus standard that supports data transfer rates of up to 400 megabits per second (Mbps) (in 1394a) and 800 Mbps (in 1394b). Products supporting the 1394 standard assume different names, depending on the company. Apple, which originally developed the technology, uses the trademarked name FireWire.

FireWire is easy to use, and a single 1394 port can be used to connect up 63 external devices. The standard defines 100-, 200-, and 400-Mbps devices and

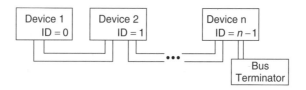

Figure 2.5 Daisy-chained devices. Daisy-chain connections are used in many kinds of devices in an embedded system (e.g., interrupt controllers), because they allow for an easy "extension" of the system bus by simply attaching to the device at the end.

can support the multiple speeds on a single bus, and is flexible – the standard supports freeform daisy chaining and branching for peer-to-peer implementations. It is also hot pluggable, that is, devices can be added and removed while the bus is active.

FireWire supports two types of data transfer: asynchronous and isochronous. For traditional computer memory-mapped, load, and store applications, asynchronous transfer is appropriate and adequate. Isochronous data transfer provides guaranteed data transport at a predetermined rate. This is especially important for multimedia applications where uninterrupted transport of time-critical data and just-in-time delivery reduce the need for costly buffering. This makes it ideal for devices that need to transfer high levels of data in real time, such as cameras, VCRs, and televisions.

2.3   CENTRAL PROCESSING UNIT

A reasonable understanding of the internal organization of the CPU is quite helpful in understanding the basic principles of real-time response; hence, those concepts are briefly reviewed here.[1]

The CPU can be thought of as containing several components connected by its own internal bus, which is distinct from the memory and address buses of the system. As shown in Figure 2.6 the CPU contains a program counter (PC), an arithmetic logic unit (ALU), internal CPU memory–scratch pad memory and

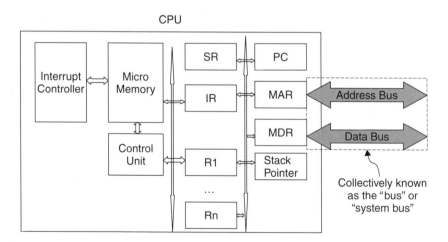

Figure 2.6   Partial, stylized, internal structure of a typical CPU. The internal paths represent connections to the internal bus structure. The connection to the system bus is shown on the right.

[1] Some of the following discussion in this section is adapted from *Computer Architecture: A Minimalist Perspective* by Gilreath and Laplante [Gilreath03].

micromemory, general registers (labelled 'R1' through 'Rn'), an instruction register (IR), and a control unit (CU). In addition, a memory address register (MAR) holds the address of the memory location to be acted on, and a memory date register (MDR) holds the data to be written to the MAR or that have been read from the memory location held in the MAR.

There is an internal clock and other signals used for timing and data transfer, and other hidden internal registers that are typically found inside the CPU, but are not shown in Figure 2.6.

2.3.1 Fetch and Execute Cycle

Programs are a sequence of macroinstructions or macrocode. These are stored in the main memory of the computer in binary form and await execution. The macroinstructions are sequentially fetched from the main memory location pointed to by the program counter, and placed in the instruction register.

Each instruction consists of an operation code (opcode) field and zero or more operand fields. The opcode is typically the starting address of a lower-level program stored in micromemory (called a microprogram), and the operand represents registers, memory, or data to be acted upon by this program.

The control unit decodes the instruction. Decoding involves determining the location of the program in micromemory and then internally executing this program, using the ALU and scratch-pad memory to perform any necessary arithmetic computations. The various control signals and other internal registers facilitate data transfer, branching, and synchronization.

After executing the instruction, the next macroinstruction is retrieved from main memory and executed. Certain macroinstructions or external conditions may cause a nonconsecutive macroinstruction to be executed. This case is discussed shortly. The process of fetching and executing an instruction is called the fetch–execute cycle. Even when "idling," the computer is fetching and executing an instruction that causes no effective change to the state of the CPU and is called a no-operation (no-op). Hence, the CPU is constantly active.

2.3.2 Microcontrollers

Not all real-time systems are based on a microprocessor. Some may involve a mainframe or minicomputers, while others are based on a microcontroller. Very large real-time systems involving mainframe or minicomputer control are unusual today unless the system requires tremendous CPU horsepower and does not need to be mobile (for example, an air traffic control system). But, microcontroller-based real-time systems abound.

A microcontroller is a computer system that is programmable via microinstructions (Figure 2.7). Because the complex and time-consuming macroinstruction decoding process does not occur, program execution tends to be very fast.

Unlike the complex instruction decoding process found in a traditional microprocessor, the microcontroller directly executes "fine grained" instructions stored

Micromemory

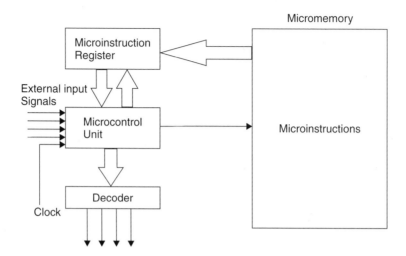

Figure 2.7 Stylized microcontroller block diagram.

in micromemory. These fine-grained instructions are wider than macroinstructions (in terms of number of bits) and directly control the internal gates of the microcontroller hardware. The microcontroller can take direct input from devices and directly control external output signals. High-level language and tool support allows for straightforward code development.

2.3.3 Instruction Forms

An instruction set constitutes the language that describes a computer's functionality. It is also a function of the computer's organization.[2] While an instruction set reflects differing underlying processor design, all instruction sets have much in common in terms of specifying functionality.

Instructions in a processor are akin to functions in procedural programming language in that both take parameters and return a result. Most instructions make reference to either memory locations, pointers to a memory location, or a register.[3] The memory locations eventually referenced contain data that are processed to produce new data. Hence, any computer processor can be viewed as a machine for taking data and transforming it, through instructions, into new information.

It is important to distinguish which operand is being referenced in describing an operation. As in arithmetic, different operations use different terms for the parameters to distinguish them. For example, addition has addend and augends,

---

[2] Traditionally, the distinction between computer organization and computer architecture is that the latter involves using only those hardware details that are visible to the programmer, while the former involves implementation details.

[3] An exception to this might be a HALT instruction. However, any other instruction, even those that are unary, will affect the program counter, accumulator, or a stack location.

subtraction has subtract and and subtrahend, multiplication has multiplicand and multiplier, and division has dividend and divisor.

In a generic sense, the two terms "operandam" and "operandum" can be used to deal with any unary or binary operation. The operandam is the first parameter, like an addend, multiplicand, or dividend. The operandum is the second parameter, like the augend, multiplier, or divisor. The following formal definitions will be helpful, as these terms will be used throughout the text.

The defining elements of instructions hint at the varying structures for organizing information contained within the instruction. In the conventional sense, instructions can be regarded as an *n*-tuple, where the *n* refers to the parameters of the instruction.

In the following sections, the instruction formats will be described beginning with the most general to the more specific. The format of an instruction provides some idea of the processor's architecture and design. However, note that most processors use a mix of instruction forms, especially if there is an implicit register. The following, self-descriptive examples illustrate this point.

2.3.3.1 1-Address and 0-Address Forms   Some processors have instructions that use a single, implicit register called an accumulator as one of the operands. Other processors have instruction sets organized around an internal stack in which the operands are found in the two uppermost stack locations (in the case of binary operations) or in the uppermost location (in the case of unary operations). These 0-address (or 0-address or stack) architectures can be found in programmable calculators that are programmed using postfix notation.

2.3.3.2 2-Address Form   A 2-address form is a simplification (or complication, depending on the point of view) of the 3-address form. The 2-address (or 2-tuple) form means that an architectural decision was made to have the resultant and operandum as the same. The 2-address instruction is of the form:

```
op-code operandam, operandum
```

As a mathematical function, the 2-address would be expressed as:

```
operandum = op-code(operandam, operandum)
```

Hence, the resultant is implicitly given as the operandum and stores the result of the instruction.

The 2-address form simplifies the information provided, and many high-level language program instructions often are self-referencing, such as the C language statement:

```
i=i+1;
```

which has the short form:

```
i++;
```

This operation could be expressed with an ADD instruction in 2-address form as:

```
ADD 0x01, &i     ; 2-address
```

where &i is the address of the i variable.[4] A 3-address instruction would redundantly state the address of the i variable twice: as the operandum and as the resultant as follows:

```
ADD 0x01, &i, &i ; 3-address
```

However, not all processor instructions map neatly into 2-address form, so this form can be inefficient. The 80×86 family of processors, including the Pentium,[®] use this instruction format.

2.3.3.3 *3-Address Form*   The 3-address instruction is of the form:

```
op-code operandam, operandum, resultant
```

This is closer to a mathematical functional form, which would be

```
resultant = op-code(operandam, operandum)
```

This form is the most convenient from a programming perspective and leads to the most compact code.

2.3.4 Core Instructions

In any processor architecture, there are many instructions, some oriented toward the architecture and others of a more general kind. In fact, all processors share a core set of common instructions.

There are generally six kinds of instructions. These can be classified as:

- Horizontal-bit operation
- Vertical-bit operation
- Control
- Data movement
- Mathematical/special processing
- Other (processor specific)

The following sections discuss these instruction types in some detail.

2.3.4.1 *Horizontal-Bit Operation*   The horizontal-bit operation is a generalization of the fact that these instructions alter bits within a memory in the horizontal direction, independent of one another. For example, the third bit in

---

[4] This convention is used throughout the book.

the operands would affect the third bit in the resultant. Usually, these instructions are the AND, IOR, XOR, NOT operations.

These operations are often called "logical" operators, but practically speaking, they are bit operations. Some processors have an instruction to specifically access and alter bits within a memory word.

2.3.4.2 *Vertical-Bit Operation*   The vertical-bit operation alters a bit within a memory word in relation to the other bits. These are the rotate-left, rotate-right, shift-right, and shift-left operations. Often shifting has an implicit bit value on the left or right, and rotating pivots through a predefined bit, often in a status register of the processor.

2.3.4.3 *Control*   Both horizontal- and vertical-bit operations can alter a word within a memory location, but a processor has to alter its state to change flow of execution and which instructions the processor executes.[5] This is the purpose of the control instructions, such as compare and jump on a condition. The compare instruction determines a condition such as equality, inequality, and magnitude. The jump instruction alters the program counter based upon the condition of the status register.

Interrupt handling instructions, such as the Intel 80×86's CLI, clears the interrupt flag in the status register, or the TRAP in the Motorola 68000 handles exceptions. Interrupt handling instructions can be viewed as asynchronous control instructions.

The enable priority interrupt (EPI) is used to enable interrupts for processing by the CPU. The disable priority interrupt (DPI) instruction prevents the CPU from processing interrupts (i.e., being interrupted). Disabling interrupts does not remove the interrupt as it is latched; rather, the CPU "holds off" the interrupt until an EPI instruction is executed.

Although these systems may have several interrupt signals, assume that the CPU honors only one interrupt signal. This has the advantage of simplifying the instruction set and off-loading certain interrupt processing. Such tasks as prioritization and masking of certain individual interrupts are handled by manipulating the interrupt controller via memory-mapped I/O or programmed I/O.

Modern microprocessors also provide a number of other instructions specifically to support the implementation of real-time systems. For example, the Intel IA-32 family provides LOCK, HLT, and BTS instructions, among others.

The LOCK instruction causes the processor's LOCK# signal to be asserted during execution of the accompanying instruction, which turns the instruction into an atomic (uninterruptible) instruction. Additionally, in a multiprocessor environment, the LOCK# signal ensures that the processor has exclusive use of any shared memory while the signal is asserted.

The HLT (halt processor) instruction stops the processor until, for example, an enabled interrupt or a debug exception is received. This can be useful for

---

[5] If this were not the case, the machine in question would be a calculator, not a computer!

debugging purposes in conjunction with a coprocessor (discussed shortly), or for use with a redundant CPU. In this case, a self-diagnosed faulty CPU could issue a signal to start the redundant CPU, then halt itself, which can be awakened if needed.

The BTS (bit test and set) can be used with a LOCK prefix to allow the instruction to be executed atomically. The test and set instructions will be discussed later in conjunction with the implementation of semaphores.

Finally, the IA-32 family provides a read performance-monitoring counter and read time-stamp counter instructions, which allow an application program to read the processor's performance-monitoring and time-stamp counters, respectively. The Pentium 4® processors have eighteen 40-bit performance-monitoring counters, and the P6® family processors have two 40-bit counters. These counters can be used to record either the occurrence or duration of events.

2.3.4.4 *Mathematical*   Most applications require that the computer be able to process data stored in both integer and floating-point representation. While integer data can usually be stored in 2 or 4 bytes, floating-point quantities typically need 4 or more bytes of memory. This necessarily increases the number of bus cycles for any instruction requiring floating-point data.

In addition, the microprograms for floating-point instructions are considerably longer. Combined with the increased number of bus cycles, this means floating-point instructions always take longer than their integer equivalents. Hence, for execution speed, instructions with integer operands are always preferred over instructions with floating-point operands.

Finally, the instruction set must be equipped with instructions to convert integer data to floating-point and vice versa. These instructions add overhead while possibly reducing accuracy. Therefore mixed-mode calculations should be avoided if possible.

The bit operation instructions can create the effects of binary arithmetic, but it is far more efficient to have the logic gates at the machine hardware level implement the mathematical operations. This is true especially in floating-point and dedicated instructions for math operations. Often these operations are the ADD, SUB, MUL, DIV, as well as more exotic instructions. For example, in the Pentium, there are built-in instructions for more efficient processing of graphics.

2.3.4.5 *Data Movement*   The I/O movement instructions are used to move data to and from registers, ports, and memory. Data must be loaded and stored often. For example in the C language, the assignment statement is

```
i = c;
```

As a 2-address instruction, it would be

```
MOVE &c, &i
```

Most processors have separate instructions to move data into a register from memory (LOAD), and to move data from a register to memory (STORE). The Intel

80×86 has dedicated IN, OUT to move data in and out of the processor through ports, but it can be considered to be a data movement instruction type.

2.3.4.6 Other Instructions The only other kinds of instructions are those specific to a particular architecture. For example, the 8086 LOCK instruction previously discussed. The 68000 has an ILLEGAL instruction, which does nothing but generate an exception. Such instructions as LOCK and ILLEGAL are highly processor architecture specific, and are rooted in the design requirements of the processor.

2.3.5 Addressing Modes

The addressing modes represent how the parameters or operands for an instruction are obtained. The addressing of data for a parameter is part of the decoding process for an instruction (along with decoding the instruction) before execution.

Although some architectures have ten or more possible addressing modes, there are really three basic types of addressing modes:

- Immediate data
- Direct memory location
- Indirect memory location

Each addressing mode has an equivalent in a higher-level language.

2.3.5.1 Immediate Data Immediate data are constant, and they are found in the memory location succeeding the instruction. Since the processor does not have to calculate an address to the data for the instruction, the data are immediately available. This is the simplest form of operand access. The high-level language equivalent of the immediate mode is a literal constant within the program code.

2.3.5.2 Direct Memory Location A direct memory location is a variable. That is, the data are stored at a location in memory, and it is accessed to obtain the data for the instruction parameter. This is much like a variable in a higher-level language – the data are referenced by a name, but the name itself is not the value.

2.3.5.3 Indirect Memory Location An indirect memory location is like a direct memory location, except that the former does not store the data for the parameter, it references or "points" to the data. The memory location contains an address that then refers to a direct memory location. A pointer in the high-level language is the equivalent in that it references where the actual data are stored in memory and not, literally, the data.

2.3.5.4 Other Addressing Modes Most modern processors employ combinations of the three basic addressing modes to create additional addressing modes. For example, there is a computed offset mode that uses indirect memory locations. Another would be a predecrement of a memory location, subtracting one from the address where the data are stored. Different processors will expand upon these basic addressing modes, depending on how the processor is oriented to getting and storing the data.

One interesting outcome is that the resultant of an operational instruction cannot be immediate data; it must be a direct memory location, or indirect memory location. In 2-address instructions, the destination, or operandum resultant, must always be a direct or indirect memory location, just as an L-value in a higher-level language cannot be a literal or named constant.

2.3.6 RISC versus CISC

Complex instruction set computers (CISC) supply relatively sophisticated functions as part of the instruction set. This gives the programmer a variety of powerful instructions with which to build applications programs and even more powerful software tools, such as assemblers and compilers. In this way, CISC processors seek to reduce the programmer's coding responsibility, increase execution speeds, and minimize memory usage.

The CISC is based on the following eight principles:

1. Complex instructions take many different cycles.
2. Any instruction can reference memory.
3. No instructions are pipelined.
4. A microprogram is executed for each native instruction.
5. Instructions are of variable format.
6. There are multiple instructions and addressing modes.
7. There is a single set of registers.
8. Complexity is in the microprogram and hardware.

In addition, program memory savings are realized because implementing complex instructions in high-order language requires many words of main memory. Finally, functions written in microcode always execute faster than those coded in the high-order language.

In a reduced instruction set computer (RISC) each instruction takes only one machine cycle. Classically, RISCs employ little or no microcode. This means that the instruction-decode procedure can be implemented as a fast combinational circuit, rather than a complicated microprogram scheme. In addition, reduced chip complexity allows for more on-chip storage (i.e., general-purpose registers). Effective use of register direct instructions can decrease unwanted memory fetch time

The RISC criteria are a complementary set of eight principles to CISC. These are:

1. Simple instructions taking one clock cycle.
2. LOAD/STORE architecture to reference memory.
3. Highly pipelined design.

4. Instructions executed directly by hardware.

5. Fixed-format instructions.

6. Few instructions and addressing modes.

7. Large multiple-register sets.

8. Complexity handled by the compiler and software.

A RISC processor can be viewed simply as a machine with a small number of vertical microinstructions, in which programs are directly executed in the hardware. Without any microcode interpreter, the instruction operations can be completed in a single microinstruction.

RISC has fewer instructions; hence, more complicated instructions are implemented by composing a sequence of simple instructions. When this is a frequently used instruction, the compiler's code generator can use a template of the instruction sequence of simpler instructions to emit code as if it were that complex instruction.

RISC needs more memory for the sequences of instructions that form a complex instruction. CISC uses more processor cycles to execute the microinstructions used to implement the complex macroinstruction within the processor instruction set.

RISCs have a major advantage in real-time systems in that, in theory, the average instruction execution time is shorter than for CISCs. The reduced instruction execution time leads to shorter interrupt latency and thus shorter response times. Moreover, RISC instruction sets tend to allow compilers to generate faster code. Because the instruction set is limited, the number of special cases that the compiler must consider is reduced, thus permitting a larger number of optimization approaches.

On the downside, RISC processors are usually associated with caches and elaborate multistage pipelines. Generally, these architectural enhancements greatly improve the average case performance of the processor by reducing the memory access times for frequently accessed instructions and data. However, in the worst case, response times are increased because low cache hit ratios and frequent pipeline flushing can degrade performance. But in many real-time systems, worst-case performance is typically based on very unusual, even pathological, conditions. Thus, greatly improving average-case performance at the expense of degraded worst-case performance is usually acceptable.

2.4 MEMORY

An understanding of certain characteristics of memory technologies is important when designing real-time systems. The most important of these characteristics is access time, which is the interval between when a datum is requested from a memory cell and when it is available to the CPU. Memory access times can have a profound effect on real-time performance and should influence the choice of instruction modes used, both when coding in assembly language and in the use of high-order language idioms.

The effective access time depends on the memory type and technology, the memory layout, and other factors; its method of determination is complicated and beyond the scope of this book. Other important memory considerations are power requirements, density (bits per unit area), and cost.

2.4.1 Memory Access

The typical microprocessor bus read cycle embodies the handshaking between the processor and the main memory store. The time to complete the handshaking is entirely dependent on the electrical characteristics of the memory device and the bus (Figure 2.8). Assume the transfer is from the CPU to main memory. The CPU places the appropriate address information on the address bus and allows the signal to settle. It then places the appropriate data onto the data bus. The CPU asserts the DST[6] signal to indicate to the memory device that the address lines have been set to the address and the data lines to the data to be accessed. Another signal (not shown) is used to indicate to the memory device whether the transfer is to be a load (from) or store (to) transfer. The reverse transfer from memory to the CPU is enacted in exactly the same way.

2.4.2 Memory Technologies

Memory can be volatile (the contents will be lost if power is removed) or non-volatile (the contents are preserved upon removing power). In addition there

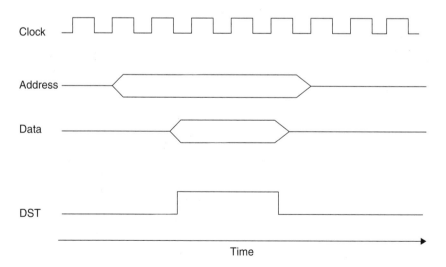

Figure 2.8 Illustration of the clock-synchronized memory-transfer process between a device and the CPU. The symbols "<>" shown in the data and address signals indicates that multiple lines are involved during this period in the transfer.

[6] The symbol names here are typical and will vary significantly from one system to another.

is RAM which is both readable and writeable, and ROM. Within these two groups are many different classes of memories. Only the more important ones will be discussed.

RAM memories may be either dynamic or static, and are denoted DRAM and SRAM, respectively. DRAM uses a capacitive charge to store logic 1s and 0s, and must be refreshed periodically due to capacitive discharge. SRAMs do not suffer from discharge problems and therefore do not need to be refreshed. SRAMs are typically faster and require less power than DRAMs, but are more expensive.

2.4.2.1 *Ferrite Core*
More for historical interest than a practical matter, consider ferrite core, a type of nonvolatile static RAM that replaced memories based on vacuum tubes in the early 1950s. Core memory consists of a doughnut-shaped magnet through which a thin drive line passes.

In a core-memory cell, the direction of flow of current through the drive lines establishes either a clockwise or counterclockwise magnetic flux through the doughnut that corresponds to either logic 1 or logic 0. A sense line is used to "read' the memory (Figure 2.9). When a current is passed through the drive line, a pulse is generated (or not) in the sense line, depending on the orientation of the magnetic flux.

Core memories are slow (10-microsecond access), bulky, and consume lots of power. Although they have been introduced here for historical interest, they do have one practical advantage – they cannot be upset by electrostatic discharge or by a charged particle in space. This consideration is important in the reliability of space-borne and military real-time systems. In addition, the new ferroelectric memories are descendents of this type of technology.

2.4.2.2 *Semiconductor Memory*
RAM devices can be constructed from semiconductor materials in a variety of ways. The basic one-bit cells are then configured in an array to form the memory store. Both static and dynamic RAM can be constructed from several types of semiconductor materials and designs.

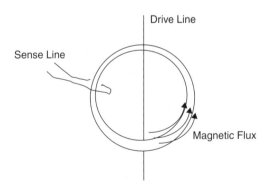

Figure 2.9 A core-memory element. The figure is approximately 15 times larger than actual size.

Static memories rely on bipolar logic to represent ones and zeros. Dynamic RAMs rely on capacitive charges, which need to be refreshed regularly due to charge leakage. Typically, dynamic memories require less power and are denser than static ones; however, they are much slower because of the need to refresh them. A SRAM with a battery back up is referred to as an NVRAM (nonvolatile RAM).

The required refresh of the dynamic RAM is accomplished by accessing each row in memory by setting the row address strobe (RAS) signal without the need to activate the column address strobe (CAS) signals. The RAM refresh can occur at a regular rate (e.g., 4 milliseconds) or in one burst.

A significant amount of bus activity can be held off during the dynamic refresh, and this must be taken into account when calculating instruction execution time (and hence system performance). When a memory access must wait for a DRAM refresh to be completed, cycle stealing occurs, that is, the CPU is stalled until the memory cycle completes. If burst mode is used to refresh the DRAM, then the timing of critical regions may be adversely affected when the entire memory is refreshed simultaneously.

Depending on the materials used and the configuration, access times of 15 nanoseconds or better can be obtained for static semiconductor RAM.

2.4.2.3 Fusible Link Fusible-link ROMs are a type of nonvolatile memory. These one-time programmable memories consist of an array of paths to ground or "fusible links." During programming these fuses are shorted or fused to represent either 1s or 0s, thus embedding the program into the memory. Just as fusible-link memories cannot be reprogrammed, they cannot be accidentally altered. They are fast and can achieve access time of around 50 nanoseconds, though they are not very dense.

Fusible-link ROM is used to store program instructions and data that are not to be altered and that require a level of immutability, such as in hardened military applications.

2.4.2.4 Ultraviolet ROM Ultraviolet ROM (UVROM) is a type of nonvolatile programmable ROM (PROM), with the special feature that it can be reprogrammed a limited number of times. For reprogramming, the memory is first erased by exposing the chip to high-intensity ultraviolet light. This reprogrammability, however render UVROMS susceptible to upset.

UVROM is typically used for the storage of program and fixed constants. UVROMs have access times similar to those of fusible-link PROMs.

2.4.2.5 Electronically Erasable PROM Electronically erasable PROM (EEPROM) is another type of PROM with the special feature that it can be reprogrammed *in situ*, without the need for a special programming device (as in UVROM or fusible-link PROM). These memories are erased by toggling signals on the chip, which can be accomplished under program control.

EEPROMs are used for long-term storage of variable information. For example, in embedded applications, "black-box" recorder information from diagnostic tests might be written to EEPROM for postmission analysis.

These memories are slower than other types of PROMs (50–200 nanosecond access times), limited rewrite cycles (e.g., 10,000), and have higher power requirements (e.g., 12 volts).

2.4.2.6 Flash Memory

Flash memory is another type of rewritable PROM that uses a single transistor per bit, whereas EEPROM uses two transistors per bit. Hence, flash memory is more cost effective and denser then EEPROM. Read times for flash memory are fast, 20 to 30 nanoseconds, but write speeds are quite slow – up to 1 microsecond. Another disadvantage of flash memory is that it can be written to and erased about 100,000 times, whereas EEPROM is approximately 1 million. Another disadvantage is that flash memory requires rather high voltages: 12 V to write; 2 V to read. Finally, flash memory can only be written to in blocks of size 8–128 kilobytes at a time.

This technology is finding its way into commercial electronics applications, but it is expected to appear increasingly in embedded real-time applications.

2.4.2.7 Ferroelectric Random-Access Memory

An emerging technology, ferroelectric RAM relies on a capacitor employing a special insulating material. Data are represented by the orientation of the ferroelectric domains in the insulting material, much like the old ferrite-core memories. This similarity also extends to relative immunity to upset. Currently, ferroelectric RAM is available in arrays of up to 64 megabytes with read/write 40 nanosecond access time and 1.5/1.5 read/write voltage

2.4.3 Memory Hierarchy

Primary and secondary memory storage forms a hierarchy involving access time, storage density, cost, and other factors. Clearly, the fastest possible memory is desired in real-time systems, but cost control generally dictates that the fastest affordable technology is used as required. In order of fastest to slowest, and considering cost, memory should be assigned as follows:

1. Internal CPU memory
2. Registers
3. Cache
4. Main memory
5. Memory on board external devices

Selection of the appropriate technology is a systems design issue. Table 2.1 summarizes the previously discussed memory technologies and some appropriate associations with the memory hierarchy.

Note that these numbers vary widely depending on many factors, such as manufacturer, model and cost, and change frequently. These figures are given for relative comparison purposes only.

Table 2.1 A summary of memory technologies

Memory Type	Typical Access Time	Density	Typical Applications
DRAM	50–100 ns	64 Mbytes	Main memory
SRAM	10 ns	1 Mbyte	μmemory, cache, fast RAM
UVROM	50 ns	32 Mbytes	Code and data storage
Fusible-link PROM	50 ns	32 Mbytes	Code and data storage
EEPROM	50–200 ns	1 Mbyte	Persistent storage of variable data
Flash	20–30 ns (read) 1 μs (write)	64 Mbytes	Code and data storage
Ferroelectric RAM	40 ns	64 Mbytes	Various
Ferrite core	10 ms	2 kbytes or less	None, possibly ultrahardened nonvolatile memory

2.4.4 Memory Organization

To the real-time systems engineer, particularly when writing code, the kind of memory and layout is of particular interest. Consider, for example, an embedded processor that supports a 32-bit address memory organized, as shown in Figure 2.10. Of course, the starting and ending addresses are entirely imaginary, but could be representative of a particular embedded system. For example, such a map might be consistent with the memory organization of the inertial measurement system.

The executable program resides in memory addresses 00000000 through E0000000 hexadecimal in some sort of programmable-only ROM, such as fusible link. It is useful to have the program in immutable memory so that an accidental write to this region will not catastrophically alter the program. Other data, possibly related to factory settings and tuned system parameters, are stored at locations E000001 through E0000F00 in EPROM, which can be rewritten only when the system is not in operation. Locations E0000F01 through FFC00000 are RAM memory used for the run-time stack, memory heap, and any other transient data storage. Addresses FFC00001 through FFFFE00 are fixed system parameters that might need to be rewritten under program control, for example, calibration constants determined during some kind of diagnostic or initialization mode. During

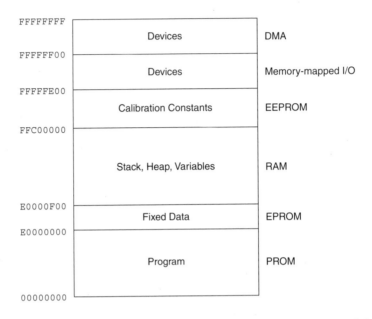

Figure 2.10 Typical memory map showing designated regions. (Not to scale.).

run time, diagnostic information or black box data might be stored here. These data are written to the nonvolatile memory rather than to RAM so that they are available after the system is shut down (or fails) for analysis. Finally, locations FFFFE00 through FFFFFFFF contain addresses associated with devices that are accessed either through DMA or memory-mapped I/O.

2.5 INPUT/OUTPUT

In real-time systems the input devices are sensors, transducers, steering mechanisms, and so forth. Output devices are typically actuators, switches, and display devices.

Input and output are accomplished through one of three different methods: programmed I/O, memory-mapped I/O, or direct memory address (DMA). Each method has advantages and disadvantages with respect to real-time performance, cost, and ease of implementation.

2.5.1 Programmed Input/Output

In programmed I/O, special data-movement instructions are used to transfer data to and from the CPU. An IN instruction will transfer data from a specified I/O device into a specified CPU register. An OUT instruction will output from a register to some I/O device. Normally, the identity of the operative CPU register is embedded in the instruction code. Both the IN and OUT instructions require the efforts of the CPU, and thus cost time that could impact real-time performance.

For example, a computer system is used to control the speed of a motor. An output port is connected to the motor, and a signed integer is written to the port to set the motor speed. The computer is configured so that when an OUT instruction is executed, the contents of register 1 are placed on the data bus and sent to the I/O port at the address contained in register 2. The following code fragment allows the program to set the motor speed.[7]

```
LOAD R1 &speed          ;motor speed into register 1
LOAD R2 &motoraddress   ;address of motor control into register 2
OUT                     ;output from register 1 to the memory-mapped I/O
                        ;port address contained in register 2
```

2.5.2 Direct Memory Access

In DMA, access to the computer's memory is given to other devices in the system without CPU intervention. That is, information is deposited directly into main memory by the external device. Here a DMA controller is required (Figure 2.11) unless the DMA circuitry is integrated into the CPU. Because CPU participation is not required, data transfer is fast.

The DMA controller prevents collisions by requiring each device to issue a DMA request signal (DMARQ) that will be acknowledged with a DMA acknowledge signal (DMACK). Until the DMACK signal is given to the requesting device, its connection to the main bus remains in a tristate condition. Any device that is tristated cannot affect the data on the memory data lines. Once the DMACK

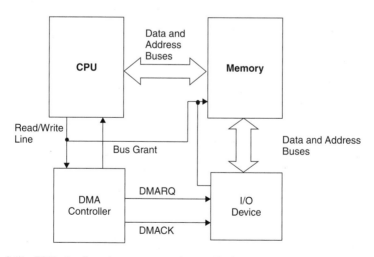

Figure 2.11 DMA circuitry where an external controller is used. This functionality can also be integrated on-chip with the CPU.

---

[7] So, for example, "R1" is register number 1.

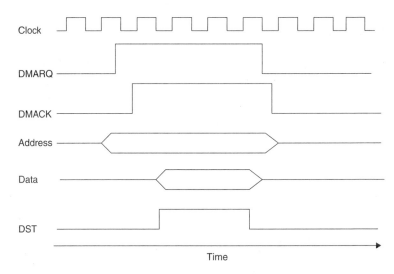

Figure 2.12 The DMA timing process. The sequence is: request transfer (DMARQ high), receive acknowledgment (DMACK high), place data on bus, and indicate data are present on bus (DST high). The signal height indicates voltage high/low.

is given to the requesting device, its memory bus lines become active, and data transfer occurs, as with the CPU (Figure 2.12).

The CPU is prevented from performing a data transfer during DMA through the use of a signal called a bus grant. Until the bus grant signal is given by the controller, no other device can obtain the bus. The DMA controller is responsible for assuring that only one device can place data on the bus at any one time through bus arbitration. If two or more devices attempt to gain control of the bus simultaneously, bus contention occurs. When a device already has control of the bus and another obtains access, an undesirable occurrence (a collision) occurs.

The device requests control of the bus by signaling the controller via the DMARQ signal. Once the DMACK signal is asserted by the controller, the device can place (or access) data to/from the bus (which is indicated by another signal, typically denoted DST).

Without the bus grant (DMACK) from the DMA controller, the normal CPU data-transfer processes cannot proceed. At this point, the CPU can proceed with non-bus-related activities (e.g., the execution phase of an arithmetic instruction) until it receives the bus grant, or until it gives up (after some predetermined time) and issues a bus time-out signal. Because of its speed, DMA is often the best method for input and output for real-time systems.

2.5.3 Memory-Mapped Input/Output

Memory-mapped I/O provides a data-transfer mechanism that is convenient because it does not require the use of special CPU I/O instructions. In memory-mapped I/O certain designated locations of memory appear as virtual I/O ports

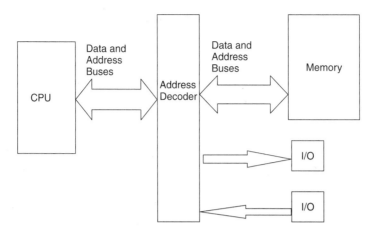

Figure 2.13 Memory-mapped I/O circuitry.

(Figure 2.13). For example, consider the control of the speed of a stepping motor. If it were to be implemented via memory-mapped I/O, the required assembly language code might look like the following:

```
LOAD    R1  &speed          ;motor speed into register 1
STORE   R1  &motoraddress   ;store to address of motor control
```

where `speed` is a bit-mapped variable and `motoraddress` is a memory-mapped location.

In many computer systems, the video display is updated via memory-mapped I/O. For example, suppose that a display consists of a 24 row by 80 column array (a total of 1920 cells). Each screen cell is associated with a specific location in memory. To update the screen, characters are stored on the address assigned to that cell on the screen.

Input from an appropriate memory-mapped location involves executing a `LOAD` instruction on a pseudomemory location connected to an input device.

2.5.3.1 Bit Maps A bit map describes a view of a set of devices that are accessed by a single (discrete) signal and organized into a word of memory for convenient access either by DMA or memory-mapped addressing. Figure 2.14

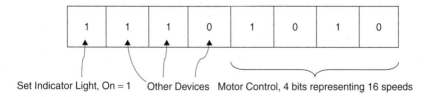

Set Indicator Light, On = 1   Other Devices   Motor Control, 4 bits representing 16 speeds

Figure 2.14 Bit map showing mappings between specific bits and the respective devices in a memory-mapped word.

illustrates a typical bit map for a set of output devices. Each bit in the bit map is associated with a particular device. For example, in the figure the high-order bit is associated with a display light. When the bit is set to one, it indicates that the indicator light is on. The low-order four bits indicate the settings for a 16-speed stepping motor. Other devices are associated with the remaining bits.

Bit maps can represent either output states, that is, the desired state of the device, or an indication of the current state of the device in questions, that is, it is an input or an output.

2.5.4 Interrupts

An interrupt is a hardware signal that initiates an event. Interrupts can be initiated by external devices, or internally if the CPU is has this capability. External interrupts are caused by other devices (e.g., clocks and switches), and in most operating systems such interrupts are required for scheduling. Internal interrupts, or traps, are generated by execution exceptions, such as a divide-by-zero. Traps do not use external hardware signals; rather, the exceptional conditions are dealt with through branching in the microcode. Some CPUs can generate true external interrupts, however.

2.5.4.1 Instruction Support for Interrupts   Processors provide two instructions, one to enable or turn on interrupts EPI, and another to disable or turn them off (DPI). These are atomic instructions that are used for many purposes, including buffering, within interrupt handlers, and during parameter passing.

2.5.4.2 Internal CPU Handling of Interrupts   Upon receipt of the interrupt signal, the processor completes the instruction that is currently being executed. Next, the contents of the program counter are saved to a designated memory location called the interrupt return location. In many cases, the CPU "flag" or condition status register (SR) is also saved so that any information about the previous instruction (for example, a test instruction whose result would indicate that a branch is required) is also saved. The contents of a memory location called the interrupt-handler location are loaded into the program counter. Execution then proceeds with the special code stored at this location, called the interrupt handler. This process is outlined in Figure 2.15.

Processors that are used in embedded systems are equipped with circuitry that enables them to handle more than one interrupt in a prioritized fashion. The overall scheme is depicted in Figure 2.16.

Upon receipt of interrupt i, the circuitry determines whether the interrupt is allowable given the current status and mask register contents. If the interrupt is allowed, the CPU completes the current instruction and then saves the program counter in interrupt-return location i. The program counter is then loaded with the contents of interrupt-handler location i. In some architectures, however, the return address is saved in the system stack, which allows for easy return from a sequence of interrupts by popping the stack. In any case, the code at the address there is used to service the interrupt.

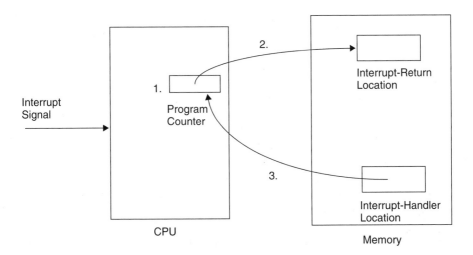

Figure 2.15 Sketch of the interrupt-handling process in a single-interrupt system. Step 1: finish the currently executing macroinstruction. Step 2: save the contents of the program counter to the interrupt-return location. Step 3: load the address held in the interrupt-handler location into the program counter. Resume the fetch and execute sequence.

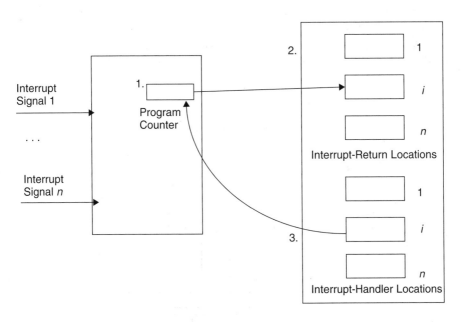

Figure 2.16 The interrupt-handling process in a multiple-interrupt system. Step 1: complete the currently executing instruction. Step 2: save the contents of the program counter to interrupt-return location i. Step 3: load the address held in interrupt-handler location i into the program counter. Resume the fetch–execute cycle.

To return from the interrupt, the saved contents of the program counter at the time of interruption are reloaded into the program counter and the usual fetch and execute sequence is resumed.

Interrupt-driven I/O is simply a variation of program I/O, memory-mapped I/O, or DMA, in which an interrupt is used to signal that an I/O transfer has completed or needs to be initiated via one of the three mechanisms.

2.5.4.3 *Programmable Interrupt Controller*   Not all CPUs have the built-in capability to prioritize and handle multiple interrupts. An external interrupt-controller device can be used to enable a CPU with a single-interrupt input to handle interrupts from several sources. These devices have the ability to prioritize and mask interrupts of different priority levels. The circuitry on board these devices is quite similar to that used by processors that can handle multiple interrupts (Figure 2.17).

This additional hardware includes special registers, such as the interrupt vector, status register, and mask register. The interrupt vector contains the identity of the highest-priority interrupt request; the status register contains the value of the lowest interrupt that will currently be honored; and the mask register contains a bit map that either enables or disables specific interrupts. Another specialized register is the interrupt register, which contains a bit map of all pending (latched) interrupts. Programmable interrupt controllers (PICs) can support a large number of devices. For example, the Intel 82093AA I/O Advanced Programmable Interrupt Controller supports 24 programmable interrupts. Each can be independently set to be edge or level triggered, depending on the needs of the attached device.

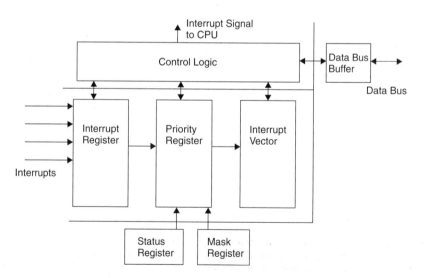

Figure 2.17 A programmable interrupt controller (PIC). The registers-interrupt, priority, vector, status, and mask-serve the same functions previously described for the interrupt control circuitry on board a similarly equipped CPU.

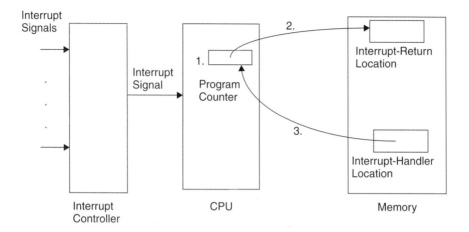

Figure 2.18 Handling multiple interrupts with an external interrupt controller. Step 1: finish the currently executing instruction. Step 2: save the contents of the program counter into the interrupt-return location. Step 3: load the address held in the interrupt-handler location into the program counter. Resume the fetch and execute cycle. The interrupt-handler routine will interrogate the PIC and take the appropriate action.

When configured as in Figure 2.18, a single-interrupt CPU in conjunction with an interrupt controller can handle multiple interrupts.

The following scenario illustrates the complexity of writing interrupt-handler software, and points out a subtle problem that can arise.

An interrupt handler executes upon receipt of a certain interrupt signal that is level triggered. The first instruction of the routine is to clear the interrupt by strobing bit 1 of the interrupt clear signal. Here, intclr is a memory-mapped location whose least significant bit is connected with the clear interrupt signal. Successively storing 0, 1, and 0 serves to strobe the bit.

Although the interrupt controller automatically disables other interrupts on receipt of an interrupt, the code immediately reenables them to detect spurious ones. The following code fragment illustrates this process for a 2-address architecture pseudoassembly code:

```
LOAD  R1,0          ;load register 1 with the constant value 0
LOAD  R2,1          ;load register 2 with the constant value 1
STORE R1, &intclr   ;set clear interrupt signal low
STORE R2, &intclr   ;set clear interrupt signal high
STORE R1, &intclr   ;set clear interrupt signal low
EPI                 ;enable interrupt
```

The timing sequence is illustrated in Figure 2.19.

Note, however, that a problem could occur if the interrupt is cleared too quickly. Suppose that the clear, LOAD, and STORE instructions take 0.75 microsecond, but the interrupt pulse is 4 microseconds long. If the clear interrupt instruction is executed immediately upon receipt of the interrupt, a total of

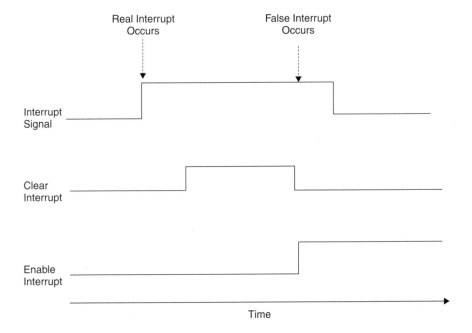

Figure 2.19 Timing sequence for interrupt clearing that could lead to a problem.

3 microseconds will elapse. Since the interrupt signal is still present, when interrupts are enabled, a spurious interrupt will be caused. This problem is insidious, because most of the time software and hard delays hold off the interrupt-handler routine until long after the interrupt signal has latched and gone away. It often manifests itself when the CPU has been replaced by a faster one.

2.5.4.4 *Interfacing Devices to the CPU via Interrupts* Most processors have at least one pin designated as an interrupt input pin, and many peripheral-device controller chips have a pin designated as an interrupt output pin. The interrupt request line (IRL) from the peripheral controller chip connects to an interrupt input pin on the CPU (Figure 2.20).

When the controller needs servicing from the CPU, the controller sends a signal down the IRL. In response, the CPU begins executing the interrupt service routine associated with the device in the manner previously described. When the CPU reads data from (or writes data to) the peripheral controller chip, the CPU first places the controller's address on the address bus. The decode logic interprets that address and enables I/O to the controller through the device-select line.

Suppose now that the system is equipped with a PIC chip that can handle multiple peripheral controllers and can support 8 or 16 peripheral devices. The interrupt request lines from the peripheral controllers connect to the interrupt controller chip. Figure 2.21 depicts a hardware arrangement to handle multiple peripheral devices.

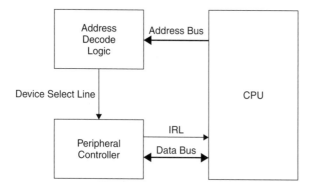

Figure 2.20 A single peripheral controller. IRL is the interrupt request line.

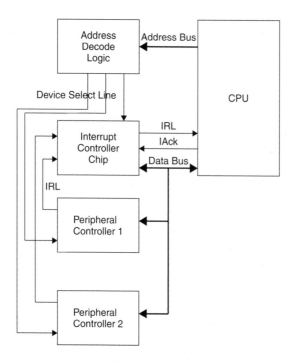

Figure 2.21 Several peripheral controllers connected to the CPU via a PIC. Notice that the devices share the common data bus, which is facilitated by tristating, nonactive devices via the device-select lines.

The interrupt controller chip demultiplexes by combining two or more IRLs into one IRL that connects to the CPU. Interrupt controllers can be cascaded in master–slave fashion. When an interrupt arrives at one of the slave interrupt controllers, the slave interrupts the master controller, which in turn interrupts the CPU. In this way, the interrupt hardware can be extended.

How does a system respond when more than one device generates an interrupt at the same time? Essentially, each hardware interrupt is assigned a unique priority. For systems that use an interrupt controller, whether the controller is on-chip or external, the priorities are programmed into the controller by software, usually when the system is initialized (though there may be times where dynamic assignment of priorities is desirable). So if two or more interrupts happen simultaneously, one of them will have the highest priority. In systems that support multiple interrupts, the interrupt controller keeps track of pending interrupts, and passes them over to the CPU in order of their priority. In most systems, the interrupt controller responds to a given interrupt by setting a bit in the interrupt vector to indicate that the interrupt is being serviced. Then at the end of processing the interrupt, the interrupt service routine (ISR) executes an instruction that informs the interrupt controller that the ISR has completed. The interrupt controller then clears the appropriate bit in the interrupt vector.

When the CPU acknowledges the interrupt request, the CPU interrogates the interrupt controller and reads the interrupt vector. The CPU uses this interrupt number to vector to the correct ISR related to the device that initiated the interrupt.

2.5.4.5 Interruptible Instructions   In rare instances certain macroinstruction may need to be interruptible. This might be the case where the instruction takes a great deal of time to complete. For example, consider a memory-to-memory instruction that moves large amounts of data. In most cases, such an instruction should be interruptible between blocks to reduce interrupt latency. However, interrupting this particular instruction could cause data integrity problems. Ultimately, it is rare that an architecture will support interruptible instructions because of precisely this kind of problem that can be averted.

2.5.4.6 Watchdog Timers   In many computer systems, the CPU or other devices are equipped with a counting register that is incremented periodically. The register must be cleared by appropriate code using memory-mapped I/O before the register overflows and generates an interrupt. This type of hardware is called a watchdog timer (WDT) (Figure 2.22).

Watchdog timers are used to ensure that certain devices are serviced at regular intervals, that certain processes execute according to their prescribed rate, and that

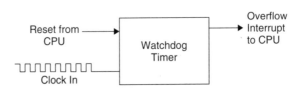

Figure 2.22 A watchdog timer. Software issues a reset signal via memory-mapped or programmed I/O to reset the timer before it can overflow, issuing a watchdog timer interrupt.

the CPU continues to function. Clearing the watchdog timer is often humorously referred to as "petting the dog."

2.6 ENHANCING PERFORMANCE

The limitations of the single fetch-and-execute cycle of the von Neumann architecture have caused various architectural enhancements to evolve. Most of these architectural enhancements benefit from a high locality of reference.

Two architectural enhancements that can improve average case performance in real-time systems are caches and pipelines. Both of these enhancements illustrate the fact that when the locality of reference is high, performance is improved. But when locality of reference is low, which often occurs in worst-case and exceptional conditions, performance is actually deteriorated.

2.6.1 Locality of Reference

Locality of reference refers to the relative "distance" in memory between consecutive code or data accesses. If the data or code fetched tend to reside relatively close in memory, then the locality of reference is high. Conversely, when programs execute instructions that are relatively widely scattered (as in "spaghetti" code), locality of reference is low, and adverse effects will be seen.

Well-written programs in procedural languages tend to execute sequentially within code modules and within the body of loops, and hence have a high locality of reference. While this is not necessarily true for object-oriented code (which tends to execute in a much more nonlinear fashion), portions of such code can be linearized. For example, arrays tend to be stored in blocks in sequence, with elements often accessed sequentially. When software is executed in a linear sequential fashion, instructions are in sequence and therefore are stored in nearby memory locations, thus yielding a high locality of reference.

2.6.2 Cache

A cache is a small store of fast memory where frequently used instructions and data are kept. The cache also contains a list of memory blocks that are currently in the cache. The list is often in the cache itself. Each block can hold a small number of instructions or data (e.g., 1 K).

The basic operation of the cache is as follows. Suppose the CPU requests contents of memory location. First, it checks the address tags to see if data are in the cache. If present, the data are retrieved from the cache, which is significantly faster than a fetch from main memory. However, if the needed data or instruction are not already in the cache one cache block must be written back and the required new block read from main memory to the cache. The needed information is then delivered from the cache to CPU and the address tags adjusted.

Cache design considerations include: cache size, mapping function, block replacement algorithm (e.g., first-in-first out; least recently used), write policy (e.g., should any altered data be written through or wait for cache replacement), block size, and number of caches (e.g., there can be separate data and instruction cache).

What performance benefit does the cache give? Consider a simple system with a single cache. Assume a noncached memory reference costs 100 ns, whereas an access from the cache only takes 30 ns. Now assume that the cache hit ratio is 60%. Then the apparent (average) access time would be

$$0.6 \cdot 30 \text{ ns} + 0.4 \cdot 100 \text{ ns} = 58 \text{ ns}$$

Because access time for the cache is faster than main memory, performance benefits are a function of the cache hit radio, that is, the percentage of time that the needed instruction or data is found in the cache. This is due to the fact that if needed data or instructions are not found in the cache, then the cache contents need to be written back (if any were altered) and overwritten by a memory block containing the needed information. This overhead can become significant when the hit ratio is low. Therefore a low hit ratio can degrade performance. Hence, if the locality of reference is low, a low number of cache hits would be expected, degrading performance.

Another drawback of using a cache is that its performance is nondeterministic; it is impossible to know a priori what the cache contents and hence the overall access time will be. Thus system performance in a system using a cache has an element of nondeterminism in that a worse-case access time (every access causes a cache replacement) theoretically should be used in any analysis.

2.6.3 Pipelining

Pipelining imparts an implicit execution parallelism in the different cycles of processing an instruction. Suppose execution of an instruction consists of the following stages:

1. Fetch – Get the instruction from memory.
2. Decode – Determine what the instruction is.
3. Execute – Perform the instruction decode.
4. Write – Store the results to memory.

In nonpipelined execution (scalar execution), one instruction is processed through a cycle at a time. With pipelining, more instructions can be processed in different cycles simultaneously, improving processor performance.

For example, consider Figure 2.23. The first picture shows the sequential execution of the fetch, decode, execute, and write components of three instructions. The sequence requires twelve clock cycles. Beneath the sequence is shown another set of the same three instructions, plus eight more instructions, with overlapping execution of the components. The first three instructions are completed

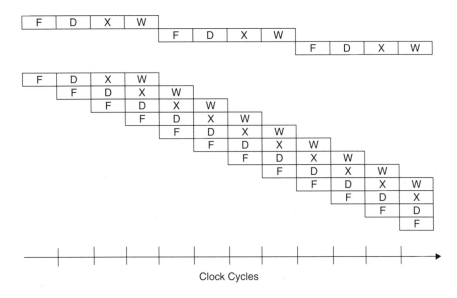

Figure 2.23 Sequential instruction execution versus pipelined instruction execution. Nine complete instructions can be completed in the pipelined approach in the same time it takes to complete three instructions in the sequential (scalar) approach [Gilreath03].

in only six clock cycles, and most of the remaining instructions are completed within the twelve clock cycles.

Pipelining is a form of speculative execution in that the instructions that are prefetched are taken to be the next sequential instructions. If any of the instructions in the pipeline are a branch instruction, the prefetched instructions further in the pipeline are no longer valid.

Higher-level pipelines, or superpipelining, can be achieved if the fetch-and-execute cycle can be decomposed further. For example, a six-stage pipeline can be achieved, consisting of a fetch stage, two decode stages (needed to support indirect addressing modes), an execute stage, a write-back stage (which finds completed operations in the buffer and frees corresponding functional units), and a commit stage (in which the validated results are written back to memory).

Another approach to pipelining is to use redundant hardware to replicate one or more stages in the pipeline. Such a design is called a superscalar architecture. Furthermore, superscalar and superpipelined architectures can be combined to obtain a superscalar, superpipelined computer.

When a branch instruction is executed, the pipeline registers and flags must all be reset, that is, the pipeline is flushed, which takes time. Data and input dependencies can also slow pipeline flowthrough. Pipelining will improve overall performance when locality of reference is high. Otherwise, it may degrade performance.

The real-time systems engineer must realize that before an interrupt can be handled, the oldest instruction in the pipeline must be completed and then the others either saved somehow or flushed (the preferred technique). Saving the intermediate steps of the other instructions requires a sophisticated processor and increases system response time.

2.6.4 Coprocessors

Many embedded systems incorporate a second specialized CPU, or coprocessor, to perform special instructions that are not part of the base instruction set. Coprocessors improve real-time performance because they extend the instruction set to support specialized instructions faster. They do not improve performance because of any inherent parallelism.

For example, in imaging and signal-processing applications a digital signal-processing (DSP) coprocessor might be used to provide for native instructions to perform convolution, fast Fourier transforms (FFT), and so forth.

The main processor loads certain registers with data for the coprocessor, issues an interrupt to the coprocessor, then halts itself. When the coprocessor finishes, it awakens the main processor via an interrupt, and then halts itself.

The coprocessor and its resources are a critical resource and need to be protected. For example, registers belonging to the coprocessor should be saved as part of the context-switching process and a separate stack must be kept.

2.7 OTHER SPECIAL DEVICES

A class of logic devices, called programmable logic devices (PLDs), are used to provide certain digital logic to the embedded system. Typical programmable devices include programmable array logics (PALs) and programmable logic arrays (PLAs) for relatively simple needs. Programmable logic is widely used in embedded real-time systems for creating fast, hard-wired logic for smart devices.

For more complex functionality, custom IC chips or applications-specific integrated circuits (ASIC) can be used. ASICs typically consist of a core processor of some kind and some simple, built-in peripheral support. Field-programmable gate arrays (FPGA) provide more sophisticated functionality, like the ASIC, but they have the added advantage of being reprogrammable in place. In all cases the real-time systems engineer should be familiar with the operation of these devices and their role in the embedded system.

2.7.1 Applications-Specific Integrated Circuits

The ASIC is a special-purpose integrated circuit designed for one application only. In essence, these devices are systems on a chip that can include a microprocessor, memory, I/O devices, and other specialized circuitry. ASICs are used in many embedded applications, including image processing, avionics systems, and medical systems, and the real-time design issues are the same for them as they are for most other systems.

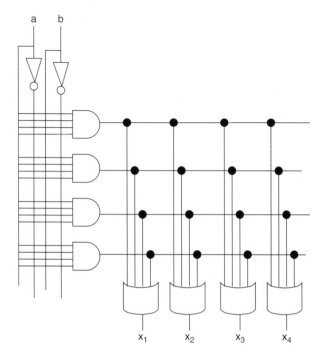

Figure 2.24 A typical PAL. Here only the external junctions are programmable. The internal junctions, marked with a dot, are fixed connections.

2.7.2 Programmable Array Logic/Programmable Logic Array

The PAL consists of a programmable AND array followed by a fixed-number of input OR elements. Each OR element has a certain number of dedicated product terms, and sharing of these product terms is not allowed (Figure 2.24).

The PLA array differs from the PAL in that the AND array is followed by a programmable-width OR array (Figure 2.25). Having a programmable OR array allows the product terms to be shared between macrocells, which effectively increases device density. Compared to the PAL, the PLA is much more flexible and yields more efficient logic, but is more expensive. On the other hand, the PAL is faster (because it uses fewer fuses) and is less expensive than the PLA.

2.7.3 Field-Programmable Gate Arrays

The FPGA technology provides for the construction of a system on a chip with an integrated processor, memory, and I/O. The FPGA differs from the ASIC in that it is reprogrammable, even while embedded in the system.

A reconfigurable architecture allows for the programmed interconnection and functionality of a conventional processor, with the added advantage that it can

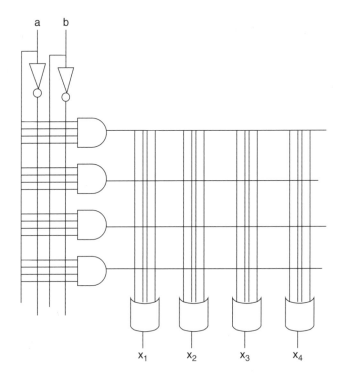

Figure 2.25 A typical PLA. Here all junctions are programmable. That is, they can be selectively fused to form the necessary product terms.

be tailored to the types of applications involved. Algorithms and functionality are moved from the software side into the hardware side.

In an FPGA, the programmable logic consists of a series of logic blocks (LBs) connected in either segmented or continuous interconnect structures (Figure 2.26). Segmented interconnections are used for short signals between adjacent configurable LBs (CLB), while continuous interconnections are used for bus-structured architectures [Xilinx98].

Each logic block uses one or more look-up tables (LUT) and several bits of memory. The contents of the LUTs are reprogrammable by changing their contents. I/O blocks are usually connected to local memories. This design allows for maximum flexibility, and FPGAs can be used to implement circuit diagrams into the logic blocks (Figure 2.27). The logic blocks can be configured to exploit parallelism.

The continuous structure is ideal for connecting large numbers of simple logic units, such as half adders, full adders, and twos-complement inverters. Moreover, these logic blocks can also be predesigned to implement higher-level functions, such as vector addition, convolution, or even an FFT. The ability to reconfigure logic blocks gives the flexibility of selecting a single instruction and use it to implement the necessary functionality.

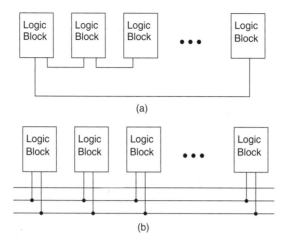

Figure 2.26 (a) Segmented and (b) continuous interconnection strategies for FPGA logic blocks.

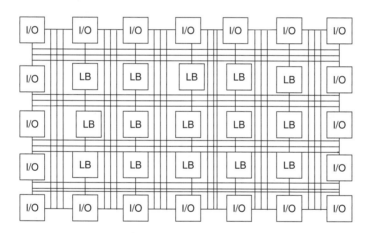

Figure 2.27 A conceptual illustration of an FPGA showing internal configurable logic blocks and periphery I/O elements [Gilreath03].

High-level language and sophisticated tool support are available for development using FPGAs. Because of standardization, implementations are portable across commercial FPGA types and further expand the set of available CLBs.

FPGAs have reached usable densities in the hundreds of thousands of gates and flip-flops that can be integrated to form system-level solutions. Clock structures can be driven using dedicated clocks that are provided within the system. FPGAs are infinitely reprogrammable (even within the system) and design modifications can be made quickly and easily [Xilinx98]. Hence, they are well adapted to many embedded real-time systems.

2.7.4 Transducers

Transducers are, generally, any device that converts energy from one form to another. In the case of the embedded system, transducers provide a usable output in response to a specified measured quantity. The input is an analog signal, which must be converted to digital form by another device (discussed shortly). The output is a scaled number or bit map that is made available to the CPU either via DMA or memory-mapped I/O.

Typical transducers in embedded systems include temperature sensors, accelerometers, gyroscopic devices, pressure and altitude sensors, mechanical and optical motion-sensing devices, and even chemical sensors. A few of these devices are discussed, particularly those that relate to the inertial measurement system already introduced.

2.7.4.1 *Temperature Sensors*   Temperature is an important control parameter of most embedded real-time systems. The most commonly used electrical temperature sensors are thermocouples, thermistors, and resistance thermometers.

Thermocouples take advantage of the junction effect, which is the voltage difference generated at a junction due to the difference in the energy distribution of two dissimilar metals. Resistance thermometers rely on the increase in resistance of a metal wire as temperature increases. Thermistors are resistive elements made of semiconductor materials that have changing resistance properties with temperature.

2.7.4.2 *Accelerometers*   Accelerometers use a simple transducing function to convert the compression or stretching of a spring or the deformation of a membrane into an electrical output. For example, one mechanism takes advantage of the fact that the capacitance associated with a capacitor is a function of the gap width, which changes according to the spring or membrane deformation.

Another kind of force sensor is a strain gauge, which takes advantage of the fact that as a wire is stretched or compressed, its resistance changes. Accelerometers can be constructed using a strain gauge.

Accelerometers can also be built to take advantage of the piezoelectric effect. The piezoelectric effect is the phenomenon that if a crystal is compressed in such a way that the lattice structure is disrupted, it causes the discharge of electrons. In this way, the compression of the device due to acceleration can be measured. Piezoelectric accelerometers are widely used where miniaturization is desirable.

2.7.4.3 *Gyroscopes*   Gyroscopes are used in the inertial navigation of aircraft, spacecraft, robots, and automotive applications. The gyroscope is based on the fact that a vertically oriented rotating mass will remain fixed with respect to two spatial directions, but not the third. Mechanical gyroscopes are not used to sense position; rather, they are used to maintain a platform in a fixed position with respect to space, that is, in an inertial reference frame (Figure 2.28).

The configuration is as follows. Two accelerometers are fixed to this platform in orthogonal directions, as shown in Figure 2.28. Because each gyroscope stays

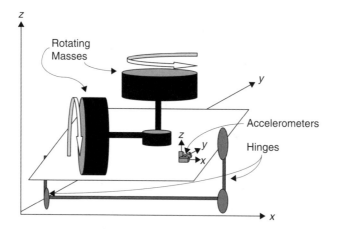

Figure 2.28 Stylized representation of two gyroscopes used to hold a hinged platform with three orthogonal accelerometers, fixed with respect to an inertial reference frame (not to scale).

fixed with respect to two directions, two gyroscopes aligned perpendicularly will stay fixed with respect to all three directions, with one redundant direction represented. Thus, accelerations in the x, y, and z directions can be accumulated over time to yield the velocity vector, and accumulating the velocity vector over time yields the position vector.

Of course, the platform must be held absolutely fixed with respect to the inertial frame, which is impossible because of gyro drift due to physical imperfections. Therefore the gyros need to be restored to their original positions by torquing them, that is, by pivoting them slightly on their axes. In addition, the accelerometers are imperfect and subject to temperature and other variations. Hence, the accelerometer readings are somewhat inaccurate, leading to less than perfect position estimates. For this reason, various compensations of the imperfections must be performed, and various filtering mechanisms used to improve the quality of the estimation in real-time.

Rotating mass gyroscopes do not need to be used to hold the platform stable. Instead, the accelerometers can be attached directly to the frame of the vehicle. But now the accelerometers do not stay fixed within the inertial reference frame of space. In these "strap-down systems" position resolvers are needed to measure the change in orientation of the airframe, and then the accelerometer readings are mathematically rotated into the reference frame.

2.7.4.4 Position Resolvers Position resolvers are sensors that provide angular measurements pertaining to the orientation or attitude of the vehicle (roll, pitch, and yaw; see Figure 1.6). Accelerometers that are mounted orthogonally can provide enough information from which orientation can be determined via geometry. Other techniques take advantage of the piezoelectric effect or magnetic induction to determine position.

Ring-laser gyros can also be used for position resolution. These devices are constructed from two concentric fiber-optic loops. A laser pulse is sent in opposite directions through each of the two loops. If the vehicle rotates in the direction of the loops, then one beam will travel faster than the other (twice the angular velocity of rotation). This difference can be measured and the amount of rotation determined. One ring-laser gyro each is needed to measure yaw, pitch, and roll angles.

2.7.5 Analog/Digital Converters

Analog-to-digital (A/D) conversion, or A/D circuitry, converts continuous (analog) signals from various transducers and devices into discrete (digital) ones. Similar circuitry can be used to convert temperature, sound, pressure, and other inputs from transducers by using a variety of sampling schemes.

The output of A/D circuitry is a discrete version of the time-varying signal being monitored. This information can be passed on to the real-time computer system using any of the three data-transfer methods, but in each case the A/D circuitry makes available an n-bit number that represents a discrete version of the signal. The discrete version of the continuous value can be treated as a scaled number.

The key factor in the service of A/D circuitry for time-varying signals is the sampling rate. In order to convert an analog signal into a digital form without loss of information, samples of the analog signal must be taken at twice the rate of the highest-frequency component of the signal (the Nyquist rate). Thus, a signal with highest-frequency component at 60 hertz must be sampled at 120 times per second. This implies that software tasks serving A/D circuitry must run at least at the same rate, or risk losing information. This consideration is an inherent part of the design process for the scheduling of tasks.

2.7.6 Digital/Analog Converters

Digital-to-analog (D/A) conversion, or D/A circuitry, performs the inverse function of A/D circuitry; that is, it converts a discrete quantity to a continuous one. D/A devices are used to allow the computer to output analog voltages based on the digital version stored internally. Communication with D/A circuitry uses one of the three input/output methods discussed.

2.8 NON-VON-NEUMANN ARCHITECTURES

The von Neumann bottleneck refers to the fact that only one instruction or one datum can occupy the bus at any one time. In addition, only one instruction can be executed at any given time. Architectural enhancements such as caching, pipelining, and coprocessing have already been discussed as workarounds to the fundamental von Neumann bottleneck and serial instruction execution constraints.[8]

---

[8] Some of the following discussion in this section is adapted from *Computer Architecture: A Minimalist Perspective* by Gilreath and Laplante [Gilreath03]

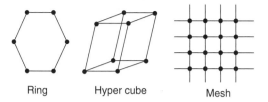

Ring   Hyper cube   Mesh

Figure 2.29 Three different multiprocessor interconnection schemes: (a) ring, (b) mesh, (c) hypercube. There are, of course, many other configurations [Gilreath03].

2.8.1 Parallel Systems

The major difference between two parallel systems is the means for which data are exchanged and interchanged between processors. The two mechanisms for data interchange are message passing and shared memory.

Message passing involves exchanging discrete messages between processors. The standard 1553B protocol is one architecture for message passing. Message passing is a software-oriented approach to parallelism, in that the actual processors can be distributed across a network.

Another parallel metric is the interconnection among the processors, measured in terms of number of connections for processors to communicate with one another. There are many different interconnection schemes. The more common ones include ring, mesh, or hypercube (Figure 2.29) and the bus interconnection topology used in 1553B (Figure 2.4).

Shared memory among the processors is a hardware-oriented solution. Shared memory uses a model where each processor can address another processor as memory. This is a nonuniform means of interchanging data, as memory for a processor is different from memory to a shared memory space, which are the other processor registers. Shared memory is often organized into different configurations, such as interleaved or distributed memory. The programmable random-access machine (PRAM) is the general form used for parallel shared-memory systems.

2.8.2 Flynn's Taxonomy for Parallelism

The generally accepted taxonomy of parallel systems was proposed by Flynn [Flynn66]. The classification is based on the notion of two streams of information flow to a processor; instructions, and data. These two streams can be either single or multiple, given four classes of machines:

1. Single instruction, single data (SISD)
2. Single instruction, multiple data (SIMD)
3. Multiple instruction, single data (MISD)
4. Multiple instruction, multiple data (MIMD)

Table 2.2   Flynn's classification for computer architectures

	Single Data Stream	Multiple Data Stream
Single instruction stream	von Neumann processors RISC	Systolic processors Wavefront Processors
Multiple-instruction stream	Pipelined architectures VLIW processors	Data flow processors Transputers Grid computers Hypercube processors

Table 2.2 shows the four primary classes and some of the architectures that fit in those classes. Most of these architectures either have been discussed or will be discussed shortly.

Besides the SISD and pipelined MISD architectures, the others tend to be found only in limited real-time applications in industry. Nevertheless, it is worth surveying these other architectures for completeness and in case they are encountered in some special-purpose application.

2.8.2.1   *Single Instruction, Single Data*   The SISD architectures encompass standard serial von Neumann architecture computers. There are a number of ways to introduce parallelism into the architecture of serial computers, such as microparallelism and using multiple functional units within the processor. Another possibility is to overlap processor operations with I/O, letting the processor concurrently execute instructions during I/O operations.

2.8.2.2   *Multiple Instruction, Single Data*   The MISD classification is a parallel computer architecture that lends itself naturally to those computations requiring an input to be subjected to several operations, each receiving the input in its original form. These applications include classification problems and digital signal processing. Pipelined and very long instruction word (VLIW) architectures are usually considered in this class.

In pipelined architectures, for example, more than one instruction can be processed simultaneously (one for each level of pipeline). However, since only one instruction can use data at any one time, it is considered MISD.

Similarly, VLIW computers tend to be implemented with microinstructions that have very long bit lengths (and hence more capability). Hence, rather than breaking down macroinstructions into numerous microinstructions, several (nonconflicting) macroinstructions can be combined into several microinstructions. For example, if object code was generated that loaded one register followed by an increment of another register, these two instructions could be executed simultaneously by the processor (or at least appear so at the macroinstruction level) with a series of long microinstructions. Since only nonconflicting instructions can be combined, any two accessing the data bus cannot. Therefore the VLIW computer is MISD.

2.8.2.3 *Single Instruction, Multiple Data* Two computer architectures that are usually classified as SIMD are systolic and wavefront-array parallel computers. In both systolic and wavefront processors, each processing element executes the same (and only) instruction, but on different data.

1. *Systolic Processors* Systolic processors consist of a large number of uniform processors connected in an array topology. Each processor usually performs only one specialized operation and has only enough local memory to perform its designated operation, and to store the inputs and outputs. The individual processors, or processing elements (PE), take inputs from the top and left, perform a specified operation, and output the results to the right and bottom. One such processing element is depicted in Figure 2.30. The processors are connected to the four nearest neighboring processors in the nearest-neighbor topology.

Processing or firing at each of the cells occurs simultaneously in synchronization with a central clock. The name comes from the fact that each cell fires on this heartbeat. Inputs to the system are from memory stores or input devices at the boundary cells at the left and top. Outputs to memory or output devices are obtained from boundary cells at the right and bottom.

Systolic processors are fast and can be implemented in VLSI. They are somewhat troublesome, however, in dealing with propagation delays in the connection buses and in the availability of inputs when the clock ticks.

2. *Wavefront Processor* Wavefront processors consist of an array of identical processors, each with its own local memory and connected in a nearest-neighbor topology. Each processor usually performs only one specialized operation. Hybrids containing two or more different type cells are possible. The cells fire asynchronously when all required inputs from the left and top are present. Outputs

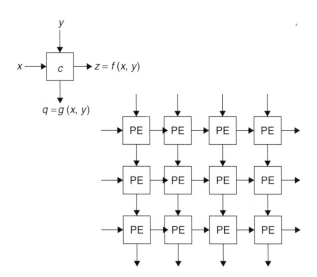

Figure 2.30 A systolic computer showing basic processing element (PE) and mesh configuration of multiple elements [Gilreath03].

then appear to the right and below. Unlike the systolic processor, the outputs are the unaltered inputs. That is, the top input is transmitted unaltered to the bottom output bus, and the left input is transmitted unaltered to the right output bus. Also, different from the systolic processor, outputs from the wavefront processor are read directly from the local memory of selected cells and output obtained form boundary cells. Inputs are still placed on the top and left input buses of boundary cells. The fact that inputs propagate through the array unaltered like a wave gives this architecture its name.

Wavefront processors combine the best of systolic architectures with data flow architectures. That is, they support an asynchronous data flow computing structure; timing in the interconnection buses and at input and output devices is not a problem. Furthermore, the structure can be implemented in VLSI.

2.8.2.4 *Multiple Instruction, Multiple Data*   MIMD computers involve large numbers of processors that are capable of executing more than one instruction and on more than one datum at any instant. Except for networks of distributed multiprocessors working on the same problem (grid computing), these are "exotic" architectures. Two paradigms that follow MIMD are data flow computers and transputers.

1. *Data Flow Architectures*   Data flow architectures use a large number of special processors in a topology in which each of the processors is connected to every other. Each of the processors has its own local memory and a counter. Special tokens are passed between the processors asynchronously. These tokens, called activity packets, contain an opcode, operand count, operands, and list of destination addresses for the result of the computation. An example of a generic activity packet is given in Figure 2.31.

Each processor's local memory is used to hold a list of activity packets for that processor, the operands needed for the current activity packet, and a counter used

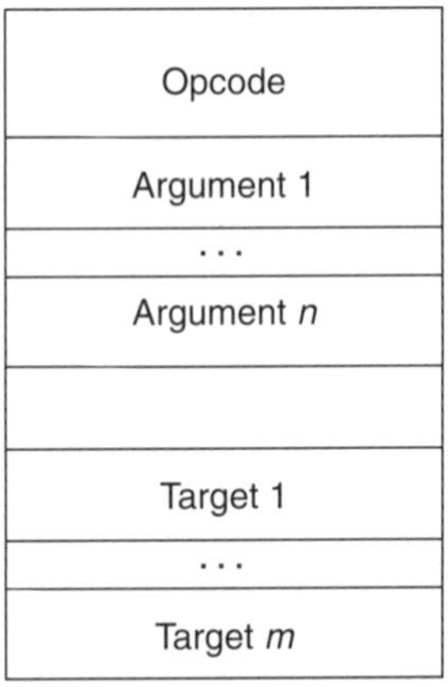

| Opcode |
| Argument 1 |
| . . . |
| Argument n |
| |
| Target 1 |
| . . . |
| Target m |

Figure 2.31   Generic activity template for data flow machine [Gilreath03].

to keep track of the number of operands received. When the number of operands stored in local memory is equivalent to that required for the operation in the current activity packet, the operation is performed and the results are sent to the specified destinations. Once an activity packet has been executed, the processor begins working on the next activity packet in its execution list.

Data flow architectures are an excellent parallel solution for signal processing, but require a cumbersome graphical programming language, and hence are rarely seen today.

2. *Transputers* Transputers are fully self-sufficient, multiple instruction set, von Neumann processors. The instruction set includes directives to send data or receive data via ports that are connected to other transputers. The transputers, though capable of acting as a uniprocessor, are best utilized when connected in a nearest-neighbor configuration. In a sense, the transputer provides a wavefront or systolic processing capability, but without the restriction of a single instruction. Indeed, by providing each transputer in a network with an appropriate stream of data and synchronization signals, wavefront, or systolic computers – which can change configurations – can be implemented.

Transputers have been used in some embedded real-time applications, and commercial implementations are available. Moreover, tool support, such as the multitasking language occam-2, has made it easier to build transputer-based applications. Nevertheless, transputer-based systems are relatively rare, especially in the United States.

3. *Transport Triggered Architecture* A special case of the MIMD computer is the distributed heterogeneous architecture in which a number of independent von Neumann CPUs communicate over a network and employ a time-driven processing model rather than an event-driven one. One example, the time-triggered architecture (TTA) developed by Kopetz and others, can be used for implementing distributed hard real-time systems [Kopetz97]. TTA models a distributed real-time system as a set of nodes interconnected by a real-time communication system (Figure 2.32).

TTA is based on fault-tolerant clock synchronization. Each node consists of a communication controller and a host computer, which are provided with a global, synchronized clock with a 1-microsecond tick duration. Each node is autonomous but communicates with other nodes over a replicated broadcast channel. Using time division multiple access (TDMA), each node is allocated a time slot in which it can send information on the bus to one or more receiving nodes, through a unique addressing scheme. It is thus possible to predict the latency of all messages on the bus, which guarantees hard real-time message delivery. Furthermore, since the messages are sent at a predetermined point in time, the latency jitter is minimal. By comparing the known point in time at which a particular message was sent and when it is received, host computers can synchronize their clocks without the overhead of messages [Kopetz98]. Each individual node in the TTA needs to be designed as a self-sufficient real-time system. But the architecture provides a very reliable and predictable mechanism for communication between

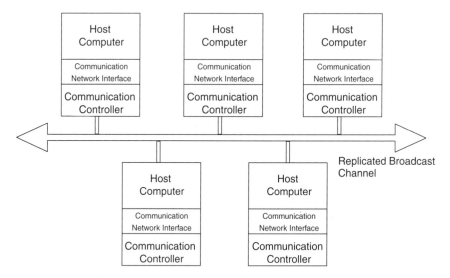

Figure 2.32   Time-triggered architecture with five nodes.

nodes and for fault tolerance, because should a node fail, it can be detected by another node that can assume the failed node's responsibilities.

Commercial products for implementing TTA are available and have been deployed for safety-critical automotive and avionics applications, mostly in Europe.

2.9 EXERCISES

2.1.   Discuss the possibility of an "*n*-address" machine, where $n > 3$.

2.2.   The instruction set of a certain processor does not have the JLE, JLT, JGE (jump less equal, less than, and greater or equal), and JGT instructions. Assume the process does not have all other arithmetic instructions nor the JNE (jump not equal) and JUA instructions. Implement the missing instructions for the generic assembly language in a

  (a)   0-Address machine

  (b)   1-Address machine

  (c)   2-Address machine

  (d)   3-Address machine

2.3.   Why is DMA controller access to main memory in most systems given higher priority than CPU access to main memory?

2.4.   Discuss the relative advantages/disadvantages of DMA, program I/O, and memory-mapped data transfer as they pertain to real-time systems.

2.5.   Describe the relationship between the main processor and coprocessor in a system with which you are familiar or one that you discover through Web research.

2.6. What special problems do pipelined architectures pose for real-time system designers? Are they any different for non-real-time systems?

2.7. Compare and contrast the different memory technologies discussed in this chapter as they pertain to real-time systems.

2.8. Should the instruction following the TEST instruction be interruptible? If so, what must the implicit BRANCH instruction (interrupt) do?

2.9. It is common practice for programmers to create continuous test and loop code in order to poll I/O devices or wait for interrupts to occur. Some processors provide an instruction (WAIT or HALT) that allows the processor to hibernate until an interrupt occurs. Why is the second form more efficient and desirable?

2.10. In general terms, suggest a possible scheme that would allow a CPU macroinstruction to be interruptible. What would be the overall effect (if any) on macroinstruction execution time and CPU throughput and response times?

2.11. Microcoded computers tend to be superior to 1-, 2-, or 3- address computers with respect to overall performance. Why?

2.12. What is the difference between coprocessing and multiprocessing? What are the advantages and disadvantages of each?

2.13. Find a processor with unique, special instructions and, considering the typical application area for that processor, discuss the need for the special instruction.

2.14. GCP, Inc., has contracted you to analyze and develop commercial off-the-shelf (COTS) processors to be phased into their existing product lines. Your objective is to select four popular commercial processor solutions, develop a series of relevant metrics for them, and make a preliminary determination of which is best suited to each of GCP's five largest product lines. Physical characteristics such as size and weight are considered unimportant and negligible in comparison to the present design of GCP's products.

2.15. It seems that there are far fewer commercial implementations of SIMD, MIMD, and MISD architectures than there were just 10 years ago. This is probably due to a variety of factors, including high design hosts, low demand, lack of support tools, and, simply, bad business decisions.

A company is considering a number of different commercial SIMD processors for use in the next generation of pasta sauce bottling systems. Your task is to prepare a report analyzing the market in commercial SIMD processors for presentation to management. You will need to generate a set of metrics for an apples-to-apples comparison of the various solutions out there, such as processing power, scalability, support tools available, support costs, and purchase price. Try to generate at least 10 relevant metrics. You might want to review existing articles surveying the market, but please cite these articles if you use them. You should be able to gather enough material from Web searches and from a literature search of IEEE sources. Your analysis should take into consideration the likely applications for the chip in the inertial measurement system; after all, the solution should fit the problem. Your market analysis should be sufficient to make a preliminary decision to invite three or four manufacturers in for a presentation of their devices.

2.16. What would be the appropriate operating system architecture for the:

(a) Inertial measurement system

(b) Nuclear monitoring system

(c) Patient monitoring system

(d) Airline reservations system

(e) Pasta sauce bottling system

(f) Traffic intersection control system

Use a Web search to select any appropriate commercial processor and support tool set.

3

REAL-TIME OPERATING SYSTEMS

3.1 REAL-TIME KERNELS

A process[1] is an abstraction of a running program and is the logical unit of work scheduled by the operating system. It is typically represented by a data structure that contains at least a state of execution, an identity (real-time), attributes (e.g., execution time), and the resources associated with it. A thread is a lightweight process that shares resources with other processes or threads. Each thread must "reside" within some process and make use of the resources of that process. Threads that reside within the same process share that processes' resources.

Real-time operating systems must provide three specific functions with respect to tasks: scheduling, dispatching, and intercommunication and synchronization. The kernel of the operating system is the smallest portion that provides for these functions. A scheduler determines which task will run next in a multitasking system, while a dispatcher performs the necessary bookkeeping to start that task. Intertask communication and synchronization assures that the tasks cooperate. Various layers of operating system functionality and an associated taxonomy are given in Figure 3.1.

A nanokernel provides simple thread (lightweight process) management. It essentially provides only one of the three services provided by a kernel, whereas a microkernel in addition provides for task scheduling. A kernel also provides for intertask synchronization and communication via semaphores, mailboxes, and other methods. A real-time executive is a kernel that includes privatized memory

[1] Synonymously called "task."

Real-Time Systems Design and Analysis, By Phillip A. Laplante
ISBN 0-471-22855-9 © 2004 Institute of Electrical and Electronics Engineers

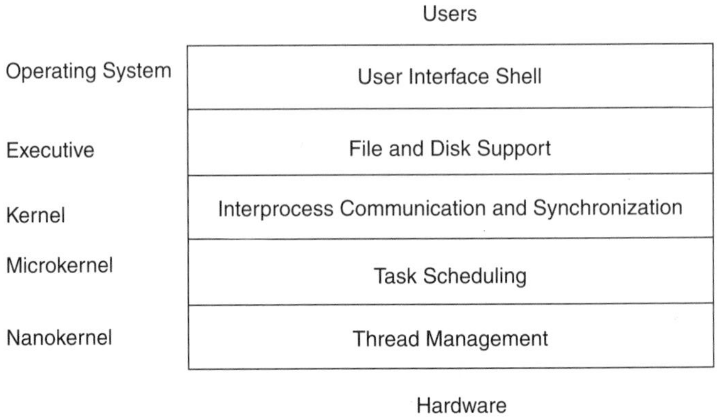

Users

Operating System	User Interface Shell
Executive	File and Disk Support
Kernel	Interprocess Communication and Synchronization
Microkernel	Task Scheduling
Nanokernel	Thread Management

Hardware

Figure 3.1 The role of the kernel in operating systems. Moving up the taxonomy from the low-level nanokernel to the full-featured operating system shows the additional functionality provided and also indicates the relative closeness to hardware versus human users.

blocks, I/O services, and other complex features. Most commercial real-time kernels are executives. Finally, an operating system is an executive that provides for a generalized user interface, security, and a file-management system.

Regardless of the operating system architecture used, the objective is to satisfy real-time behavioral requirements and provide a seamless multitasking environment that is flexible and robust.

3.1.1 Pseudokernels

Real-time multitasking can be achieved without interrupts and even without an operating system per se. When feasible, these approaches are preferred because resultant systems are easier to analyze.

3.1.1.1 Polled Loop   Polled loops are used for fast response to single devices. In a polled-loop system, a single and a repetitive instruction is used to test a flag that indicates whether or not some event has occurred. If the event has not occurred, then the polling continues.

For example, suppose a software system is needed to handle packets of data that arrive at a rate of no more than 1 per second. A flag named `packet_here` is set by the network, which writes the data into the CPU's memory via direct memory access (DMA). The data are available when `packet_here` = 1. Using a C code fragment, such a polled loop to handle such a system is:

```
for(;;)    {                        /* do forever    */
           if (packet_here)         /* check flag       */
           {
               process_data();      /* process data */
               packet_here=0;       /* reset flag    */
           }
       }
```

Polled-loop schemes work well when a single processor is dedicated to handling the I/O for some fast device and when overlapping of events is not allowed or minimized. Polled loops are ordinarily implemented as a background task in an interrupt-driven system, or as a task in a cyclic executive. In the latter case, the polled loop polls each cycle for a finite number of times to allow other tasks to run. Other tasks handle the nonevent-driven processing.

3.1.1.2 *Synchronized Polled Loop*   A variation on the polled loop uses a fixed clock interrupt to pause between the time when the signaling event is triggered and then reset. Such a system is used to treat events that exhibit switch bounce. Switch bounce is a phenomenon that occurs because it is impossible to build a switch, whether mechanical or electrical, that can change state instantaneously. A typical response for such a switch is given in Figure 3.2. Events triggered by switches, levers, and buttons all exhibit this phenomenon. If, however, a sufficient delay occurs between the initial triggering of the event and the reset, the system will avoid interpreting the settling oscillations as events. These are, of course, spurious events that would surely overwhelm any polled-loop service. For instance, suppose a polled-loop system is used to handle an event that occurs randomly, but no more than once per second. The event is known to exhibit a switch-bounce effect that disappears after 20 milliseconds. A 10-millisecond fixed-rate interrupt is available for synchronization. The event is signaled by an external device that sets a memory location via DMA. The C code looks like the following:

```
for(;;)       {                      /* do forever    */
              if(flag)               /* check flag    */
              {
                  pause(20);         /* wait 20 ms    */
                  process_event();   /* process event */
                  flag=0;            /* reset flag    */
              }
}
```

where pause(20) is a system timer call that provides a delay in increments of 1 millisecond. Since there is overhead in the system call and return, the wait time will always be greater than the needed 20 milliseconds, which avoids the interpretation of spurious events.

Assuming the pause system call is available, polled-loop systems are simple to write and debug, and the response time is easy to determine.

Polled loops are excellent for handling high-speed data channels, especially when the events occur at widely dispersed intervals and the processor is dedicated to handling the data channel. Polled-loop systems most often fail, however, because bursts are not taken into account. Furthermore, polled loops by themselves are generally not sufficient to handle complex systems. Finally, polled loops inherently waste CPU time, especially if the event being polled occurs infrequently.

3.1.1.3 *Cyclic Executives*   Cyclic executives are noninterrupt-driven systems that can provide the illusion of simultaneity by taking advantage of relatively

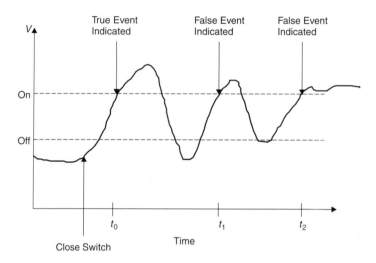

Figure 3.2 Switch-bounce phenomenon. The switch is closed at time t_0, signaling the event; however, due to the ringing of the signal and the edge-triggered logic several false events could be indicated at times t_1, and t_2.

short processes on a fast processor in a continuous loop. For example, consider the set of self-contained processes Process1 through Process_N in a continuous loop as depicted below:

```
for(;;)     {                          /* do forever    */
            Process_1();
            Process_2();
            ...
            Process_N();
            }
        }
```

In this case it can be seen that a de facto cycle rate is established, which is the same for each task, as they execute in "round-robin" fashion. Different rate structures can be achieved by repeating a task in the list. For example, in the following code:

```
for(;;)     {                          /* do forever    */
            Process_1();
            Process_2();
            Process_3();
            Process_3();
            }
        }
```

Process_3 runs twice as frequently as Process_1 and Process_2.

The task list can be made adjustable by keeping a list of pointers to processes that are managed by the "operating system" as tasks are created and completed. Intertask synchronization and communication can be achieved through global variables or parameter lists.

Consider a more pedestrian example, such as a Space Invaders® game. The game involves, essentially, servicing three button events (left movement of the tank, right movement of the tank, and fire missile), moving the aliens, computing collisions, and updating the screen. Surprisingly, it is possible to build a simple version of the game on a personal computer with essentially the following cyclic executive:

```
for(;;)      {                                      /* do forever    */
             check_for_keypressed();
             move_aliens();
             check_for_keypressed();
             check_for_collison()
             check_for_keypressed();
             update_screen();
             }
      }
```

Note that the process check_for_keypressed, which services the three button pressings, contains the processes move_tank and fire_missile, and is executed three times as frequently as the others in order to provide faster response to user input. If each process is relatively short and uniform in size, then reactivity and simultaneity can be achieved without interrupts. Moreover, if each process is carefully constructed including synchronization through messaging or global variables, complete determinism and schedulability can be achieved. Cyclic executives are, however, inadequate for all but the simplest of real-time systems because of the difficulties in uniformly dividing the processes and in the long response times that are created, as will be seen later.

3.1.1.4 *State-Driven Code*
State-driven code uses nested if–then statements, case statements, or finite state automata to break up the processing of functions into code segments. The separation of processes allows each to be temporarily suspended before completion, without loss of critical data. This, in turn, facilitates multitasking via a scheme such as coroutines, which we will discuss shortly. State-driven code works well in conjunction with cyclic executives when the processes are too long or nonuniform in size.

Finally, because mathematical techniques for reducing the number of states exist, programs based on finite state machines (FSMs) can be formally optimized. A rich theory surrounds FSMs, and this can be exploited in the development of system specifications.

Not all processes lend themselves naturally to division into states; some processes are therefore unsuitable for this technique. In addition, the tables needed to implement the code can become quite large. Finally, the manual translation process from the finite state automaton to tabular form is error-prone.

3.1.1.5 *Coroutines*
Coroutines or cooperative multitasking systems require disciplined programming and an appropriate application. These types of kernels are employed in conjunction with code-driven by finite state automata. In this

scheme, two or more processes are coded in the state-driven fashion just discussed, and after each phase is complete, a call is made to a central dispatcher. The dispatcher holds the program counter for a list of processes that are executed in round-robin fashion; that is, it selects the next process to execute. This process then executes until the next phase is complete, and the central dispatcher is called again. Note that if there is only one coroutine, then it will be repeated cyclically. Such a system is called a cycle executive.

Communication between the processes is achieved via global variables. Any data that need to be preserved between dispatches must be deposited to global variables. Consider a system in which two processes are executing "in parallel" and in isolation. After executing phase_a1, process_a returns control to the central dispatcher by executing break. The dispatcher initiates process_b, which executes phase_b1 to completion before returning control to the dispatcher. The dispatcher then starts process_a, which begins phase_a2, and so on. The C code is depicted below, for process_a and process_b:

```
void process_a(void)
{
for(;;)
    {
    switch(state_a)
        {
        case 1: phase_a1();
             break;
        case 2: phase_a2();
             break;
        case 3: phase_a3();
             break;
        case 4: phase_a4();
             break;
        case 5: phase_a5();
             break;
        }
    }
}

void process_b(void)
{
for(;;)
    {
        switch(state_b)
        {
            case 1:     phase_b1();
                    break;
            case 2:     phase_b2();
                    break;
            case 3:     phase_b3();
                    break;
            case 4:     phase_b4();
                    break;
            case 5:     phase_b5();
                    break;
        }
    }
}
```

Note that state_a and state_b are state counters that are global variables managed by the dispatcher. Indeed, for simplicity, synchronization and communication are maintained entirely via global variables and managed by the dispatcher (forget about the dangers of this approach for now). The coroutine approach can be extended to any number of processes, each divided into arbitrary phases. If each programmer provides calls to the dispatcher at known intervals, the response time is easy to determine because this system is written without hardware interrupts.

A variation of this scheme occurs when a polled loop must wait for a particular event while other processing can continue. Such a scheme reduces the amount of time wasted polling the event flag, and allows for processing time for other tasks. In short, coroutines are the easiest type of "fairness scheduling" that can be implemented. In addition, the processes can be written by independent parties, and the number of processes need not be known beforehand. Finally, certain languages such as Ada have built-in constructs for implementing coroutines in a way that is superior to the somewhat artificial (but more general) implementation.

Some surprisingly large and complex applications have been implemented using coroutines; for example, IBM's old transaction processing system, Customer Information Control System (CICS), was originally constructed entirely via coroutines. In addition, IBM's OS/2 Presentation Manager used coroutines to coordinate the activities within the various user windows. Unfortunately, any use of coroutines assumes that each task can relinquish the CPU at regular intervals. It also requires a communication scheme involving global variables, which is undesirable. Finally, processes cannot always be easily decomposed uniformly, which can adversely affect response time since the minimum size is a function of the longest phase.

3.1.2 Interrupt-Driven Systems

In interrupt-driven systems, the main program is a single jump-to-self instruction. The various tasks in the system are scheduled via either hardware or software interrupts, whereas dispatching is performed by the interrupt-handling routines.

When hardware scheduling is used, a clock or other external device issues interrupt signals that are directed to an interrupt controller. The interrupt controller issues interrupt signals, depending on the order of arrival and priority of the interrupts involved. If the computer architecture supports multiple interrupts, then the hardware handles dispatching as well. If only a single interrupt level is available, then the interrupt-handling routine will have to read the interrupt vector on the interrupt controller, determine which interrupts occurred, and dispatch the appropriate tasks. Some processors implement this in microcode, and so the operating systems designer is relieved of this duty.

3.1.2.1 Interrupt Service Routines
When writing embedded applications, it is important to understand how interrupts work because often the real-time

program needs to service interrupts from one or more special-purpose devices. Commonly, the software engineer needs to write a device driver from scratch or adapt a generic device driver. In any system, there are two kinds of interrupts:

Hardware Interrupt A signal generated by a peripheral device and sent to the CPU. In turn, the CPU executes an interrupt service routine (ISR), which takes action in response to the interrupt.

Software Interrupt Similar to the hardware interrupt, in that it causes one code module to pass control to another.

The difference between hardware and software interrupts is the trigger mechanism. The trigger of a hardware interrupt is an electrical signal from an external device, whereas, the trigger of a software interrupt is the execution of a machine language instruction.

A common concept found in programming languages is that of an exception, which is akin to an internal interrupt that is triggered by a program's attempt to perform an unexpected or illegal operation. All three situations cause the CPU to transfer execution to a known location and then execute code associated with that situation.

Hardware interrupts are asynchronous in nature, that is, an interrupt can happen at any time. When interrupted, the program is suspended while the CPU invokes the ISR. Often an application developer is required to write an ISR for a specific type of hardware interrupt. In this case it is important to understand what constitutes the CPU state, and whether ISRs must preserve anything in addition to general registers.

Access to resources shared with an ISR is usually controlled by disabling interrupts in the application around any code that reads or writes to the resource. Synchronization mechanisms cannot be used in an ISR because it is not possible for an ISR to wait indefinitely for a resource to be available. When interrupts are disabled, the system's ability to receive stimuli from the outside world is minimal. It is important to keep the critical sections of code in which the interrupts are disabled as short as possible. If the ISR takes too long to process an interrupt, the external device may be kept waiting too long before its next interrupt is serviced.

Reentrant code can execute simultaneously in two or more contexts. An ISR is said to be reentrant if, while the ISR is handling an interrupt, the same interrupt can occur again and the ISR can process the second occurrence of the interrupt before it has finished processing the first.

Regardless of the type of ISR to be written, a snapshot of the machine – called the context – must be preserved upon switching tasks so that it can be restored upon resuming the interrupted process.

3.1.2.2 Context Switching
Context switching is the process of saving and restoring sufficient information for a real-time task so that it can be resumed after being interrupted. The context is ordinarily saved to a stack data structure. Context-switching time is a major contributor to response time and therefore

must be minimized. The rule for saving context is simple: save the minimum amount of information necessary to safely restore any process after it has been interrupted. This information ordinarily includes

- Contents of general registers
- Contents of the program counter
- Contents of coprocessor registers (if present)
- Memory page register
- Images of memory-mapped I/O locations (mirror images)

Normally, within the interrupt handlers, interrupts are disabled during the critical context-switching period. Sometimes, however, after sufficient context has been saved, interrupts can be enabled after a partial context switch in order to handle a burst of interrupts, to detect spurious interrupts, or to handle a time-overloaded condition.

The stack model for context switching is used mostly in embedded systems where the number of real-time or interrupt-driven tasks is fixed. In the stack model, each interrupt handler is associated with a hardware interrupt and is invoked by the CPU, which vectors to the instruction stored at the appropriate interrupt-handler location. The context is then saved to a specially designated memory area that can be static, in the case of a single-interrupt system, or a stack, in the case of a multiple-interrupt system.

Consider the following pseudocode for a partial real-time system, written in C, and consisting of a simple jump-to-self and three interrupt handlers. Each saves context using the stack model. The interrupt handlers' starting addresses should be loaded into the appropriate interrupt vector location upon initialization. Alternatively, this procedure can be performed at link time through the link editor or linker control file.

```
void main(void)
/*initialize system, load interrupt handlers */
{
    init();
    while(TRUE);          /* infinite wait loop        */
}

void intl (void)         /* interrupt handler 1        */
{
    save(context);       /* save context on stack      */
    task1();             /* execute task 1             */
    restore(context);    /* restore context from stack */
}

void int2(void)          /* interrupt handler 2        */
{
    save(context);       /* save context on stack      */
    task2();             /* execute task 2             */
    restore(context);    /* restore context from stack */
}
```

```
void  int3(void)         /* interrupt handler 3        */
{
    save(context);       /* save context on stack      */
    task3();             /* execute task 3             */
    restore(context);    /* restore context from stack */
}
```

Procedure `save` saves certain registers to a stack area, whereas `restore` restores those registers from the stack. In practice, `save` and `restore` would actually take two arguments; a pointer to data structure representing the context information and a pointer to the stack data structure, which will be discussed later. In the case of the context data structure, the programming language compiler must provide a mechanism to extract the current contents of the general registers, PCs, and so forth.[2] Finally, both `save` and `restore` must adjust the stack pointer, which is illustrated later.

3.1.3   Preemptive-Priority Systems

A higher-priority task is said to preempt a lower-priority task if it interrupts the lower-priority task. Systems that use preemption schemes instead of round-robin or first-come-first-served scheduling are called preemptive-priority systems. The priorities assigned to each interrupt are based on the urgency of the task associated with the interrupt. For example, the nuclear power station monitoring system is best designed as a preemptive-priority system. While the handling of intruder events is critical, for example, nothing is more important than processing the core over-temperature alert.

Prioritized interrupts can be either fixed priority or dynamic priority. Fixed-priority systems are less flexible, since the task priorities cannot be changed. Dynamic-priority systems can allow the priority of tasks to be adjusted at run-time to meet changing process demands.

Preemptive-priority schemes can suffer from resource hogging by higher-priority tasks. This can lead to a lack of available resources for lower-priority tasks. In this case, the lower-priority tasks are said to be facing a problem called starvation.

A special class of fixed-rate preemptive-priority interrupt-driven systems, called rate-monotonic systems, comprises those real-time systems where the priorities are assigned so that the higher the execution frequency, the higher the priority. This scheme is common in embedded applications, particularly avionics systems, and has been studied extensively. For example, in the aircraft navigation system, the task that gathers accelerometer data every 10 milliseconds has the highest priority. The task that collect gyro data, and compensates these data and the accelerometer data every 40 milliseconds, has the second highest priority. Finally, the task that updates the pilot's display every second has lowest priority.

The theoretical aspects of rate-monotonic systems will be studied shortly.

---

[2] This is not a trivial thing because the PC and registers are needed to affect the call.

3.1.4 Hybrid Systems

Hybrid systems include interrupts that occur at both fixed rates and sporadically. The sporadic interrupts can be used to handle a critical error that requires immediate attention, and thus have highest priority. This type of system is common in embedded applications.

Another type of hybrid system found in commercial operating systems is a combination of round-robin and preemptive systems. In these systems, tasks of higher priority can always preempt those of lower priority. However, if two or more tasks of the same priority are ready to run simultaneously, then they run in round-robin fashion, which will be described shortly.

To summarize, interrupt-only systems are easy to write and typically have fast response times because process scheduling can be done via hardware. Interrupt-only systems are a special case of foreground/background systems, which are widely used in embedded systems.

One weakness of interrupt-only systems, however, is the time wasted in the `jump-to-self` loop and the difficulty in providing advanced services. These services include device drivers and interfaces to multiple layered networks. Another weakness is vulnerability to malfunctions owing to timing variations, unanticipated race conditions, hardware failure, and so on. Some companies avoid designs based on interrupts for these reasons.

3.1.4.1 Foreground/Background Systems

Foreground/background systems are an improvement over the interrupt-only systems in that the polled loop is replaced by code that performs useful processing. Foreground/background systems are the most common architecture for embedded applications. They involve a set of interrupt-driven or real-time processes called the foreground and a collection of noninterrupt-driven processes called the background (Figure 3.3). The foreground tasks run in round-robin, preemptive priority, or hybrid fashion. The background task is fully preemptable by any foreground task and, in a sense, represents the lowest priority task in the system.

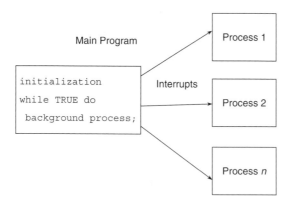

Figure 3.3 A foreground/background system.

All real-time solutions are just special cases of the foreground/background systems. For example, the polled loop is simply a foreground/background system with no foreground, and a polled loop as a background. Adding interrupts for synchronization yields a full foreground/background system. State-driven code is a foreground/background system with no foreground and phase-driven code for a background. Coroutine systems are just a complicated background process. Finally, interrupt-only systems are foreground/background systems without background processing.

3.1.4.2 *Background Processing*
As a noninterrupt-driven task, the background processing should include anything that is not time critical. While the background process is the process with the lowest priority, it should always execute to completion provided the system utilization is less than 100% and no deadlocking occurs. It is common, for instance, to increment a counter in the background in order to provide a measure of time loading or to detect if any foreground process has hung up. It might also be desirable to provide individual counters for each of the foreground processes, which are reset in those processes. If the background process detects that one of the counters is not being reset often enough, it can be assumed that the corresponding task is not being executed and, that some kind of failure is indicated. This is a form of software watchdog timer.

Certain types of low-priority self-testing can also be performed in the background. For example, in many systems, a complete test of the CPU instruction set could be performed. This kind of test should never be performed in foreground, but should be part of a robust system design. The design and coding of these CPU instruction tests require careful planning. Finally, low-priority display updates, logging to printers, or other interfaces to slow devices can be performed in the background.

3.1.4.3 *Initialization*
Initialization of the foreground/background system consists of the following steps:

1. Disable interrupts
2. Set up interrupt vectors and stacks
3. Perform self-test
4. Perform system initialization
5. Enable interrupts

Initialization is actually the first part of the background process. It is important to disable interrupts because many systems start up with interrupts enabled while time is still needed to set things up. This setup consists of initializing the appropriate interrupt vector addresses, setting up stacks if it is a multiple-level interrupt system, and initializing any data, counters, arrays, and so on. In addition, it is necessary to perform any self-diagnostic tests before enabling any interrupts. Finally, real-time processing can begin.

3.1.4.4 Real-Time Operation The real-time or foreground operation for the foreground/background system is the same as that for the interrupt-only system. For example, suppose it is desired to implement an interrupt handler for a 2-address computer architecture with a single interrupt. That is, one real-time task and the background process. The EPI and DPI instructions can be used to enable and disable the interrupt explicitly, and it is assumed that upon receiving an interrupt, the CPU will hold off all other interrupts until explicitly reenabled with an EPI instruction.

For context-switching purposes, it is necessary to save the eight general registers, R0-R7, on the stack. Note that context switching involves saving the status of the machine as it is used by the background process. The foreground process will run to completion so its context is never saved. Further, assume that the CPU will have the PC in memory location 6 at the time of interruption, and the address of the interrupt-handler routine (the interrupt vector) is stored in memory location 5.

The following assembly code could be used to trivially initialize the simple foreground/background system.

```
DPI                     ; disable interrupts
STORE &handler,5        ; put interrupt handler address in location 5
EPI                     ; enable interrupts
```

Of course, other initialization, such as initializing flags and other data, should be performed before enabling interrupts.

If symbolic memory locations reg0 through reg7 are used to save the registers, then the interrupt handler, coded in 2-address code, might look as follows:

```
DPI                     ; redundantly disable interrupts
STORE R0,&reg0          ; save register 0
STORE R1,&reg1          ; save register 1
STORE R2,&reg2          ; save register 2
STORE R3,&reg3          ; save register 3
STORE R4,&reg4          ; save register 4
STORE R5,&reg5          ; save register 5
STORE R6,&reg6          ; save register 6
STORE R7,&reg7          ; save register 7

JU @APP                 ; execute real-time application program
LOAD R7,&reg7           ; restore register 7
LOAD R6,&reg6           ; restore register 6
LOAD R5,&reg5           ; restore register 5
LOAD R4,&reg4           ; restore register 4
LOAD R3,&reg3           ; restore register 3
LOAD R2,&reg2           ; restore register 2
LOAD R1,&reg1           ; restore register 1
LOAD R0,&reg0           ; restore register 0

EPI                     ; re-enable interrupts
RI                      ; return from interrupt
```

In many computers, block save and restore instructions are available to save and restore a set of registers to consecutive memory locations. Also note that this interrupt handler does not permit the interrupt itself. If this is to be accomplished, or if more than one interrupt routine existed, a stack rather than just static memory would be needed to save context.

The background program would include the initialization procedure and any processing that was not time critical, and would be written in a high-order language. If the program were to be written in C, it might appear as:

```
void main (void)
/*allocate space for context variable */
    int reg0, reg1, reg2, reg3, reg4, reg5, reg6, reg7;
/*declare other global variables here */
{
    init();          /*initialize system */

    while (TRUE)     /*background loop */
     background();   /* non-real-time processing here */
}
```

Foreground/background systems typically have good response times, since they rely on hardware to perform scheduling. They are the solution of choice for many embedded real-time systems. But "home-grown" foreground/background systems have at least one major drawback: interfaces to complicated devices and networks must be written. This procedure can be tedious and error-prone. In addition, these types of systems are best implemented when the number of foreground tasks is fixed and known a priori. Although languages that support dynamic allocation of memory could handle a variable number of tasks, this can be tricky. Finally, as with the interrupt-only system, the foreground/background system is vulnerable to timing variations, unanticipated race conditions, hardware failures, and so on.

3.1.4.5 Full-Featured Real-Time Operating Systems
The foreground/background solution can be extended into an operating system by adding additional functions such as network interfaces, device drivers, and complex debugging tools. These types of systems are readily available as commercial products. Such systems rely on a complex operating system using round-robin, preemptive-priority, or a combination of both schemes to provide scheduling; the operating system represents the highest priority task, kernel, or supervisor.

3.1.5   The Task-Control Block Model

The task-control block model is the most popular method for implementing commercial, full-featured, real-time operating systems because the number of real-time tasks can vary. This architecture is used in interactive on-line systems where tasks (associated with users) come and go. This technique can be used in round-robin, preemptive-priority, or combination systems, although it is generally associated with round-robin systems with a single clock. In preemptive systems, however, it can be used to facilitate dynamic task prioritization. The main drawback of the task-control block model is that when a large number of tasks are created, the overhead of the scheduler can become significant.

In the task-control block (TCB) model each task is associated with a data structure, called a task control block. This data structure contains at least a PC, register contents, an identification string or number, a status, and a priority if applicable. The system stores these TCBs in one or more data structures, such as a linked list.

3.1.5.1 *Task States*
The operating system manages the TCBs by keeping track of the status or state of each task. A task typically can be in any one of the four following states:

1. Executing
2. Ready
3. Suspended (or blocked)
4. Dormant (or sleeping)

The executing task is the one that is running, and in a single-processing system there can be only one. A task can enter the executing state when it is created (if no other tasks are ready), or from the ready state (if it is eligible to run based on its priority or its position in the round-robin ready list). When a task is completed it returns to the suspended state.

Tasks in the ready state are those that are ready to run but are not running. A task enters the ready state if it was executing and its time slice runs out, or it was preempted. If it was in the suspended state, then it can enter the ready state if an event that initiates it occurs. If the task was in the dormant state, then it enters the ready state upon creation of another task. Tasks that are waiting on a particular resource, and thus are not ready, are said to be suspended or blocked.

The dormant state is used only in systems where the number of TCBs is fixed. This state allows for determining memory requirements beforehand, but limits available system memory. This state is best described as a task that exists but is unavailable to the operating system. Once a task has been created, it can become dormant by deleting it.

3.1.5.2 *Task Management*
The operating system is in essence the highest priority task. Every hardware interrupt and every system-level call (such as a request on a resource) invokes the real-time operating system. The operating system is responsible for maintaining a linked list containing the TCBs of all the ready tasks, and a second linked list of those in the suspended state. It also keeps a table of resources and a table of resource requests. Each TCB contains the essential information normally tracked by the interrupt service routine (Figure 3.4).

Task ID
Priority
Status
Register 1
...
Register n
Program Counter
Status Register(s)
Pointer to Next TCB

Figure 3.4   A typical task-control block.

The difference between the TCB model and the interrupt-service-routine model is that the resources are managed by the operating systems in the latter, while in the TCB model, tasks track their own resources. The TCB model is useful when the number of tasks is indeterminate at design time or can change while the system is in operation. That is, the TCB model is very flexible.

When it is invoked, the operating system checks the ready list to see if the next task is eligible for execution. If it is eligible, then the TCB of the currently executing task is moved to the end of the ready list, and the eligible task is removed from the ready list and its execution begins.

Task management can be achieved simply by manipulating the status word. For example, if all of the TCBs are set up in the list with the status word initially set to "dormant," then tasks can be added by changing the status to "ready" when the TCB has been initialized. During run time the status words of tasks are set accordingly, either to "executing" in the case of the next eligible task or back to "ready" in the case of the interrupted task. Blocked tasks have their status word changed to "suspended." Completed tasks can be "removed" from the task list by resetting the status word to dormant. This approach reduces overhead because it eliminates the need for dynamic memory management of the TCBs. It also provides deterministic performance because the TCB list is of constant size.

3.1.5.3 Resource Management   In addition to scheduling, the operating system checks the status of all resources in the suspended list. If a task is suspended due to a wait for a resource, then that task can enter the ready state only upon availability of the resource. The list structure is used to arbitrate two tasks that are suspended on the same resource. If a resource becomes available to a suspended task, then the resource tables are updated and the eligible task is moved from the suspended list to the ready list.

3.2   THEORETICAL FOUNDATIONS OF REAL-TIME OPERATING SYSTEMS

In order to take advantage of some of the more theoretical results in real-time operating systems (RTOS), a fairly rigorous formulation is necessary. Most real-time systems are inherently concurrent, that is, their natural interaction with external events typically requires multiple simultaneous tasks to cope with multiple threads of control. A process is the active object of a system and is the basic unit of work handled by the scheduler. As a process executes, it changes its state and at any time, and it may be in one, but only one, of the following states at any instant:

- *Dormant* (or sleeping) The task has been created and initialized. It is not yet ready to execute, that is, in this state, the process is not eligible to execute.
- *Ready* Processes in this state are those that are released and eligible for execution, but are not executing. A process enters the ready state if it was executing and its time-slice runs out, or if it was preempted. If a process was in the suspended or blocked state, then it enters the ready state if an event that initiates it occurs.
- *Executing* When a process is executing its instructions are being executed.
- *Suspended (or blocked)* Processes that are waiting for a particular resource, and thus are not ready, are said to be in the suspended or blocked state.
- *Terminated* The process has finished execution, or has self-terminated or aborted, or is no longer needed.

Similar to processes, threads can be in only one of these states at any instant.

A partial state diagram corresponding to process or thread states is depicted in Figure 3.5. It should be noted that the different operating systems have different naming conventions, but the states represented in this arbitrary nomenclature exist in one form or another in all RTOS. Many modern operating systems allow processes created within the same program to have unrestricted access to the shared memory through a thread facility.

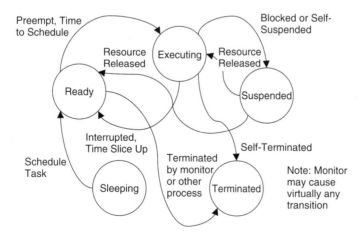

Figure 3.5 A process state diagram as a partially defined finite state machine.

3.2.1   Process Scheduling

Scheduling is a fundamental operating system function. In order to meet a program's temporal requirements in real-time systems a strategy is needed for ordering the use of system resources, and a mechanism needed for predicting the worst-case performance (or response time) when a particular scheduling policy is applied. There are two general classes of scheduling policies: pre-run-time and run-time scheduling. The goal of both types of scheduling is to satisfy time constraints.

In pre-run-time scheduling, the objective is to create a feasible schedule off-line, which guarantees the execution order of processes and prevents simultaneous access to shared resources. Pre-run-time scheduling also takes into account and reduces the cost of context switching overhead, increasing the chance that a feasible schedule can be found.

In run-time scheduling, static priorities are assigned and resources are allocated on a priority basis. Run-time scheduling relies on a complex run-time mechanism for process synchronization and communication. This approach allows events to interrupt processes and demand resources randomly. In terms of performance analysis, engineers must rely on stochastic simulations to verify these types of system designs.

3.2.1.1   *Task Characteristics of a Real Workload*   The workload on processors consists of tasks each of which is a unit of work to be allocated CPU time and other resources. Every processor is assigned to at most one task at any time. Every task is assigned to at most one processor at any time. No job is scheduled before its release time. Each task, τ_i, is typically characterized by the following temporal parameters:

- *Precedence Constraints*   Specify if any task(s) needs to precede other tasks.
- *Release or Arrival Time $r_{i,j}$*   The release time of the jth instance of task τ_i.
- *Phase ϕ_i*   The release time of the first instant of task τ_i.
- *Response Time*   Time span between the task activation and its completion.
- *Absolute Deadline d_i*   The instant by which the task must complete.
- *Relative Deadline D_i*   The maximum allowable response time of the task.
- *Laxity Type*   Notion of urgency or leeway in a task's execution.
- *Period p_i*   The minimum length of intervals between the release times of consecutive tasks.
- *Execution Time e_i*   The (maximum) amount of time required to complete the execution of a task i when it executes alone and has all the resources it requires.

Mathematically, some of the parameters just listed are related as follows:

$$\phi_i = r_{i,1} \qquad \text{and} \qquad r_{i,k} = \phi_i + (k-1) * p_i \qquad (3.1)$$

$d_{i,j}$: the absolute deadline of the jth instance of task τ_i is as follows:

$$d_{i,j} = \phi_i + (j - 1) * p_i + D_i \qquad (3.2)$$

If the relative deadline of a periodic task is equal to its period p_i, then

$$d_{i,k} = r_{i,k} + p_i = \phi_i + k * p_i \qquad (3.3)$$

where k is some positive integer greater than or equal to one, corresponding to the kth instance of that task.

3.2.1.2   *Typical Task Model*   A simple task model is presented in order to describe some standard scheduling policies used in real-time systems. The task model has the following simplifying assumptions:

- All tasks in the task set are strictly periodic.
- The relative deadline of a task is equal to its period/frame.
- All tasks are independent; there are no precedence constraints.
- No task has any nonpreemptible section, and the cost of preemption is negligible.
- Only processing requirements are significant; memory and I/O requirements are negligible.

For real-time systems, it is of the utmost importance that the scheduling algorithm produces a predictable schedule, that is, at all times it is known which task is going to execute next. Many RTOS use a round-robin scheduling policy because it is simple and predictable. Therefore, it is natural to describe that algorithm more rigorously.

3.2.2   Round-Robin Scheduling

In a round-robin system several processes are executed sequentially to completion, often in conjunction with a cyclic executive. In round-robin systems with time slicing, each executable task is assigned a fixed-time quantum called a time slice in which to execute. A fixed-rate clock is used to initiate an interrupt at a rate corresponding to the time slice. The task executes until it completes or its execution time expires, as indicated by the clock interrupt. If the task does not execute to completion, its context must be saved and the task is placed at the end of the executable list. The context of the next executable task in the list is restored, and it resumes execution. Essentially, round-robin scheduling achieves fair allocation of the CPU to tasks of the same priority by time multiplexing.

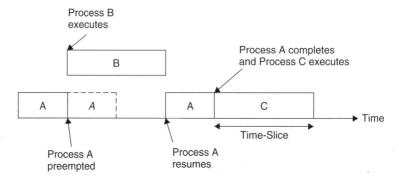

Figure 3.6   Mixed scheduling of three tasks.

Round-robin systems can be combined with preemptive priority systems, yielding a kind of mixed system. Figure 3.6 illustrates the process. Here processes A and C are of the same priority, whereas process B is of higher priority. Process A is executing for some time when it is preempted by task B, which executes until completion. When process A resumes, it continues until its time slice expires, at which time context is switched to process C, which begins executing.

3.2.3   Cyclic Executives

The cyclic-executive (CE) approach is very popular, as it is simple and generates a complete and highly predictable schedule. The CE refers to a scheduler that deterministically interleaves and sequentializes the execution of periodic tasks on a processor according to a pre-run-time schedule. In general terms, the CE is a table of procedure calls, where each task is a procedure, within a single do loop.

In the CE approach, scheduling decisions are made periodically, rather than at arbitrary times. Time intervals during scheduling decision points are referred to as frames or minor cycles, and every frame has a length, f, called the frame size. The major cycle is the minimum time required to execute tasks allocated to the processor, ensuring that the deadlines and periods of all processes are met. The major cycle or the hyperperiod is equal to the least common multiple (lcm) of the periods, that is, $\text{lcm}(p_1, \ldots p_n)$.

As scheduling decisions are made only at the beginning of every frame, there is no preemption within each frame. The phase of each periodic task is a non-negative integer multiple of the frame size. Furthermore, it is assumed that the scheduler carries out monitoring and enforcement actions at the beginning of each frame (see Figure 3.7).

Frames must be sufficiently long so that every task can start and complete with a single frame. This implies that the frame size, f, is to be larger than the execution time, e_i, of every task, T_i, that is,

$$C_1 : f \geq \max_{1 \leq i \leq n} (e_i)$$

(3.4)

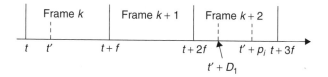

Figure 3.7 Constraints on the value of frame size.

In order to keep the length of the cyclic schedule as short as possible, the frame size, f, should be chosen so that the hyperperiod has an integer number of frames:

$$C_2 : \lfloor p_i/f \rfloor - p_i/f = 0 \tag{3.5}$$

In order to ensure that every task completes by its deadline, frames must be small so that between the release time and deadline of every task, there is at least one frame. The following relation is derived for a worst-case scenario, which occurs when the period of a process starts just after the beginning of a frame and, consequently, the process cannot be released until the next frame.

$$C_3 : 2f - gcd(p_i, f) \le D_i \tag{3.6}$$

where gcd is the greatest common divisor and D_i is the relative deadline of task i.

To illustrate the calculation of the framesize, consider the set of tasks shown in Table 3.1. The hyperperiod is equal to 660, since the least common multiple of 15, 20, and 22 is 660. The three conditions, C_1, C_2 and C_3 are evaluated as follows:

$$C_1 : \forall i f \ge e_i \Rightarrow f \ge 3$$
$$C_2 : \lfloor p_i/f \rfloor - p_i/f = 0 \Rightarrow f = 2, 3, 4, 5, 10, \dots$$
$$C_3 : 2f - gcd(p_i, f) \le D_i \Rightarrow f = 2, 3, 4, 5$$

From these three conditions, it can be inferred that a possible value for f could be any one of the values of 3, 4, or 5.

Table 3.1 Example task set for framesize calculation

τ_i	p_i	e_i	D_i
τ_2	15	1	14
τ_3	20	2	26
τ_4	22	3	22

3.2.4   Fixed-Priority Scheduling–Rate-Monotonic Approach

In the fixed-priority scheduling policy, the priority of each periodic task is fixed relative to other tasks. A seminal fixed-priority algorithm is the rate-monotonic (RM) algorithm [Liu73]. It is an optimal static priority algorithm for the task model previously described, in which a task with a shorter period is given a higher priority than a task with a longer period. The theorem, known as the rate-monotonic theorem is the most important (and useful) result of real-time systems theory. It can be stated as follows.

Theorem (Rate-monotonic) [Liu73]   Given a set of periodic tasks and preemptive priority scheduling, then assigning priorities such that the tasks with shorter periods have higher priorities (rate-monotonic), yields an optimal scheduling algorithm.

In other words, optimality of RM implies that if a schedule that meets all the deadlines exists with fixed priorities, then RM will produce a feasible schedule. A critical instant of a task is defined to be an instant at which a request for that task will have the largest response time. Liu and Layland proved that a critical instant for any task occurs whenever the task is requested simultaneously with requests for all higher-priority tasks. It is then shown that to check for RM schedulability it suffices to check the case where all tasks phasings are zero [Liu73].

The formal proof of the theorem is rather involved. However, a nice sketch of the proof due to Shaw uses an inductive argument [Shaw01].

Basis Step   Consider two fixed but non-RM priority tasks $\tau_1 = (e_1, p_1, d_1)$ and $\tau_2 = (e_2, p_2, d_2)$ where τ_2 has the highest priority, and $p_1 < p_2$. Suppose both processes are released at the same time. It is clear that this leads to the worst-case response time for τ_1. However, at this point, in order for both processes to be schedulable, it is necessary that $e_1 + e_2 \leq p_i$; otherwise, τ_1 could not meet its period or deadlines. Because of this relation between the compute times and the period (deadline) of τ_2, we can obtain a feasible schedule by reversing priorities, thereby scheduling τ_1 first, that is with RM assignment.

Induction Step   Suppose that $\tau_1, \ldots, \tau_n$ are schedulable according to RM, with priorities in ascending order, but the assignment is not RM. Let τ_i and τ_{i+1}, $1 \leq i < n$, be the first two tasks with non-RM priorities. That is, $p_i < p_{i+1}$. The "proof" proceeds by interchanging the priorities of these two processes and showing the set is still schedulable using the $n = 2$ result. The proof continues by interchanging non-RM pairs in this fashion until the assignment is RM. Therefore if a fixed-priority assignment can produce a feasible schedule, so can RM assignment. □

To illustrate rate-monotonic scheduling, consider the task set shown in Table 3.2.

Figure 3.8 illustrates the RM-schedule for the task set. All tasks are released at time 0. Since task τ_1 has the smallest period, it is the highest priority task and is scheduled first. Note that at time 4 the second instance of task τ_1 is released and it preempts the currently running task τ_3, which has the lowest priority.

Table 3.2 Sample task set for utilization calculation

τ_i	e_i	p_i	$u_i = e_i/p_i$
τ_1	1	4	0.25
τ_2	2	5	0.4
τ_3	5	20	0.25

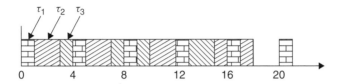

Figure 3.8 Rate-monotonic task schedule.

Here utilization, u_i, is equal to the fraction of time a task with period p_i and execution time e_i keeps a processor busy. Recall that the processor utilization of n tasks is given by Equation 1.2, that is $U = \sum_{i=1}^{n} e_i/p_i$.

3.2.4.1 Basic Results of Rate-Monotonic Algorithm Policy
From a practical point of view, it is important to know under what conditions a feasible schedule exists in the static-priority case. The following theorem [Liu73] yields a schedulable utilization of the rate-monotonic algorithm (RMA). Note that the relative deadline of every task is equal to its period.

Theorem (RMA Bound) Any set of n periodic tasks is RM schedulable if the processor utilization, U, is no greater than $n(2^{1/n} - 1)$.

This means that whenever U is at or below the given utilization bound, a schedule can be constructed with RM. In the limit when the number of tasks $n = \infty$, the maximum utilization limit is

$$\lim_{n \to \infty} n(2^{1/n} - 1) = \ln 2 \approx 0.69 \qquad (3.7)$$

The calculation of the limit in Equation 3.7 is straightforward but worth documenting. First recall that

$$\frac{d}{dx}a^x = (\ln a)a^x dx$$

Hence,

$$\frac{d}{dx}2^{n^{-1}} = (\ln 2)2^{n^{-1}}(-n^{-2})$$

Now

$$\lim_{n \to \infty} n(2^{n^{-1}} - 1) = \lim_{n \to \infty} \frac{(2^{n^{-1}} - 1)}{n^{-1}}$$

And, by L'Hôpital's rule

$$\lim_{n \to \infty} \frac{(2^{n^{-1}} - 1)}{n^{-1}} = \lim_{n \to \infty} \frac{\ln 2(2^{n^{-1}})(-n^{-2})}{-n^{-2}} = \lim_{n \to \infty} \frac{\ln 2(2^{1/n})}{1}$$

passing to the limit, it can be seen that this is just $\ln 2$.

To illustrate, the value of RMA bound for various values of n is given in Table 3.3 and illustrated in Figure 3.9. Note that the RMA utilization bound is sufficient, but not necessary. That is, it is not uncommon in practice to construct a periodic task set with total processor utilization greater than the RMA bound but still RM-schedulable. For example, the task set shown in Table 3.2 has a total utilization of 0.9, which is greater than the RM utilization bound of 0.69, but it is still schedulable using the RM policy as illustrated in Figure 3.8. Recall from Chapter 1, Table 1.3, the advice regarding utilization zones and recommendations. Indeed, many complex real-time systems are constructed with a utilization greater than 80%, with no problems.

3.2.5   Dynamic-Priority Scheduling: Earliest-Deadline–First Approach

In contrast to fixed-priority algorithms, in dynamic-priority schemes the priority of the task with respect to that of the other tasks changes as tasks are released and

Table 3.3   Upper bound on utilization *U* for *n* tasks scheduled using the rate-monotonic discipline

n	1	2	3	4	5	6	...	∞
RMA bound	1.0	0.83	0.78	0.76	0.74	0.73	...	0.69

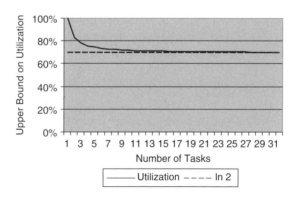

Figure 3.9 Upper bound on utilization in a rate-monotonic system as a function of the number of tasks. Notice how it rapidly converges to 0.69.

completed. One of the most well-known dynamic algorithms, earliest-deadline-first (EDF), deals with deadlines rather than execution times. The ready task with the earliest deadline has the highest priority at any point of time.

The following theorem gives the condition under which a feasible schedule exists under the EDF priority scheme [Liu73].

Theorem [EDF Bound]   A set of n periodic tasks, each of whose relative deadline equals its period, can be feasibly scheduled by EDF if and only if

$$\sum_{i=1}^{n}(e_i/p_i) \leq 1 \tag{3.8}$$

Figure 3.10 illustrates the EDF scheduling policy. The schedule is produced for the task set shown in Table 3.4.

Although τ_1 and τ_2 release simultaneously, τ_1 executes first because its deadline is earliest. At $t = 2$, τ_2 can execute. Even though τ_1 releases again at $t = 5$, its deadline is not earlier than τ_3. This sequence continues until time $t = 15$ when τ_2 is preempted, as its deadline is later ($t = 21$) than $\tau_1(t = 20)$; τ_2 resumes when τ_1 completes.

3.2.5.1 *Basic Results of EDF Policy*   EDF is optimal for a uniprocessor, with task preemption being allowed. In other words, if a feasible schedule exists, then the EDF policy will also produce a feasible schedule. There is never a processor idling prior to a missed deadline.

3.2.5.2 *Comparison of RMA and EDF Policies*   Schedulable utilization is a measure of performance of algorithms used to schedule periodic tasks. It is desired that a scheduling algorithm yield a highly schedulable utilization. By

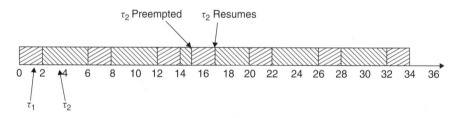

Figure 3.10   EDF task schedule for task set in Table 3.4.

Table 3.4   Task set for example of EDF scheduling

τ_i	e_i	p_i
τ_1	2	5
τ_2	4	7

Table 3.5   Task set that illustrates the advantage of RM over EDF in the presence of missed deadlines[a]

τ_i	r_i	e_i	p_i
τ_1	0	2	5
τ_2	0	4	6

[a]The reader is encouraged to draw the execution trace timeline as in Figure 3.10.

this criterion, dynamic-priority algorithms are evidently better than fixed-priority scheduling algorithms. EDF is more flexible and achieves better utilization. However, the timing behavior of a system scheduled according to a fixed-priority algorithm is more predictable than that of a system scheduled according to a dynamic-priority algorithm. In case of overloads, RM is stable in the presence of missed deadlines; the same lower-priority tasks miss deadlines every time. There is no effect on higher-priority tasks. In contrast, when tasks are scheduled using EDF, it is difficult to predict which tasks will miss their deadlines during overloads. Also, note that a late task that has already missed its deadline has a higher priority than a task whose deadline is still in the future. If the execution of a late task is allowed to continue, this may cause many other tasks to be late. A good overrun management scheme is thus needed for such dynamic-priority algorithms employed in systems where overload conditions cannot be avoided.

The following set of tasks illustrate this phenomenon (Table 3.5). It is easy to check that for this case, EDF misses deadlines for both τ_1 and τ_2 (assuming that the late task is allowed to complete at its current assigned priority). RM misses the deadline of τ_2 only (every time).

As a general comment, RM tends to need more preemption; EDF only preempts when an earlier-deadline task arrives.

3.3   INTERTASK COMMUNICATION AND SYNCHRONIZATION

The task model being considered so far assumes that all tasks are independent and that all tasks can be preempted at any point of their execution. However, from a practical viewpoint, this assumption is unreasonable, as task interaction is needed in most common applications. In this section, the effect of task synchronization to maintain the consistency/integrity of the shared data/resources is examined. The main concern is how to minimize blocking that may arise in a uniprocessor system when concurrent tasks use shared resources. Related to these issues is the problem of sharing certain resources that can only be used by one task at a time.

In the previous section, techniques for multitasking were discussed in a way that each task operated in isolation from the others. In practice, strictly controlled mechanisms are needed that allow tasks to communicate, share resources, and synchronize activity. Most of the mechanisms discussed in this section are easy to understand casually, but a deep understanding is harder to attain. Misuse of these techniques, semaphores in particular, can have a disastrous effect.

3.3.1 Buffering Data

Several mechanisms can be employed to pass data between tasks in a multitasking system. The simplest and fastest among these is the use of global variables. Global variables, though considered contrary to good software engineering practices, are often used in high-speed operations.

One of the problems related to using global variables is that tasks of higher priority can preempt lower-priority routines at inopportune times, corrupting the global data. For example, one task may produce data at a constant 100 units per second, whereas another may consume these data at a rate less than 100 units per second. Assuming that the production interval is finite (and relatively short), the slower consumption rate can be accommodated if the producer fills a storage buffer with the data. The buffer holds the excess data until the consumer task can catch up. The buffer can be a queue or other data structure, including an unorganized mass of variables. Of course, if the consumer task consumes this information faster than it can be produced, or if the consumer cannot keep up with the producer, problems occur. Selection of the appropriate size buffer is critical in reducing or eliminating these problems.

3.3.2 Time-Relative Buffering

A common use of global variables is in double buffering or Ping-Pong buffering. This technique is used when time-relative (correlated) data need to be transferred between cycles of different rates, or when a full set of data is needed by one process, but can only be supplied slowly by another process. This situation is simply a variant of the classic bounded-buffer problem in which a block of memory is used as a repository for data produced by "writers" and consumed by "readers." A further generalization is the readers and writers problem in which there are multiple readers and multiple writers of a shared resource, as shown in Figure 3.11.

Many telemetry systems, which transmit blocks of data from one device to another, use double-buffering schemes with a hardware or software switch to alternate the buffers. This strategy is also used in disk controllers, graphical interfaces, navigation equipment, robot controls, and many other places. For example, in the operator display for the pasta sauce factory, suppose, lines are drawn on the screen one by one until the image is completed. In an animated system, it is undesirable to see this drawing process. If, however, the software draws on one screen image while displaying the other and then flips the screens when the new drawing is complete, the individual line drawing commands will not be seen (Figure 3.12). If the screens can be updated at about 30 screens per second, the operator's display will appear fully animated.

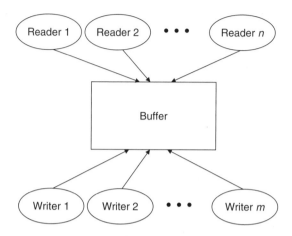

Figure 3.11 Readers and writers problem, with *n* readers and *m* writers. The shared resource is a bounded buffer. The buffer can only be written to or read from by one reader or writer at a time.

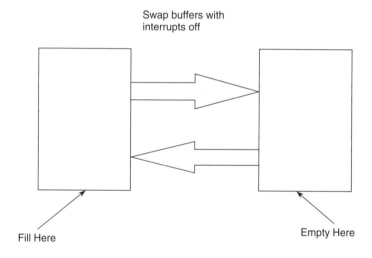

Figure 3.12 Double-buffering configuration. Two identical buffers are filled and emptied by alternating tasks. Switching is accomplished either by a software pointer or hardware discrete.

As an example of time-correlated buffering, consider the inertial measurement unit. It reads $x, y,$ and z accelerometer pulses in a 10-millisecond cycle. These data are to be processed in a 40-millisecond cycle, which has lower priority than the 10-millisecond cycle (i.e., it can be preempted). The accelerometer data processed in the 40-millisecond cycle must be time-relative; that is, it is undesirable to process x and y accelerometer pulses from time t along with z accelerometer pulses from time $t + 1$. This scenario could occur if the 40-millisecond cycle has completed processing the *x* and *y* data, but gets interrupted by the 10-millisecond cycle. To avoid this problem, use buffered variables *xb, yb,* and *zb* in the 40-millisecond cycle and buffer them, with interrupts disabled. The 40-millisecond routine might contain the following C code to handle the buffering:

```
introf();            /* disable interrupts */
xb=x;                /* buffer data        */
yb=y;
zb=z;
intron();            /* enable interrupts  */

process(xb,yb,zb);   /* use buffered data  */
```

In practice, the first procedure in any cycle would be a buffering routine to buffer all data from tasks of higher priority into the current task ("buffer in" routine). The last procedure in the cycle is a routine to buffer out data to any tasks of lower priority ("buffer out" routine).

3.3.3 Ring Buffers

A special data structure called a circular queue or ring buffer is used in the same way as a queue and can be used to solve the problem of synchronizing multiple reader and writer tasks. Ring buffers, however, are easier to manage than double buffers or queues when there are more than two readers or writers.

In the ring buffer, simultaneous input and output to the list are achieved by keeping head and tail indices. Data are loaded at the tail and read from the head. Figure 3.13 illustrates this. Suppose the ring buffer is a structure

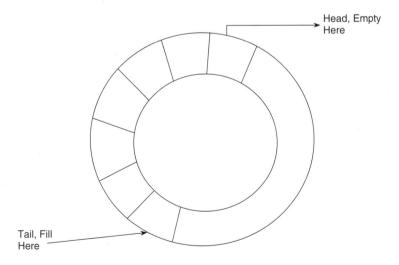

Figure 3.13 A ring buffer. Processes write to the buffer at the tail index and read data from the head index. Data access is synchronized with a counting semaphore set to size of ring buffer, to be discussed later.

of type `ring_buffer` that includes an integer array of size of N called contents, namely,

```
typedef struct ring_buffer
{
    int contents[N];
    int head;
    int tail;
}
```

It is further assumed that the `head` and `tail` indices have been initialized to 0, that is, the start of the buffer.

An implementation of the `read(data,S)` and `write(data,S)` operations, which reads from and writes to ring buffer S, respectively, are given below in C code.[3]

```
void read (int data, ring_buffer *s)
{
    if (s->head==s->tail)
      data=NULL;                    /* underflow */
    else
      {
        data=s->contents +head;    /* retrieve data from buffer */
        s->head=(s->head+1) % N;   /* decrement head index */
      }
}
void write (int data, ring_buffer *s)
{
    if ((s->tail+1) %N==head)
      error();                       /* overflow, invoke error handler */
    else
    {
        s->contents+tail=data;
        tail=(tail+1) % N;    /*take care of wrap-around */
    }
}
```

Additional code is needed to test for the overflow condition in the ring buffer, and the task using the ring buffer needs to test the data for the underflow (NULL) value. An overflow occurs when an attempt is made to write data to a full queue. Underflow is the condition when a task attempts to retrieve data from an empty buffer. Implementation of these exception handlers is left as Exercise 3.31.

Ring buffers can be used in conjunction with a counting or binary semaphore to control multiple requests for a single resource such as memory blocks, modems, and printers.

3.3.4 Mailboxes

Mailboxes or message exchanges are an intertask communication device available in many commercial, full-featured operating systems. A mailbox is a mutually

---

[3] For those unfamiliar with C, the notation "->" indicates accessing a particular field of the structure that is referenced by the pointer.

agreed upon memory location that one or more tasks can use to pass data, or more generally for synchronization. The tasks rely on the kernel to allow them to write to the location via a post operation or to read from it via a pend operation.

The mailbox operations, pend and post can be described with the following interfaces:

```
void pend (int data, s);

void post (int data, s);
```

The difference between the pend operation and simply polling the mailbox is that the pending task is suspended while waiting for data to appear. Thus, no time is wasted continually checking the mailbox; that is, the busy waiting condition is eliminated.

The datum that is passed can be a flag used to protect a critical resource (called a key), a single piece of data, or a pointer to a data structure. In most implementations, when the key is taken from the mailbox, the mailbox is emptied. Thus, although several tasks can pend on the same mailbox, only one task can receive the key. Since the key represents access to a critical resource, simultaneous access is precluded.

3.3.4.1 Mailbox Implementation
Mailboxes are best implemented in systems based on the task control block model with a supervisor task. A table containing a list of tasks and needed resources (e.g., mailboxes, printers, etc.) is kept along with a second table containing a list of resources and their states. For example, in Tables 3.6 and 3.7, three resources currently exist; a printer and two mailboxes. Here, the printer is being used by tasks #100, while mailbox #1 is being used (currently being read from or written to) by task #102. Task #104 is pending on mailbox #1 and is suspended because it is not available. Mailbox #2 is currently not being used or pended on by any task.

When the supervisor is invoked by a system call or hardware interrupt, it checks the tables to see if some task is pending on a mailbox. If the key is available (key status is "full"), then that task must be restarted. Similarly, if a task posts to a mailbox, then the operating system must ensure that the key is placed in the mailbox and its status updated to "full."

There are often other operations on the mailbox. For example, in some implementations, an accept operation is permitted. accept allows tasks to read the

Table 3.6 Task resource request table

Task ID #	Resource	Status
100	Printer	Has it
102	Mailbox 1	Has it
104	Mailbox 1	Pending

Table 3.7   Resource table used in conjunction with task resource request table

Resource	Status	Owner
Printer 1	Busy	100
Mailbox 1	Busy	102
Mailbox 2	Empty	None

key if it is available, or immediately return an error code if the key is not available. In other implementations, the pend operation is equipped with a timeout, to prevent deadlocks.

3.3.5   Queues

Some operating systems support a type of mailbox that can queue multiple pend requests. These systems provide qpost, qpend, and qaccept operations to post, pend, and accept a data to/from the queue. In this case, the queue can be regarded as any array of mailboxes, and its implementation is facilitated through the same resource tables already discussed.

Queues should not be used to pass arrays of data; pointers should be used instead. Queues are useful in implementing device servers where a pool of devices is involved. Here the ring buffer holds requests for a device, and queues can be used at both the head and the tail to control access to the ring buffer. Such a scheme is useful in the construction of device-controlling software.

3.3.6   Critical Regions

Multitasking systems are concerned with resource sharing. In most cases, these resources can only be used by one task at a time, and use of the resource cannot be interrupted. Such resources are said to be serially reusable and they include certain peripherals, shared memory, and the CPU. While the CPU protects itself against simultaneous use, the code that interacts with the other serially reusable resources cannot. Such code is called a critical region. If two tasks enter the same critical region simultaneously, a catastrophic error can occur. For example, consider two C programs, Task_A and Task_B, which are running in a round-robin system. Task_B outputs the message "I am task_B" and Task_A outputs the message "I am Task_A." In the midst of printing, Task_B is interrupted by Task_A, which begins printing. The result is the incorrect output:

```
I am I am Task_A Task_B
```

The emphasis is placed on the middle text to show that it interrupted the output of the Task_B. More serious complications could arise if both tasks were controlling devices in an embedded system. Simultaneous use of a serial reusable resource results in a collision. The concern, then, is to provide a mechanism for preventing collisions.

3.3.7 Semaphores

The most common method for protecting critical regions involves a special variable called a semaphore. A semaphore S is a memory location that acts as a lock to protect critical regions. Two operations, wait and signal are used either to set or to reset the semaphore. Traditionally, one denotes the wait operation as P(S) and the signal operations V(S).[4] The primitive operations are defined by the following C code:

```
void P(int S)
{
    while (S == TRUE);
    S=TRUE;
}

void V(int S)
{
    S=FALSE;
}
```

The wait operation suspends any program calling until the semaphore S is FALSE, whereas the signal operation sets the semaphore S to FALSE. Code that enters a critical region is bracketed by calls to wait and signal. This prevents more than one process from entering the critical region. Incidentally, recall that C passes by value unless forced to pass by reference by passing a pointer; therefore, when calling functions the dereferencing operator "&" should be used. However, for convenience of notation, when a parameter is passed, it is as if the address of the parameter is passed to the function. Alternatively, the parameter can be viewed as a global variable.

Now consider two concurrent processes in a multitasking system illustrated by the pseudocode shown side-by-side:

```
Process_1
.
.
P(S)
critical region
V(S)
.
.
```

```
Process_2
.
.
P(S)
critical region
V(S)
.
.
```

Both processes can access the same critical region, so semaphores are used to protect the critical region. Note that the semaphore S should be initialized to FALSE before either process is started.

---

[4] P and V are the first letters of the Dutch "to test" – *proberen* – and "to increment" – *verhogen*. They were first suggested by Dijkstra [Dijkstra65]. P and wait, and V and signal will be used synonymously throughout the text.

Again, for example, consider the C code for Task_A and Task_B, mentioned before. The problem can be solved by bracketing the output statements with semaphore operations as follows:

```
void Task_A(void)
{
    P(S);
    printf("I am Task_A");
    V(S);
}
```

```
void Task_B(void)
{
    P(S);
    printf("I am Task_B");
    V(S);
{
```

Assume that S is within the scope of both Task_A and Task_B and that it is initialized to FALSE by the system.

A process will spend much of its time in wait semaphore operation (busy–wait) if a large amount of contention for the resource is protected by it. Because the wait operation involves a repeated test of a while loop condition, semaphore protection is sometimes called a spin lock. Furthermore, in many books the semaphore variable of choice is mutex, emphasizing the fact that mutual exclusion is enforced.

Semaphores appear to be simple, yet they are subtly complex. A thorough understanding of their nuances is essential to avoid implanting logic errors that lead to insidious problems. Some of these will be discussed later.

3.3.7.1 *Mailboxes and Semaphores*

Mailboxes can be used to implement semaphores if semaphore primitives are not provided by the operating system. In this case, there is the added advantage that the pend instruction suspends the waiting process rather than actually waiting for the semaphore. For example, using the dummy data, KEY, the operations are implemented as follows

```
void P(int S)
{
   int KEY=0;

   pend(KEY,S);
}
```

The accompanying signal operation utilizes the mailbox post operation.

```
void V(int S)
{
   int KEY=0;

   post(KEY,S);
}
```

A binary semaphore can be used to implement a mailbox where initially mutex=1 and proc_sem=0. A sketch of the implementation in C follows:

```
bool mutex, proc_sem, full_slots, empty_slots;
void post(int mailbox, int message)
{
     wait(mutex);
```

```
if (empty_slots)
      {
            insert(mailbox,message);
            update();
            signal(mutex);
            signal(proc_sem);
      }
else
      {

            :
            :
            signal(mutex);
            wait(proc_sem);
            wait(mutex);
            insert(mailbox,message);
            update();
            signal(mutex);
            signal(proc_sem);
      }
};

void pend(int *mailbox,int *message)
{
      wait(mutex);
      if((full_slots))
            {
                  extract(mailbox,message);
                  update();
                  signal(mutex);
            }
      else
            {

                  :
                  :
                  signal(mutex);
                  wait(proc_sem);
                  wait(mutex);
                  extract(mailbox,message);
                  update();
                  signal(mutex);
            }
};
```

As an example application, the driver–controller interface is readily accomplished with semaphores (Figure 3.14).

A device driver is software that communicates directly with the firmware of a particular hardware device. The firmware is implemented on board the device either via hard-coded logic, a microcontroller, a field-programmable gate array, (FPGA) or even an on-board processor. Firmware provides the handshaking and processing at the interface. Here the driver signals the controller with a V(busy) then waits for completion with a P(done). The controller waits for work with a P(busy), then indicates completion with V(done).

3.3.7.2 Counting Semaphores
The P and V semaphores are called binary semaphores because they can take one of two values. Alternatively, a counting semaphore or general semaphore can be used to protect pools of resources, or to

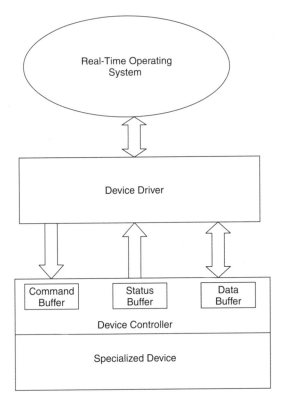

Figure 3.14   The device–controller interface.

keep track of the number of free resources. The semaphore must be initialized to the total number of free resources before real-time processing can commence.

The new `wait` and `signal` semaphore primitives, MP and MV, are designed to present access to a semaphore-protected region when the semaphore is less than or equal to zero. The semaphore is released or signaled by incrementing it. The counting `wait` becomes

```
void MP(int S)
{
    S=S-1;
    while (S < 0);
}
```

and `signal` becomes

```
void MV(int S)
{
    S=S+1
}
```

Some real-time kernels provide only binary semaphores, but the counting semaphore can be simulated with binary semaphores in the following way. Suppose S and T are binary semaphores and P(S) and V(S) are the wait and signal operations, respectively, on binary semaphores. The counting semaphore operations MP(R) and MV(R) can be created on multiple semaphores R using global binary semaphores S and T and integer R as follows:

```
void MP (int R)               /* multiple wait */
{
    P(S);                     /* lock counter       */
    R=R-1;                    /* request a resource */
    if(R < 0)                 /* none available?    */
    {
        V(S);                 /* release counter    */
        P(T);                 /* wait for free resource   */
    };
    V(S);                     /* release counter    */
}

void MV(int R)                /* multiple signal    */
{
    P(S);                     /* lock counter       */
    R=R+1;                    /* free resource      */
    if (R <= 0)               /* give that task the go ahead */
        V(T);
    else
        V(S);                 /* release counter}
}
```

The integer R keeps track of the number of free resources. Binary semaphore S protects R, and binary semaphore T is used to protect the pool of resources. The initial values of S are set to FALSE, T to TRUE, and R to the number of available resources in the kernel.

The operation of the code is subtly intricate. In the multiple wait routine, MP, the counting variable is locked and decremented by 1 to indicate a resource request. If enough resources are available (R >0), then the counting variable is released and processing proceeds. If no resources are free, then the counting variable is released and the process waits until there is a free resource using the P(T) call. Upon return from the P(T) call the counting variable is freed.

In the multiple-signal routine, the counting variable is locked and incremented. If any process was pending on the resource pool (R ≤ 0), then that process is signaled to go ahead using the V(T) call. If no process was waiting, then the counter is released. Note that when the waiting process gets the go ahead, it releases the counting variable with the final V(S) call in the procedure MP. Finally, it is easy to see that a binary semaphore can be simulated with a counting semaphore simply by initializing the counting semaphore to 1. It is helpful to study this code thoroughly by working out examples for different initial values of R and different resource request scenarios.

3.3.7.3 Problems with Semaphores   Certain problems arise if the operation of testing and subsequent setting of a semaphore are not atomic – that is, uninterruptible. To illustrate the problem, consider the following example.

Suppose two tasks in a round-robin system with time slicing are using a resource protected by a binary semaphore S. The wait operation discussed in the previous section,

```
void P (int S)
{
    while (S == TRUE);
    S = TRUE;
}
```

would generate assembly instructions, in 2-address code, similar to

```
@1    LOAD    R1,&S
      TEST    R1,1
      JEQ     @1       ; S = TRUE?
      STORE   &S,1     ; S := TRUE
```

where "1" is TRUE and "0" is FALSE. Suppose the process using the semaphore primitive were interrupted between the TEST and STORE instructions. The interrupting routine, which might use the same resource, finds S to be available and begins using it. If this task then suspends (because, say, its time slice ran out) and the interrupted task resumes, it will still see the device as free because the old contents of S are still in register 1. Thus, two tasks attempt to use the same resource and a collision occurs. This problem may occur infrequently, and so it may be difficult to test and detect.

3.3.7.4 The Test-and-Set Instruction   To solve the problem of atomic operation between testing a memory location and storing a specific value in it, most instruction sets provide a test-and-set macroinstruction. The instruction fetches a word from memory and tests the high-order (or other) bit. If the bit is 0, it is set to 1 and stored again, and a condition code of 0 is returned. If the bit is 1, a condition code of 1 is returned and no store is performed. The fetch, test, and store are indivisible.

The wait and signal operations can be implemented easily with a test-and-set instruction.

```
void P(int S);
{
    while (test_and_set(S) == TRUE); /* wait */
}

void V(int S);
{
    S=FALSE
}
```

Procedure P would generate assembly language code, that may look like

```
@loop   TANDS &S
        JNE @loop
```

where TANDS is a test-and-set instruction.

If a machine does not support the TANDS instruction, a semaphore can still be implemented, for example, by using the solution first proposed by Dijkstra [Dijkstra68b], shown here in C.

```
void P(int S)
{
int temp=TRUE;

    while (temp<>TRUE)
    {
        disable();              /*disable interrupts */
        temp=S;
        S=TRUE;
        enable();               /* enable interrupts */
    };
}
```

Of course, disable() and enable() must be uninterruptible procedures or in-line assembly code.

3.3.8 Other Synchronization Mechanisms

Monitors are abstract data types that encapsulate the implementation details of the serial reusable resource and provides a public interface. Instances of the monitor type can only be executed by one process at a time. Monitors can be used to implement any critical section.

Certain languages provide for synchronization mechanisms called event flags. These constructs allow for the specification of an event that causes the setting of some flag. A second process is designed to react to this flag. Event flags in essence represent simulated interrupts created by the programmer. Raising the event flag transfers the flow of control to the operating system, which can then invoke the appropriate handler. Tasks that are waiting for the occurrence of an event are said to be blocked.

3.3.9 Deadlock

When tasks are competing for the same set of two or more serially reusable resources, then a deadlock situation or deadly embrace may ensue. The notion of deadlock is best illustrated by example.

For example, TASK_A requires resources 1 and 2, as does Task_B. Task_A is in possession of resource 1 but is waiting on resource 2. Task_B is in possession of resource 2 but is waiting on resource 1. Neither Task_A nor Task_B will relinquish the resource until its other request is satisfied. The situation is illustrated as follows where two semaphores, S and R, are used to protect resource 1 and resource 2, respectively, by the side-by-side pseudocode:

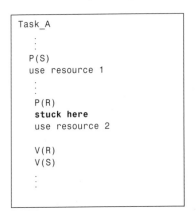
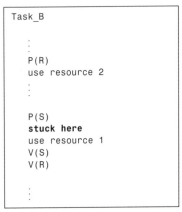

```
Task_A                          Task_B

   :                               :
   :                               :
P(S)                            P(R)
use resource 1                  use resource 2
   :                               :
   :                               :
P(R)                            P(S)
stuck here                      stuck here
use resource 2                  use resource 1
                                V(S)
V(R)                            V(R)
V(S)
   :                               :
   :                               :
```

Pictorially, if semaphore S guards device 1 and semaphore R guards device 2, then the realization of the two might appear as the resource diagram in Figure 3.15.

Deadlock is a serious problem because it cannot always be detected through testing. In addition, it may occur very infrequently, making the pursuit of a known deadlock problem difficult. Finally, the solution of the deadlock problem is by no means straightforward and is not without consequences, including significant impact on real-time performance.

While it is unlikely that such obvious race and deadlock scenarios as those just described are going to be created, bad designs might be masked by complex structures. For example, if the system resource diagram contains subgraphs that look like Figure 3.15, that is, it contains cycles then deadlock can occur. Petri net analysis can also be helpful in identifying such situations (see Chapter 4).

Starvation differs from deadlock in that at least one process is satisfying its requirements but one or more are not. In deadlock, two or more processes cannot advance due to mutual exclusion. Livelock is a related condition in which two or more processes are stuck in their busy wait loops and hence are "alive" but unable to progress.

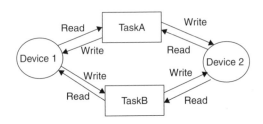

Figure 3.15   Deadlock realization in a resource diagram.

Four conditions are necessary for deadlock:

1. Mutual exclusion
2. Circular wait
3. Hold and wait
4. No preemption

Eliminating any one of the four necessary conditions will prevent deadlock from occurring.

Mutual exclusion applies to those resources that cannot be shared, such as printers, disk devices, and output channels. Mutual exclusion can be removed through the use of programs that allow these resources to appear to be shareable by applications (spoolers and daemons).

The circular-wait condition occurs when a sequential chain of processes exist that hold resources needed by other processes further down the chain (such as in cyclic processing). One way to eliminate circular wait is to impose an explicit ordering on the resources and to force all processes to request all resources above the number of the lowest one needed. For example, suppose that a collection of devices is ranked as shown in Table 3.8. Now if a process wishes to use just the printer, it will be assigned the printer, motor-control channel, and the monitor. If another process requires the monitor only, it will have to wait until the processes releases the resources. It is easy to see that such an approach eliminates the circular wait at the potential cost of starvation.

The hold-and-wait condition occurs when processes request a resource and then lock that resource until subsequent resource requests are filled. One solution to this problem is to allocate to a process all potentially required resources at the same time, as in the previous example. This approach can, however, lead to starvation in other processes. Another solution is never to allow a process to lock more than one resource at a time. For example, when writing one semaphore-protected disk file to another, lock one file and copy a record, unlock that file, lock the other file, write the record, and so on. This, of course, can lead to poor resource utilization as well as windows of opportunity for other processes to interrupt and interfere with resource utilization.

Table 3.8 Device numbering scheme to eliminate the circular wait condition

Device	Number
Disk	1
Printer	2
Motor control	3
Monitor	4

Finally, eliminating preemption will preclude deadlock. Namely, if a low-priority task holds a resource protected by semaphore S, and if a higher-priority task interrupts and then waits for semaphore S, the priority inversion will cause the high-priority task to wait forever, since the lower-priority task can never run to release the resource and signal the semaphore. If the higher-priority task is allowed to preempt the lower one, then the deadlock will be avoided. However, this solution can lead to starvation in the low-priority process, as well as to nasty interference problems. For example, what if the low-priority task had locked the printer for output, and now the high-priority task starts printing?

Two other ways of combating deadlock are to avoid it completely by identifying unsafe states using the Banker's algorithm, or to detect it and recover from it. Detection of deadlock is not always easy, although watchdog timers or system monitors can be used for this purpose.

3.3.9.1 *Deadlock Avoidance*   The best way to deal with deadlock is to avoid it altogether. Several techniques for avoiding deadlock are available. For example, if the semaphores protecting critical resources are implemented by mailboxes with time-outs, then deadlocking cannot occur, but starvation of one or more tasks is possible.

Suppose a lock refers to any semaphore used to protect a critical region. Then the following resource-management approach is recommended to help avoid deadlock.

1. Minimize the number of critical regions as well as minimizing their size.
2. All processes must release any lock before returning to the calling function.
3. Do not suspend any task while it controls a critical region.
4. All critical regions must be error free.
5. Do not lock devices in interrupt handlers.
6. Always perform validity checks on pointers used within critical regions. Pointer errors are common in certain languages, like C, and can lead to serious problems within the critical regions.

Nevertheless items 1 through 6 are difficult to achieve and other means are often necessary to avoid deadlock.

1. *The Banker's Algorithm*   The Banker's Algorithm can be used to avoid unsafe situations that can lead to deadlock. The technique, suggested by Dijkstra, uses the analogy of a small-town bank, its depositors, and cash reserve [Dijkstra68b]. In the analogy, depositors have placed money in the bank and could potentially withdraw it all. The bank never keeps all of its deposits on hand as cash (it invests about 95% of it). If too many depositors were to withdraw their savings simultaneously, the bank could not fulfill the requests. The Banker's Algorithm was originally formulated for a single resource type, but was soon extended for multiple resource types by Habermann [Habermann69].

The algorithm ensures that the number of resources attached to all processes and potentially needed for at least one to complete, can never exceed the number of resources for the system. These unsafe states might lead to deadlock, but do not necessarily. As an example, consider a system with three processes, A, B, and C, and a pool of 10 resources of a certain type (e.g., memory blocks). It is known that process A will never need more than 6 blocks at any one time. For processes B and C the totals are 5 and 7, respectively. A table such as the one that follows is constructed to keep track of the resource needs and availability.

Process	Max Requirement	Used	Possibly Needed
A	6	0	6
B	5	0	5
C	7	0	7
		Total available	10

When resources are requested, the operating system updates the table, ensuring that a possible deadlock state is not reached. An example of a "safe state" is:

Process	Max Requirement	Used	Possibly Needed
A	6	2	4
B	5	3	2
C	7	1	6
		Total available	4

Here, the requirements of process A or B can be satisfied by the total available, so the state is safe. An example of an "unsafe state" is:

Process	Max Requirement	Used	Possibly Needed
A	6	4	2
B	5	3	2
C	7	2	5
		Total available	1

In this case, the total requirements of no task can be met with the total available resources, so deadlock could ensue.

2. *Generalized Banker's Algorithm*   The Banker's Algorithm can be extended to a set of two or more resources by expanding the table corresponding to one resource type to one that tracks multiple resources. Formally, consider a set of processes $p_1 \cdots p_n$ and a set of resources $r_1 \cdots r_m$. Form the matrix $max[i, j]$ that represents the maximum claim of resource type j by process i. Let the matrix $alloc[i, j]$ represent the number of units of resource j held by process i. Upon each request of resources of type j, c_j, the resulting number of available resources of type j if the resource is granted $avail[j]$, is computed

$$avail[j] = c_j - \sum_{0 \leq i < n} alloc[i, j] \qquad (3.9)$$

From this information, it can be determined whether or not an unsafe state will be entered.

The procedure is as follows.

1. Update $alloc[i, j]$, yielding $alloc[i, j]'$, the new alloc table.
2. Given c, max, and alloc', compute the new avail vector.
3. Determine if there exists a p_i such that $max[i, j] - alloc[i, j]' \leq avail[j]$ for $0 \leq j < m$ and $0 \leq i < n$.

    a. If no such p_i exists, then the state is unsafe.

    b. If $alloc[i, j]'$ is for all i and j, the state is safe.

Finally, set $alloc[i, j]'$ to 0 and deallocate all resources held by process i. For example, in the previous scenario suppose that the three processes A, B, C share resources of type R_1, R_2, and R_3. Resource R_1, is the memory block resource described in the previous example. Resources R_2 and R_3 represent other resources, such as printers and disk drives, respectively. The initial resource table now becomes

Process	Max Requirement			Used			Possibly Needed		
	R_1	R_2	R_3	R_1	R_2	R_3	R_1	R_2	R_3
A	6	3	4	0	0	0	6	3	4
B	5	3	5	0	0	0	5	3	5
C	7	2	1	0	0	0	7	2	1
				Total available		10	4	5	

Again, when a process requests a resource, the table is inspected to ensure that a safe state will result in granting the request. An example of a safe state is shown below. It is safe because at least process A can satisfy all of its resource needs with available resources.

Process	Max Requirement			Used			Possibly Needed		
	R_1	R_2	R_3	R_1	R_2	R_3	R_1	R_2	R_3
A	6	3	4	2	2	2	4	1	2
B	5	3	5	3	1	0	2	2	5
C	7	2	1	1	0	1	6	2	0
				Total available			4	1	2

However, the following state is unsafe:

Process	Max Requirement			Used			Possibly Needed		
	R_1	R_2	R_3	R_1	R_2	R_3	R_1	R_2	R_3
A	6	3	4	2	2	2	4	1	2
B	5	3	5	3	1	1	2	2	4
C	7	2	1	1	0	1	6	2	0
				Total available			4	1	1

because none of the processes can be fully satisfied in a potential request for all their maximum needs.

Unfortunately, the Banker's Algorithm adds excessive overhead in real-time systems. Moreover, the resource needs for each task may not be known a priori, which is necessary for proper implementation of the algorithm.

3.3.9.2 Detect and Recover

Assuming that a deadlock situation can be detected (for example, by using a watchdog timer), what can be done about it? If the deadlock situation is known to occur infrequently, for example, once per year, and the system is not a critical one, this approach may be acceptable. For example, if in a video game this problem is known to occur rarely, the effort needed to detect and correct the problem may not be justified, given the cost and function of the system. For any of the example systems introduced in Chapter 1, however, ignoring this problem is unacceptable.

What about handling the deadlock by resetting the system? Again, this may be unacceptable for most critical systems.

Finally, if a deadlock is detected, some form of rollback to a predeadlock state can be performed, although this may lead to a recurrent deadlock, and operations such as writing to certain files or devices cannot be rolled back easily.

3.3.10 Priority Inversion

When a low-priority task blocks a higher-priority one, a priority inversion is said to occur. Consider the following example. Let three tasks τ_1, τ_2, and τ_3 have decreasing priorities, and τ_1 and τ_3 share some data or resource that requires

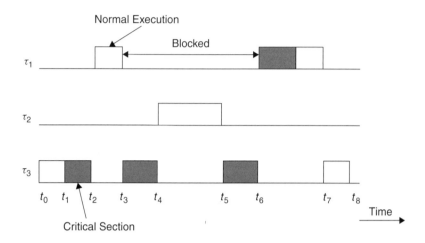

Figure 3.16   A priority inversion problem.

exclusive access, while τ_2 does not interact with either of the other tasks. Access to the critical section is done through the P and V operations on semaphore S.

Now consider the following execution scenario, illustrated in Figure 3.16. Task τ_3 starts at time t_0, and locks semaphore S at time t_1. At time t_2, τ_1 arrives and preempts τ_3 inside its critical section. At the same time, τ_1 attempts to use the shared resource by locking S, but it gets blocked, as τ_3 is currently using it. At time t_3, τ_3 continues to execute inside its critical section. Now if τ_2 arrives at time, t_4, it preempts τ_3, as it has a higher priority and does not interact with either τ_1 or τ_3. The execution of τ_2 increases the blocking time of τ_1, as it is no longer dependent only on the length of the critical section executed by τ_3. This event can take place with other intermediate priority tasks, and thereby can lead to an unbounded or an excessive blocking. Task τ_1 resumes its execution at time t_6 when τ_3 completes its critical section. A priority inversion is said to occur between time interval $[t_3, t_6]$ during which the highest priority task τ_1 has been unduly prevented from execution by a medium-priority task. Note that the blocking of τ_1 during the periods $[t_3, t_4]$ and $[t_5, t_6]$ by τ_3, which has the lock, is preferable to maintain the integrity of the shared resources.

3.3.10.1   *The Priority Inheritance Protocol*   The problem of priority inversion in real-time systems has been studied intensively for both fixed-priority and dynamic-priority scheduling. One result, the Priority Inheritance Protocol, offers a simple solution to the problem of unbounded priority inversion.

In the Priority Inheritance Protocol the priority of tasks are dynamically changed so that the priority of any task in a critical region gets the priority of the highest task pending on that same critical region. In particular, when a task, τ_i, blocks one or more higher-priority tasks, it temporarily inherits the highest priority of the blocked tasks. The highlights of the protocol are:

- The highest-priority task, τ, relinquishes the processor whenever it seeks to lock the semaphore guarding a critical section that is already locked by some other job.
- If a task, τ_1, is blocked by τ_2 and $\tau_1 \succ \tau_2$, (i.e., τ_1 has precedence over τ_2), task τ_2 inherits the priority of τ_1 as long as it blocks τ_1. When τ_2 exits the critical section that caused the block, it reverts to the priority it had when it entered that section.
- Priority inheritance is transitive. If τ_3 blocks τ_2, which blocks τ_1 (with $\tau_1 \succ \tau_2 \succ \tau_3$), then τ_3 inherits the priority of τ_1 via τ_2.

Thus, in the example just discussed, τ_3 priority would be temporarily raised to that of τ_1 at time t_3, thereby preventing τ_2 from preempting it at time t_4. The resulting schedule incorporating the Priority Inheritance Protocol is shown in Figure 3.17. Here the priority of τ_3 reverts back to its original at time t_5, and τ_2 gets to execute only after τ_1 completes its computation, as desired.

It is important to point out that the Priority Inheritance Protocol does not prevent deadlock. In fact, Priority Inheritance can cause deadlock or multiple blocking. Nor can it prevent other problems induced by semaphores. For example, consider the following sequence (with $\tau_1 \succ \tau_2$):

$$\tau_1: \text{Lock } S_1; \text{ Lock } S_2; \text{ Unlock } S_2; \text{ Unlock } S_1$$
$$\tau_2: \text{Lock } S_2; \text{ Lock } S_1; \text{ Unlock } S_1; \text{ Unlock } S_2$$

Here two tasks use two semaphores in a nested fashion, but in reverse order. The explanation of the resulting problem is left as Exercise 3.30. Although the deadlock does not depend on the Priority Inheritance Protocol (it is caused by an erroneous use of a semaphore), the Priority Inheritance Protocol does not prevent the problem. To get around these kinds of problems, it is necessary to use the

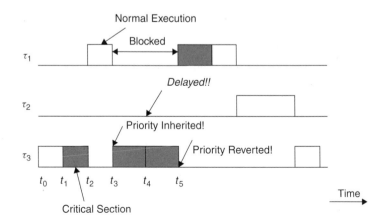

Figure 3.17 Priority Inheritance Protocol illustration.

Priority Ceiling Protocol, which imposes a total ordering on the semaphore access. This will be discussed momentarily.

A popular commercial operating system implements the Priority Inheritance Protocol as follows (in this case it is termed "pseudopriority inheritance"). When process A sees that process B has the resource locked, it suspend waiting for either (1) the resource to become unlocked, or (2) a minimal time-delay event. When it does suspend, process A gives the remainder of its time slice to B (to allow it to run immediately at a pseudohigher priority). If it is ready to run, B then runs in place of A. If it is not ready to run (i.e., suspended), B and A both are suspended to allow all other tasks to run. When A wakes up, the process is repeated until the resource becomes unlocked.

Finally, a notorious incident of the priority inversion problem occurred in 1997 in NASA's Mars Pathfinder Space mission's *Sojourner* rover vehicle, which was used to explore the surface of Mars. In this case the Mil-std-1553B information bus manager was synchronized with mutexes. Accordingly a meteorological data-gathering task that was of low priority and low frequency blocked a communications task that was of higher priority and higher frequency. This infrequent scenario caused the system to reset. The problem would have been avoided if the priority inheritance mechanism provided by the commercial real-time operating system (just mentioned) had been used. But it had been disabled. Fortunately, the problem was diagnosed in ground-based testing and remotely corrected by reenabling the priority inheritance mechanism [Cottet02].

3.3.10.2 *Priority Ceiling Protocol*   The Priority Ceiling Protocol extends to the Priority Inheritance Protocol through chained blocking in such a way that no task can enter a critical section in a way that leads to blocking it. To achieve this, each resource is assigned a priority (the priority ceiling) equal to the priority of the highest priority task that can use it.

The Priority Ceiling Protocol is the same as the Priority Inheritance Protocol, except that a task, T, can also be blocked from entering a critical section if there exists any semaphore currently held by some other task whose priority ceiling is greater than or equal to the priority of T. For example, consider the scenario illustrated in Table 3.9. Suppose that τ_2 currently holds a lock on S_2, and τ_1 is initiated. Task τ_1 will be blocked from entering S_1 because its priority is not greater than the priority ceiling of S_2.

As a further example, consider the three tasks with the following sequence of operations, and having decreasing priorities:

τ_1: Lock S_1; Unlock S_1
τ_2: Lock S_1; Lock S_2; Unlock S_2; Unlock S_1
τ_3: Lock S_2; Unlock S_2

Following the rules of assigning priority ceiling to semaphores, the priority ceilings of S_1 and S_2 are $P(\tau_1)$ and $P(\tau_2)$, respectively. The following description

Table 3.9 Data for the Priority Ceiling Protocol illustration

Critical Section	Accessed by	Priority Ceiling
S_1	τ_1, τ_2	$P(\tau_1)$
S_2	τ_1, τ_2, τ_3	$P(\tau_1)$
S_3	τ_3	$P(\tau_3)$
S_4	τ_2, τ_3	$P(\tau_2)$

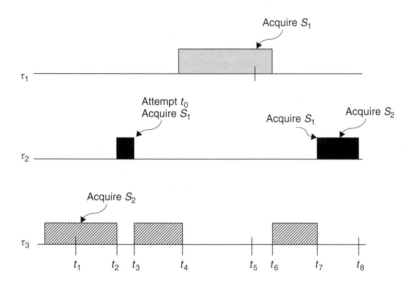

Figure 3.18 Illustration of the Priority Ceiling Protocol in action.

and Figure 3.18 illustrate the execution of the Priority Ceiling Protocol. Suppose that τ_3 starts executing first, locks the semaphore S_2 at time t_1 and enters the critical section. At time t_2, τ_2 starts executing, preempts τ_3, and attempts to lock semaphore S_1 at time t_3. At this time, τ_2 is suspended because its priority is not higher than priority ceiling of semaphore S_2, currently locked by τ_3. Task τ_3 temporarily inherits the priority of τ_2 and resumes execution. At time t_4, τ_1 enters, preempts τ_3, and executes until time t_5, where it tries to lock S_1. Note that τ_1 is allowed to lock S_1 at time t_5, as its priority is greater than the priority ceiling of all the semaphores currently being locked (in this case, it is compared with S_2). Task τ_1 completes its execution at t_6 and lets τ_3 execute to completion at t_7. Task τ_3 is then allowed to lock S_1, and subsequently S_2, and completes at t_8.

3.3.10.3 Basic Results of Resource Access Protocols   A task can be blocked by a lower-priority task only once, and at most the duration of one critical section. The following is a sufficient condition to test for feasibility of RM scheduling of n periodic tasks under all task phasings:

$$\sum_{i=1}^{n}(e_i/p_i) + \max(B_1/p_1, \ldots, B_{n-1}/P_{n-1}) \leq n(2^{1/n} - 1) \qquad (3.10)$$

where B_i is the blocking time that task τ_i can experience from a lower-priority task and p_i is the period of task τ_i.

3.4 MEMORY MANAGEMENT

An often-neglected discussion, dynamic memory allocation, is important in terms of both the use of on-demand memory by applications tasks and the requirements of the operating system. Applications tasks use memory explicitly, for example, through requests for heap memory, and implicitly through the maintenance of the run-time memory needed to support sophisticated high-order languages. The operating system needs to perform extensive memory management in order to keep the tasks isolated.

Dangerous allocation of memory is any allocation that can preclude system determinism. Dangerous allocation can destroy event determinism, for example, by overflowing the stack, or it can destroy temporal determinism by causing a deadlock situation. Therefore, it is important to avoid dangerous allocation of memory while at the same time reducing the overhead incurred by memory allocation. This overhead is a standard component of the context switch time and must be minimized.

3.4.1 Process Stack Management

In a multitasking system, context for each task needs to be saved and restored in order to switch processes. This can be done by using one or more run-time stacks or the task-control block model. Run-time stacks work best for interrupt-only systems and foreground/background systems, whereas the task-control block model works best with full-featured real-time operating systems.

3.4.1.1 Managing the Stack   If a run-time stack is to be used to handle the run-time saving and restoring of context, two simple routines – save and restore – are necessary. The save routine is called by an interrupt handler to save the current context of the machine into a stack area. To present disaster, this call should be made immediately after interrupts have been disabled. The restore routine should be called just before interrupts are enabled and before returning from the interrupt handler.

Consider, for example, the implementation of the save routine. Assume the global variable stack is to point to the top of the stack and that eight general registers (R0–R7) are to be saved on a stack. The memory location "PC" corresponds to the interrupt return vector location, and so it contains the program counter value at the time of interruption. It is necessary to save this on the stack to allow stacking of interrupts. The pseudocode for a 2-address architecture works by saving a pointer to the top of the stack and uses an "indirect: mode represented by ",I" to sequentially store the other registers.

```
save(context)              ;context is pseudo-argument.

DPI                        ;disable interrupts
STORE R0,&stack,I          ;save contents of register 0 onto stack
LOAD  R0,&stack            ;load index register with address of stack
ADD   R0,1
STORE R1,R0,I              ;save register 1
ADD   R0,1
STORE R2,R0,I              ;save register 2
ADD   R0,1
STORE R3,R0,I              ;save register 3
ADD   R0,1
STORE R4,R0,I              ;save register 4
ADD   R0,1
STORE R5,R0, I             ;save register 5
ADD   R0,1
STORE R6,R0,I              ;save register 6
ADD   R0,1
STORE R7, R0,I             ;save register 7
ADD   R0,1
STORE PC, R0,I             ;save return location
ADD   R0,1
STORE R0,&stack            ;save new stack pointer

EPI                        ;enable interrupts
RETURN                     ;return from interrupt
```

The argument context, is a pseudoargument, in that as discussed before, it is really a pointer to a data structure that contains the contents of the general register set and PC at a given instant in time. A high-order-language-callable program, written in assemble language, would be needed to save the PC, general registers, and any other context to the context data structure. Another assembly program is needed to restore the registers. Fortunately, most high-order language compilers provide these in a run-time library.

Further, while stack is shown here as a global variable, in practice it is probably better to make it an argument of the save and restore routines so that multiple run-time stacks can be maintained. Throughout this chapter, however, the representation of save and restore have been kept simple for clarity of discussion.

Next consider the restore routine, written in 2-address code, which restores context in precisely the reverse way as the save routine using an index register.

```
restore(context):          ;context is a pseudo-argument

DPI                        ;disable interrupts
LOAD  R0,&stack
```

```
SUB    RO,1
LOAD   PC,RO,I        ;restore return location
SUB    RO, 1
LOAD   R7, RO, I      ;restore register 7
SUB    RO,1
LOAD   R6, RO,I       ;restore register 6
SUB    RO,1
LOAD   R5,RO,I        ;restore register 5
SUB    RO,1
LOAD   R4,RO,I        ;restore register 4
SUB    RO,1
LOAD   R3,RO,I        ;restore register 3
SUB    RO,1
LOAD   R2,RO,I        ;restore register 2
SUB    RO,1
LOAD   R1, RO, I      ;restore register 1
STORE  RO,&stack      ;reset stack pointer
SUB    RO,1
LOAD   RO,RO,I        ;restore register 0
EPI                   ;enable interrupts
RETURN                ;return from interrupt
```

The individual interrupt–handler routines to save to a main stack written in C follow:

```
void int_handler(void)
{
  save(mainstack);
  switch(interrupt)
  {

    case 1:  int1();
             break;
    case 2:  int2();
             break;
    case 3:  int3();
             break;

  }
   restore(mainstack);

}

void int1(void)            /* interrupt handler 1 */
{
  save(stack);        .    /* save context on stack */
  task1();                 /* execute task 1*/
  restore(stack);          /* restore context from stack */
}

void int2(void)            /* interrupt handler 2 */
{
  save(stack);             /* save context on stack */
  task2();                 /* execute task2 */
  restore(stack);          /* restore context from stack */
}

void int3(void)            /* interrupt handler 3*/
{
```

```
    save(stack);          /* save context on stack */
    task3();              /* execute task 3 */
    restore(stack);       /* restore context from stack */
}
```

For example, suppose three processes, task1, task2, and task3, are running in an interrupt-only system where a signal interrupt based on three prioritized interrupts is generated. Suppose task1 is running when it is interrupted by task2. Later task2 is interrupted by task3. The run-time stack evolves as in Figure 3.19.

Certain machine architectures allow block save and block restore instructions to store and load n general registers in n consecutive memory locations. These instructions greatly simplify the implementation of the save and restore routines. Be aware that such macroinstructions may be designed to be interruptible (to reduce context switch time), so that if interrupts have not already been disabled, they should be.

3.4.1.2 Task-Control Block Model
If the task-control block model is used, then a list of task-control blocks is kept. This list can be either fixed or dynamic. In the fixed case, n task-control blocks are allocated at system generation time, all in the dormant state. As tasks are created, the task-control block enters the ready state. Prioritization or time slicing will then move the task to the execute state. If a task is to be deleted, its task-control block is simply placed in the dormant state. In the case of a fixed number of task-control blocks, no real-time memory management is necessary.

In the dynamic case, task-control blocks are added to a linked list or some other dynamic data structure as tasks are created. Again, the tasks are in the suspended state upon creation and enter the ready state via an operating system

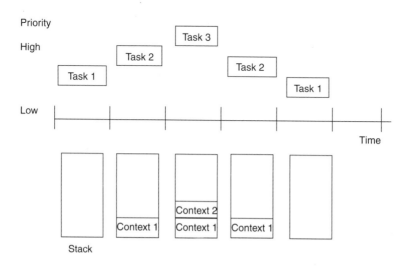

Figure 3.19 Activity on the run-time stack as tasks are interrupted.

call or event. The tasks enter the execute state owing to priority or time slicing. When a task is deleted, its task-control block is removed from the linked list, and its heap memory allocation is returned to the unoccupied or available status. In this scheme, real-time memory management consists of managing the heap needed to supply the task-control blocks; however, other data structures such as a list or sequence can be used.

3.4.2   Run-Time Ring Buffer

A run-time stack cannot be used in a round-robin system because of the first-in, first-out (FIFO) nature of the scheduling. In this case a ring buffer or circular queue can be used to save context. The context is saved to the tail of the list and restored from the list and restored from the head. The save and restore routines can be easily modified to accomplish this operation.

3.4.3   Maximum Stack Size

The maximum amount of space needed for the run-time stack needs to be known a priori. In general stack size can be determined if recursion is not used and heap data structures are avoided. If maximum stack memory requirements are not known, then a catastrophic memory allocation can occur, and the system will fail to satisfy event determinism. Ideally, provision for at least one more task than anticipated should be allocated to the stack to allow for spurious interrupts and time overloading.

3.4.4   Multiple-Stack Arrangements

Often a single run-time stack is inadequate to manage several processes in, say, a foreground/background system. A multiple-stack scheme uses a single run-time stack and several application stacks. Using multiple stacks in embedded real-time systems has several advantages:

- It permits tasks to interrupt themselves, thus allowing for handling transient overload conditions or for detecting spurious interrupts.
- The system can be written in a language that supports reentrancy and recursion, such as C or Ada. Individual run-time stacks can be kept for each process, which contains the appropriate activation records with dynamic links needed to support recursion. Or two stacks for each process can be kept, one for the activation records and the other for the display (a stack of pointers used to keep track of variable and procedure scope). In either case, a pointer to these stacks needs to be saved in the context or task-control block associated with that task.
- Only nonreentrant languages such as older versions of Fortran or assembly language are recommended with a single-stack model.

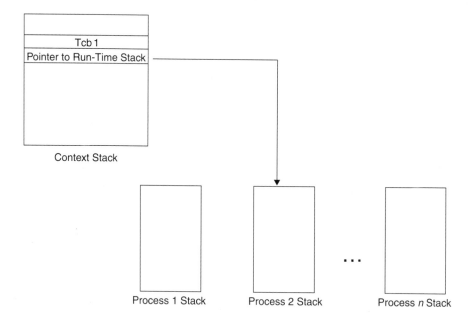

Figure 3.20 Multiple-stack management.

A multiple-stack arrangement is illustrated in Figure 3.20. The process stacks are not the same as the context stack. The stacks shown in Figure 3.19 are those needed to support the run-time allocation and deallocation of storage for the high-level language.

3.4.5 Memory Management in the Task-Control-Block Model

When implementing the TCB model of real-time multitasking, the chief memory-management issue is the maintenance of the linked lists for the ready and suspended tasks. As shown in Figure 3.21, when the currently executing task completes, is preempted, or is suspended while waiting for a resource, the next highest priority task in the ready list is removed and is made the executing one. If the executing task needs to be added to the suspended list, that is done. (If the executing task has completed, then its TCB is no longer needed.) Hence, by properly managing the linked lists, updating the status word in the TCBs, and adhering to the appropriate scheduling policy by checking the priority word in the TCBs, round-robin, preemptive priority, or both kinds of scheduling can be induced. Other memory management can include the maintenance of several blocks of memory that are allocated to individual applications as requested.

An alternative to multiple lists involves a single list in which only the status variable in the TCB is modified rather than moving the block. Thus, for example, when a task is moved from suspended to ready state, or from the ready to executing state, only the status word is changed. This approach has the advantage

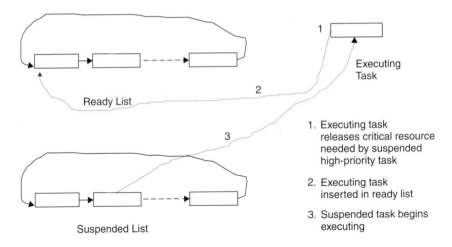

Figure 3.21   Memory management in the task-control-block model.

of less list management but slower traversal times, since the entire list must be traversed during each context switch to identify the next highest priority task that is ready to run.

3.4.6   Swapping

The simplest scheme that allows the operating system to allocate memory to two processes "simultaneously" is swapping. In this case, the operating system is always memory resident, and one process can coreside in the memory space not required by the operating system, called the user space. When a second process needs to run, the first process is suspended and then swapped, along with its context, to a secondary storage device, usually a disk. The second process, along with its context, is then loaded into the user space and initiated by the dispatcher. This type of scheme can be used along with round-robin or preemptive priority systems, but it is desirable to have the execution time of each process to be long relative to the swap time. The access time to the secondary store is the principal contributor to the context switch overhead and real-time response delays.

3.4.7   Overlays

Overlaying is a simple technique that allows a single program to be larger than the allowable memory. In this case the program is broken up into dependent code and data sections called overlays, which can fit into available memory. Special program code must be included that permits new overlays to be swapped into memory as needed (over the existing overlays), and care must be exercised in the design of such systems.

This technique has negative real-time implications because the overlays must be swapped from secondary storage devices. Nevertheless, overlaying can be used to extend the available address space. Some commercial real-time operating systems support overlaying in conjunction with commonly used programming languages and machines.

Note that in both swapping and overlaying a portion of memory is never swapped to disk or overlaid. This memory contains the swap or overlays manager (and in the case of overlaying any code that is common to all overlays is called the root).

3.4.8 Block or Page Management

A more elegant scheme than simple swapping allows more than one process to be memory-resident at any one time by dividing the user space into a number of fixed-size partitions. This scheme is useful in systems where the number of tasks to be executed is known and fixed, as in many embedded applications. Partition swapping to disk can occur when a task is preempted. Tasks, however, must reside in continuous partitions, and the dynamic allocation and deallocation of memory can cause problems.

In some cases main memory can become checkered with unused but available partitions, as in Figure 3.22. In this case, the memory space is said to be externally fragmented. This type of fragmentation causes problems when memory requests cannot be satisfied because a contiguous block of the size requested does not exist, even though the actual memory is available.

Another problem, internal fragmentation, occurs in fixed-partition schemes when, for example, a process requires 1 megabyte of memory when only 2-megabyte partitions are available. The amount of wasted memory or internal fragmentation can be reduced by creating fixed partitions of several sizes and then allocating the smallest partition greater than the required amount. Both

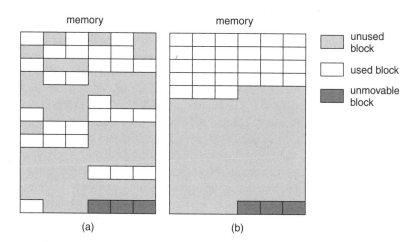

Figure 3.22 Fragmented memory (a) before and (b) after compaction, including unmovable blocks (for example, representing the operating system root program).

internal and external fragmentation hamper efficient memory usage and ultimately degrade real-time performance because of the overhead associated with their correction.

This type of dynamic memory allocation uses memory inefficiently as a result of the overhead associated with fitting a process to available memory and disk swapping. However, in some implementations, particularly in commercial real-time executives, memory can be divided into regions in which each region contains a collection of different-sized, fixed-sized partitions. For example, one region of memory might consist of 10 blocks of size 16 Mb, while another region might contain 5 blocks of 32 Mb, and so on. The operating system then tries to satisfy a memory request (either directly from the program via a system call or through the operating system in the assignment of that process to memory), so that the smallest available partitions are used. This approach tends to reduce internal fragmentation.

In an alternative scheme, memory is allocated in amounts that are not fixed, but rather are determined by the requirements of the process to be loaded into memory. This technique is more appropriate when the number of real-time tasks is unknown or varies. In addition, memory utilization is better for this technique than for fixed-block schemes because little or no internal fragmentation can occur, as the memory is allocated in the amount needed for each process. External fragmentation can still occur because of the dynamic nature of memory allocation and deallocation, and because memory must still be allocated to a process contiguously.

Compressing fragmented memory or compaction can be used to mitigate internal fragmentation (see Figure 3.22). Compaction is a CPU-intensive process and is not encouraged in hard real-time systems. If compaction must be performed, it should be done in the background, and it is imperative that interrupts be disabled while memory is being shuffled.

In demand page systems, program segments are permitted to be loaded in noncontiguous memory as they are requested in fixed-size chunks called pages. This scheme helps to eliminate external fragmentation. Program code that is not held in main memory is "swapped" to secondary storage, usually a disk. When a memory reference is made to a location within a page not loaded in main memory, a page fault exception is raised. The interrupt handler for this exception checks for a free page slot in memory. If none is found, a page block must be selected and swapped to disk (if it has been altered), a process called page stealing. Paging, which is provided by most commercial operating systems, is advantageous because it allows nonconsecutive references to pages via a page table. In addition, paging can be used in conjunction with bank-switching hardware to extend the virtual address space. In either case, pointers are used to access the desired page. These pointers may represent memory-mapped locations to map into the desired hard-wired memory bank, may be implemented through associative memory, or may be simple offsets into memory, in which case the actual address in main memory needs to be calculated with each memory reference.

Paging can lead to problems, including very high paging activity called thrashing, internal fragmentation, and the more serious deadlock. But it is unlikely that a system would use so complex a scheme as paging in an embedded real-time system, where the overhead would be too great and the associated hardware support is not usually available.

3.4.9 Replacement Algorithms

Several methods can be used to decide which page should be swapped out of memory to disk and the same techniques are applicable to cache block replacement. The most straightforward algorithm is FIFO. Its overhead is only the recording of the loading sequence of the pages.

The best nonclairvoyant algorithm is the least recently used (LRU) rule, which simply states that the least recently used page will be swapped out if a page fault occurs. To illustrate the method, consider the following example. A paged memory system is divided into sixteen 256-kilobyte pages, of which any four can be loaded at the same time. Each page is tagged (1, 2, etc). The operating system keeps track of the usage of each page. For example, the page reference string:

    2   3   4   5

indicates that pages 2, 3, 4, and 5 have been used in that order. If a request is made for page 7, then page 2 will be swapped out in order to make room for page 7, because it was the least recently used. The loaded pages would then be 3, 4, 5, and 7 with reference string:

    2   3   4   5   7

Note that references to pages already loaded in memory cause no page fault. For instance, if a reference is now made to page 3, no pages need to be swapped, because page 3 is loaded in memory. If this reference is followed by one to page 6, page 4 would have to be swapped out because it had the least recent reference. The loaded pages would then be 3, 5, 7, and with reference string:

    2   3   4   5   7   3   6

Moreover, in a paging memory scheme, the worst possible scenario involves page stealing for each request of memory. This occurs, for example, in a four-page system when five pages are requested cyclically, as in the page reference string:

    2   4   6   8   9   2   4   6   8   9

The performance of LRU is the same in this case as FIFO (in terms of number of page faults). In FIFO page-replacement schemes, it makes sense that by increasing the number of pages in memory the number of page faults would be reduced. Often this is the case, but occasionally an anomalous condition occurs whereby increasing the number of pages actually increases the number of page faults. This is Belady's Anomaly, which fortunately, does not occur in LRU replacement schemes.

Finally, the overhead for the LRU scheme rests in recording the access sequence to all pages, which can be quite substantial. Therefore, the benefits of using LRU need to be weighed against the effort in implementing it vis-à-vis FIFO.

3.4.10 Memory Locking

In addition to thrashing, the chief disadvantage of page swapping in real-time systems is the lack of predictable execution times. In a real-time system, it is often desirable to lock all or a certain part of a process into memory in order to reduce the overhead involved in paging and to make the execution times more predictable. Certain commercial real-time kernels provide this feature, called memory locking. These kernels typically allow code or data segments, or both, for a particular process, as well as the run-time stack segment, to be locked into main memory. Any process with one or more locked pages is then prevented from being swapped out to disk. Memory locking decreases execution times for the locked modules and, more importantly, can be used to guarantee execution times. At the same time, it makes fewer pages available for the application, encouraging contention.

3.4.11 Working Sets

Working sets are based on the model of locality-of-reference. The idea is if a list of recently executed program instructions is observed on a logic analyzer, it can be noted that most of the instructions are localized to within a small number of instructions in most cases. For example, in the absence of interrupts and branching, the program is executed sequentially, or the body of a loop may be executed a large number of times. However, when interrupts, procedure calls, or branching occurs, the locality-of-reference is altered. The idea in working sets is that a set of local-code windows is maintained in the cache and that upon accessing a memory location not contained in one of the working sets, one of the windows in the working set is replaced (using a replacement rule such as FIFO or LRU). The performance of the scheme is based entirely on the size of the working-set window, the number of windows in the working set, and the locality-of-reference of the code being executed.

3.4.12 Real-Time Garbage Collection

Garbage is memory that has been allocated but is no longer being used by a task, that is, the task has abandoned it. Garbage can accumulate when processes terminate abnormally without releasing memory resources. In C, for example, if memory is allocated using the `malloc` procedure and the pointer for that memory block is lost, than that block cannot be used or properly freed. Garbage can also develop in object-oriented systems and as a normal byproduct of non-procedural languages such as C++.

Real-time garbage collection is an important function that must be performed either by the language's run-time support (e.g., in Java) or by the operating system where garbage collection is not part of the language. Garbage collection techniques are discussed further in Chapter 6.

3.4.13 Contiguous File Systems

Disk I/O is a problem in many real-time systems that can be exacerbated by file fragmentation. File fragmentation is analogous to memory fragmentation and has the same associated problems, only worse. In addition to the logical overhead incurred in finding the next allocation unit in the file, the physical overhead of the disk mechanism is a factor. For example, physical overhead involved in moving the disk's read/write head to the desired sector can be significant. To reduce or eliminate this problem, many commercial real-time systems, such as real-time UNIX, force all allocated sectors to follow one another on the disk. This technique is called contiguous file allocation.

3.4.14 Building versus Buying Real-Time Operating Systems

A common question that is asked at the time of systems requirements specification is "Should a commercial real-time solution be used or should one be built from scratch?" While the answer depends on the situation, commercial kernels are frequently chosen because they generally provide robust services, are easy to use, and may be portable.

Commercially available real-time operating systems (RTOS) are wide-ranging in features and performance, and can support many standard devices and network protocols. Often these systems come equipped with useful development and debugging tools, and they can run on a variety of hardware and environments. In short, commercial real-time systems are best used when they can satisfy response requirements at a reasonable price, and if the system must run on a variety of platforms.

While commercial RTOS provide flexibility in scheduling discipline and the number of tasks supported, there are drawbacks in their use. For example, they are usually slower than using the hard-wired interrupt-driven model, because tremendous overhead is incurred in implementing the task-control block model, which is the typical architecture for commercial RTOS. Furthermore, commercial solutions can include many unneeded features, which are incorporated in order for the product to have the widest appeal. The run-time and storage costs of these features may be excessive. Finally, manufacturers may be tempted to make misleading claims, or give best-case performance figures. The worst-case response times, which are the most informative, can generally not be known. If they are known, they are typically not published because they would place the product in an unfavorable light.

For embedded systems, when the per-unit charge for commercial products is too high, or when desired features are unavailable, or when the overhead is too high, the only alternative is to write the real-time kernel. But this is not a trivial task. Therefore, commercial RTOS should be considered wherever possible.

There are many commercial solutions available for real-time systems, but deciding which one is most suitable for a given application is difficult. Many features of embedded real-time operating systems must be considered, including cost, reliability, and speed. But there are many other characteristics that may be as important or more important, depending on the application. For example, the RTOS largely resides in some form of ROM and usually controls hardware that will not tolerate many faults; therefore, the RTOS should be fault-tolerant. Also, the hardware needs to be able to react to different events in the system very quickly; therefore, the operating system should be able to handle multiple processes in an efficient manner. Finally, because of the hardware on which the operating system has limited memory, the operating system must also require a reasonable amount of memory in which to run.

In fact, there are so many functional and nonfunctional attributes of any commercial RTOS that evaluation and comparison must be a subjective endeavor. Some structure, however, should be used in the heuristic decision-making. Using a standard set of criteria provides such structure [Laplante04].

3.4.15   Selecting Real-Time Kernels

From a business and technical perspective the selection of a commercial real-time operating system represents a potential make-or-break decision. It is therefore imperative that a rigorous set of selection criteria be used. The following are desirable characteristics for real-time systems:[5]

- Timeliness
- Design for survival under peak load
- Predictability
- Fault-tolerance
- Maintainability [Buttazzo00]

Therefore the selection criteria should reflect these desiderata. Unfortunately, unless a comprehensive experience base exists using several alternative commercial RTOS in multiple, identical application domains, there are only two ways to objectively determine the fitness of a product for a given application. The first is to rely on third-party reports of success or failure. These abound and are published widely on the Web. The second is to compare alternatives based on the manufacturer's published information from brochures, technical reports, and Web sites. The following discussion presents an objective apples-to-apples technique for comparing commercial RTOS based on marketing information. This

---

[5] This discussion is adapted from [Laplante04].

technique can be used, however, in conjunction with supplemental information from actual experience and third-party reports.

Consider thirteen selection criteria, $m_1 \cdots m_{13}$, each having a range $m_i \in [0, 1]$ where unity represents the highest possible satisfaction of the criterion and zero represents complete nonsatisfaction.

1. The minimum interrupt latency, m_1, measures the time between the occurrences of hardware interrupt and when the interrupt's service routine begins executing. A low value represents relatively high interrupt latency, while a high value represents a low latency. This criterion is important because, if the minimum latency is greater than that required by the embedded system, a different operating system must be selected.

2. This criterion, m_2, defines the most processes the operating system can simultaneously support. Even though the operating system can support a large number of tasks, this metric may be further limited by available memory. This criterion is important for systems that need numerous simultaneous processes. A relatively high number of tasks supported would result in $m_2 = 1$, while few task supported would suggest a lower value for m_2.

3. Criterion m_3 specifies the system memory required to support the operating system. It does not include the amount of additional memory required to run the system's application software. Criterion $m_3 = 1$ suggests a minimal memory requirement, while $m_3 = 0$ would represent a larger memory requirement.

4. The scheduling mechanism criterion, m_4, enumerates whether preemptive, round-robin, or some other task-scheduling mechanism is used by the operating system. If many mechanisms were supported, then a high value would be assigned to m_4.

5. Criterion m_5 refers to the available methods the operating system has to allow processes to communicate with each other. Among possible choices are mutual exclusion (mutexes), binary and counting semaphores, POSIX pipes, message queues, shared memory, FIFO buffers, control sockets, and signals and scheduling. Each mechanism has advantages and disadvantages, and they have been discussed. Let $m_5 = 1$ if the RTOS provides all desired scheduling mechanisms. A lower value for m_5 implies that fewer scheduling mechanisms are available.

6. Criterion m_6 refers to the after-sale support a company puts behind its product. Most vendors offer some sort of free technical support for a short period of time after the sale, with the option of purchasing additional support if required. Some even offer on-site consultation. A high value might be assigned to a strong support program, while $m_6 = 0$ if no support is provided.

7. Application availability, m_7, refers to the amount of software available (either that ships with the operating system or is available elsewhere)

to develop applications to run on the operating system. For example, RTLinux is supported by the GNU's suite of software, which includes the gcc C compiler and many freely available software debuggers, and other supporting software. This is an important consideration, especially when using an unfamiliar operating system. Let $m_7 = 1$ if a large amount of software were available, while 0 would mean that little or none was available.

8. Criterion m_8 refers to the different processors that the operating system supports. This is important in terms of portability and compatibility with off-the-shelf hardware and software. This criterion also encompasses the range of peripherals that the operating system can support, such as video, audio, SCSI, and such. A high value for the criterion represents a highly portable and compatible RTOS.

9. Criterion m_9 refers to whether the code of the operating system will be available to the developer, for tweaking or changes. The source also gives insight to the RTOS architecture, which is quite useful for debugging purposes and systems integration. Setting $m_9 = 1$ would suggest open source code or free source code, while a lower value might be assigned in proportion to the purchase price of the source code. Let $m_9 = 0$ if the source code were unavailable.

10. Criterion m_{10} refers to the time it takes for the kernel to save the context when it needs to switch from one task to another. A relatively fast context switch time would result in a higher value for m_{10}.

11. This criterion is directly related to the cost of the RTOS alone. This is critical because for some systems, the RTOS cost may be disproportionately high. In any case, a relatively high cost would be assigned a very low value, while a low cost would merit a higher value for m_{11}.

12. This criterion, m_{12}, rates which development platforms are available. In other words, it is a listing of the other operating systems that are compatible with the given RTOS. A high value for m_{12} would represent wide compatibility, while a lower m_{12} would indicate compatibility with only one platform.

13. This criterion, m_{13}, is based on a listing of what networks and network protocols are supported by the given RTOS. This would be useful to know because it rates what communication methods the software running on this operating system would be able to use to communicate to other computers within the same computer network. A high value for the criterion represents a relatively large number of networks supported.

Recognizing that the importance of individual criteria will differ depending on the application, a weighting factor, $w_i \in [0, 1]$, will be used for each criterion m_i, where unity is assigned if the criterion has highest importance, and zero if the criterion is unimportant for a particular application. Then a fitness metric,

$M \in [0, 13]$, is formed as

$$M = \sum_{i=1}^{13} w_i m_i \qquad (3.10)$$

Clearly, a higher value of M means that the RTOS is well suited to the application, while a lower value means that the RTOS is not well suited for the application.

While selection of the values for m_i and w_i will be subjective for any given RTOS and any given application, the availability of this heuristic metric provides a handle for objective comparison, historical perspective, and other uses.

3.4.15.1 A Case Study in Selecting a Commercial Real-Time Operating System

A typical commercial RTOS is now examined based on the criteria introduced. Although the data are real, the manufacturer name is omitted, as the intention is not to imply a recommendation of any product.

The following assumptions are made:

- For all the sample RTOS, assume that the calculations for the number of interrupt, the minimum time that it takes, and other system analysis based on the metrics chosen are performed under the same conditions, that is, sampling, time constraints, and number of processors.
- Maximum or minimum of tasks refers to the operating system object, such as the memory management unit (MMU), device drivers, and other system tasks.
- Assume that interrupt refers to "hardware interrupt." "Software interrupts," together with hardware interrupts and other vectoring mechanisms provided by the processor, are referred to as "exception handling."
- Thread switching latency time is equivalent to the measurement of context switching latency time.

In the cases where a criterion value can be assigned, this is done. Where the criteria are "processor dependent" or indeterminate, absent a real application, assignment of a rating is postponed, and a value of * is given. This "uncertain" value is fixed at the time of application analysis. Note too that the values between tables need to be consistent. So, for example, if a 6-microsecond latency yields $m_1 = 1$ for RTOS X, the same 6-microsecond latency should yield $m_1 = 1$ for RTOS Y.

Consider commercial RTOS A. Table 3.10 summarizes the criteria and ratings, which were based on the following rationale. The product literature indicated that the minimum interrupt latency is CPU dependent, therefore a * value is assigned here (which will be later resolved as 0.5 for the purposes of evaluating the metric). Context switch time is not given, and so a * is also indicated. The RTOS supports 32-thread priority levels, but it is not known if there is a limit on the total number of tasks, so a value of 0.5 is assigned. The RTOS

Table 3.10 Summary data for real-time operating system A[a]

Criterion	Description	Rating	Comment
m_1	Minimum interrupt latency	*	CPU dependent
m_2	Number of tasks supported	0.5	32-Thread priority levels
m_3	Memory requirements	0.7	ROM: 60 K
m_4	Scheduling mechanism	0.25	Preemptive
m_5	Intertask synchronization mechanism	0.5	Direct message passing
m_6	Software support (warranty)	0.5	Paid phone support
m_7	Software support (compiler)	1	Various
m_8	Hardware compatibility	0.8	Various
m_9	Royalty free	0	No
m_{10}	Source available	1	Yes
m_{11}	Context switch time	*	NA
m_{12}	Cost	0.7	Approximately $2500
m_{13}	Supported network protocols	1	Various

[a]Some of the specific details have been deliberately omitted to preserve the identity of the product.

itself requires 60 K of memory, which is somewhat more than some of the alternatives, so a value of 0.7 is assigned. The system provides only one form of scheduling, preemptive priority, so a lower value, 0.25, is assigned here than if other forms, such as round-robin, were available. Intertask synchronization and communication is available only through message passing, so a relative low $m_5 = 0.5$ is assigned.

The company provides paid phone support, which is not as "generous" as other companies, so a value of $m_6 = 0.5$ is assigned. There is a royalty cost for each unit, so a zero was assigned. Finally, there is a wide range of software support for the product, including network protocols, and the source is available, so values of one are given for these three criteria.

3.4.15.2 Matching an Operating System to an Application

Consider the following application and a set of RTOS, including A just described and RTOS B–E, whose criteria were determined in a similar manner.

The software controlling an inertial measurement system requires substantial input/output processing, which inherently causes large amounts of system interrupts. This is a highly reactive and mission-critical system that requires fast context switching, minimal interrupt latency, a high degree of synchronization and a well-supported and reliable system. Therefore $m_1 = m_2 = m_5 = m_6 = m_7 = m_{10} = m_{11} = 1$. Hardware compatibility is not critical because there is little need to port the system and the number of tasks supported is relatively low, therefore, $m_2 = m_8 = 0.1$. The other criteria are set to 0.4 or 0.5 because they are only moderately important. The ratings assigned are summarized in Table 3.11. The metric suggests that RTOS D is the best match for the internal measurement system.

Table 3.11 Decision table for inertial measurement system

Criterion	Description	Weight, w_1	A	B	C	D	E
m_1	Minimum interrupt latency	1	0.5	0.8	1	0.5	1
m_2	Number of tasks supported	0.1	0.5	0.5	0.5	1	1
m_3	Memory requirements	1	0.7	0.2	0.5	1	0.9
m_4	Scheduling mechanism	0.5	0.25	0.5	0.25	1	0.25
m_5	Intertask synchronization mechanism	1	0.5	1	0.5	1	1
m_6	Software support (warranty)	1	0.5	0.5	1	0.8	1
m_7	Software support (compiler)	1	1	0.75	1	1	0.5
m_8	Hardware compatibility	0.1	0.8	0.5	0.2	1	0.2
m_9	Royalty free		0	1	1	1	1
m_{10}	Source available	1	1	1	0	0.4	1
m_{11}	Context switch time	1	0.5	0.5	0.5	1	0.5
m_{12}	Cost	0.4	0.5	0.5	0.1	0.1	0.7
m_{13}	Supported network protocols	0.5	1	1	1	1	0.6
M			5.66	5.80	5.24	6.94	6.73

Here the metric M in equation 3.10 is computed for five candidate RTOS (A through E). From the last row it can be seen that RTOS D is the best fit in this case.

3.5 CASE STUDY: POSIX

POSIX is the IEEE's Portable Operating System Interface for Computer Environments. The standard provides compliance criteria for operating system services and is designed to allow applications programs to write applications that can easily port across operating systems. POSIX compliant systems are used widely in real-time applications. It is the intention here to describe the POSIX standard for the purposes of further explicating the concepts of RTOS with a robust and useful example.

3.5.1 Threads

Real-time POSIX provides a mechanism for creating concurrent activities by allowing for each process to contain several threads of execution. POSIX threads are very similar to Java's threads and Ada's tasks models.

POSIX threads (or Pthreads) are defined as a set of C language programming types and procedure calls, implemented with a pthread.h header/include file and a thread library, though this library may be part of another library, such as libc. The following is the C interface for thread management in POSIX.[6]

```
Types are defined in #include <sys/types.h>
pthread_t        /* Used to identify a thread. */
phread_attr_t    /* Used to identify a thread attribute object. */
size_t           /* Used for sizes of objects. */

/* initialize and destroy threads attribute object */
    int   pthread_attr_init(pthread_attr_t *);
    int   pthread_attr_destroy(pthread_attr_t *);
```

---

[6] Refer to the UNIX man (online manual) pages to get further details on C interfaces to various POSIX functions.

```
/* cancel execution of a thread */
   int    pthread_cancel(pthread_t);
/* detach a thread */
   int    pthread_detach(pthread_t);
/* compare thread IDs */
   int    pthread_equal(pthread_t, pthread_t);
/* thread termination */
   void   pthread_exit(void *);
/* wait for thread termination */
   int    pthread_join(pthread_t, void **);
/* get calling thread's ID */
   pthread_t pthread_self(void);
/** Stack and scheduling related **/
/* set and get detachstate attribute */
   int    pthread_attr_setdetachstate(pthread_attr_t *, int);
   int    pthread_attr_getdetachstate(const pthread_attr_t *, int *);
/* set and get inheritsched attribute */
   int    pthread_attr_setinheritsched(pthread_attr_t *, int);
   int    pthread_attr_getinheritsched(const pthread_attr_t *, int *);
/* set and get schedparam attribute */
   int    pthread_attr_setschedparam(pthread_attr_t *, const struct sched_param
*);
   int    pthread_attr_getschedparam(const pthread_attr_t *, struct sched_param
*);
/* dynamic thread scheduling parameters access */
   int    pthread_getschedparam(pthread_t, int *, struct sched_param *);
   int    pthread_setschedparam(pthread_t, int , const struct sched_param *);
/* set and get schedpolicy attribute */
   int    pthread_attr_setschedpolicy(pthread_attr_t *, int);
   int    pthread_attr_getschedpolicy(const pthread_attr_t *, int *);
/* set and get stackaddr attribute */
   int    pthread_attr_setstackaddr(pthread_attr_t *, void *);
   int    pthread_attr_getstackaddr(const pthread_attr_t *, void **);
/* set and get stacksize attribute */
   int    pthread_attr_setstacksize(pthread_attr_t *, size_t);
   int    pthread_attr_getstacksize(const pthread_attr_t *, size_t *);

   int    pthread_getconcurrency(void);
   void  *pthread_getspecific(pthread_key_t);
```

The naming conventions being followed in POSIX are shown in the Table 3.12. All identifiers in the threads library begin with pthread_.

The following example illustrates how to create multiple threads (five in this example) with the pthread_create() routine. Each thread does a simple print, and then terminates with a call to pthread_exit(). The example also demonstrates how to "wait" for thread completions by using the Pthread join routine.

```
#include <pthread.h>
#include <stdio.h>

void message_printer_function(void *ptr)
{
  char *message;
  message = (char*) ptr;
  printf("%s\n",message);
}

void main()
```

Table 3.12   POSIX naming scheme

Routine Prefix	Functional Group
pthread_	Threads themselves and miscellaneous subroutines
pthread_attr_	Thread attributes objects
pthread_mutex_	Mutexes
pthread_mutexattr_	Mutex attributes objects
pthread_cond_	Condition variables
pthread_condattr_	Condition attributes objects
pthread_key_	Thread-specific data keys

```
{
    pthread_t thread[5];
    pthread_attr_t attribute;
    int errorcode,counter, status;
    char *message="TestPrint";
/* Initialize and set thread detached attribute */
    pthread_attr_init(&attribute);
    pthread_attr_setdetachstate(&attribute, PTHREAD_CREATE_JOINABLE);

    for(counter=0;counter<5;counter++)
    {
        printf("I am creating thread %d\n", counter);
        errorcode = pthread_create(&thread[counter],
&attribute,(void*)&message_printer_function,(void*)message);
        if (errorcode)
        {
            printf("ERROR happened in thread creation");
            exit(-1);
        }
    }
/* Free attribute and wait for the other threads */
    pthread_attr_destroy(&attribute);

    for(counter=0;counter<5;counter++)
    {
        errorcode = pthread_join(thread[counter], (void **)&status);
        if (errorcode)
        {
            printf("ERROR happened in thread join");
            exit(-1);
        }
        printf("Completed join with thread %d\n",counter);
        /*printf("Completed join with thread %d status= %d\n",counter, status);*/
    }
    pthread_exit(NULL);
}
```

3.5.2   POSIX Mutexes and Condition Variables

Mutex variables are one of the primary means of implementing thread synchronization. The basic concept of a mutex as used in Pthreads is that only one thread is allowed to lock (or own) a mutex variable at any given time. Thus, even if several threads try to lock a mutex, only one thread will be successful. No other thread can own/lock that mutex until the owning thread unlocks that mutex, and only the owner can unlock it. POSIX mutexes application program interfaces (APIs) are given below.

```
/** POSIX Mutexes **/
/* Creating/Destroying Mutexes */
   pthread_mutex_init(mutex, attr)
   pthread_mutex_destroy(mutex)
   pthread_mutexattr_init(attr)
   pthread_mutexattr_destroy(attr)

/* Locking/Unlocking Mutexes */
   pthread_mutex_lock(mutex)
   pthread_mutex_trylock(mutex)
   pthread_mutex_unlock(mutex)
```

As compared to mutexes, condition variables provide an alternative for threads to synchronize. The basic difference between mutexes and condition variables is that while mutexes implement synchronization by controlling thread access to data, condition variables allow threads to synchronize based upon the actual value of data.

Without condition variables, threads need to continually poll (possibly in a critical section) to check if the condition is met. This could lead to unnecessary resource consumption, as the thread would be continuously busy in this activity. A condition variable facilitates to achieve the same goal without polling.

```
/** POSIX Condition Variables **/
/* Creating/Destroying Condition Variables */
   pthread_cond_init(condition, attr)
   pthread_cond_destroy(condition)
   pthread_condattr_init(attr)
   pthread_condattr_destroy(attr)

/* Waiting/Signalling On Condition Variables */
   pthread_cond_wait(condition, mutex)
   pthread_cond_signal(condition)
   pthread_cond_broadcast(condition)
```

The following example demonstrates the use of a mutex where a single reader and a single writer communicate via a shared memory.

```
#include <stdio.h>
#include <pthread.h>
#define SET 1
#define NOTSET 0
int info_in_buffer=NOTSET;
```

```
pthread_mutex_t lock=PTHREAD_MUTEX_INITIALIZER;
void read_user(void)
{
  while(1)
  {
     /* check whether buffer is written and read data*/
     pthread_mutex_lock(&lock);
     if (info_in_buffer==SET)
     {
       printf("In read user \n");
       /* simulation the read operation by a wait (sleep(2)) */
       sleep(2);
       info_in_buffer=NOTSET;
     }
     pthread_mutex_unlock(&lock);
     /* giving the writer an opportunity to write to the buffer*/
     sleep(2);
  }
}

void write_user(void)
{
  while(1)
  {
     /* check whether buffer is free and write data*/
     pthread_mutex_lock(&lock);
     if (info_in_buffer==NOTSET)
     {
     printf("In write user \n");
     /* simulation the write operation by a wait (sleep(2)) */
     sleep(2);
       info_in_buffer=SET;
     }
     pthread_mutex_unlock(&lock);
     /* giving the reader an opportunity to read from the buffer*/
     sleep(2);
  }
}

void main()
{
  pthread_t Readthread;
  pthread_attr_t attribute;
  pthread_attr_init(&attribute);
  pthread_create(&Readthread,&attribute,(void*)&read_user,NULL);
  write_user();
}
```

3.5.3 POSIX Semaphores

POSIX provides counting semaphores and binary semaphores to enable processes running in different address spaces, or threads within the same address space, to synchronize and communicate using shared memory. The following prototypes are self-describing examples of their use.

```
int sem_init(sem_t *sem, int pshared, unsigned int value);
/* Initializes the semaphore object pointed by 'sem' */

int sem_destroy(sem_t *sem);
/* Destroys a semaphore object and frees up the resources it might hold */

/* The following three functions are used in conjunction with other
   processes. See man pages for more details.
*/
sem_t *sem_open(const char *name, int oflag, ...);
int sem_close(sem_t *sem);
int sem_unlink(const char *name);

int sem_wait(sem_t *sem);
/* Suspends the calling thread until the semaphore pointed to by 'sem' has
   non-zero count. Decreases the semaphore count. */

int sem_trywait(sem_t *sem);
/* A non-blocking variant of sem_wait. */

int sem_post(sem_t *sem);
/* Increases the count of the semaphore pointed to by 'sem'. */

int sem_getvalue(sem_t *sem, int *sval);
/* Stores the current count of the semaphore 'sem' in 'sval'. */
```

3.5.4 Using Semaphores and Shared Memory

It is important that two processes not write to the same area of shared-memory at the same time, and this is where the semaphores are useful. Before writing to a shared memory region, a process can lock the semaphore to prevent another process from accessing the region until the write operation is completed. When the process is finished with the shared-memory region, the process unlocks the semaphore and frees the shared-memory region for use by another process.

```
#include<stdio.h>
#include<pthread.h>
#include<semaphore.h>
#include<sys/sem.h>

sem_t writer_lock;
sem_t reader_lock;

void read_user(void)
{
  while(1) {
    sem_wait(&reader_lock);
    /* simulate read operation by a delay*/
    printf("in reader task \n");
    sleep(2);
    sem_post(&writer_lock);
  }
}

void write_user(void)
{
```

```
while(1) {
    sem_wait(&writer_lock);
    /* simulate read operation by a delay*/
    printf("in writer task \n");
    sleep(2);
    sem_post(&reader_lock);
    }
}

void main()
{
    pthread_t read_thread;
    pthread_attr_t attribute;
    sem_init(&writer_lock,0,1);
    sem_init(&reader_lock,0,1);
    sem_wait(&reader_lock);
    pthread_attr_init(&attribute);
    pthread_create(&read_thread,&attribute,(void*)&read_user,NULL);
    write_user();
}
```

3.5.5 POSIX Messages

Message queues work by exchanging data in buffers. Any number of processes can communicate through message queues. Message notification can be synchronous or asynchronous. The POSIX message passing through message-queue facilities provide a deterministic, efficient means for interprocess communication. Real-time message passing is designed to work with shared memory in order to accommodate the needs of real-time applications with an efficient, deterministic mechanism to pass arbitrary amounts of data between cooperating processes. The following prototypes describe the POSIX messaging capabilities.

```
mqd_t mq_open(const char *name, int oflag, ...);
/* Connects to, and optionally creates, a named message queue. */

int mq_send(mqd_t mqdes, const char *msg_ptr, oskit_size_t msg_len,
            unsigned int msg_prio);
/* Places a message in the queue. */

int mq_receive(mqd_t mqdes, char *msg_ptr, oskit_size_t msg_len,
            unsigned int *msg_prio);
/* Receives (removes) the oldest, highest priority message from the queue. */

int mq_close(mqd_t mqdes);
/* Ends the connection to an open message queue. */

int mq_unlink(const char *name);
/* Ends the connection to an open message queue and causes the
 queue to be removed when the last process closes it. */

int mq_setattr(mqd_t mqdes, const struct mq_attr *mqstat, struct mq_attr
*omqstat);
int mq_getattr(mqd_t mqdes, struct mq_attr *mqstat);
/* Set or get message queue attributes. */
```

```
int mq_notify(mqd_t mqdes, const struct sigevent *notification);
/* Notifies a process or thread that a message is available in the queue. */
```

The following example illustrates sending and receiving messages between two processes using a message queue [Marshall96]. The following two programs should be compiled and run at the same time to illustrate the basic principle of message passing:

message_send.c   Creates a message queue and sends one message to the queue.

message_rec.c   Reads the message from the queue.

The full code listing for message_send.c is as follows:

```c
#include <sys/types.h>
#include <sys/ipc.h>
#include <sys/msg.h>
#include <stdio.h>
#include <string.h>
#define MSGSZ     128
/* Declare the message structure.  */
typedef struct msgbuf {
        long    mtype;
        char    mtext[MSGSZ];
        } message_buf;

main()
{
    int msqid;
    int msgflg = IPC_CREAT | 0666;
    key_t key;
    message_buf sbuf;
    size_t buf_length;

    /* Get the message queue id for the  "name" 1234, which was created by the
server.  */
    key = 1234;

(void) fprintf(stderr, "\nmsgget: Calling msgget(%#lx,\ %#o)\n", key, msgflg);

    if ((msqid = msgget(key, msgflg )) < 0)
    {
        perror("msgget");
        exit(1);
    }
    else
     (void) fprintf(stderr,"msgget: msgget succeeded: msqid = %d\n", msqid);

     /* We'll send message type 1 */

    sbuf.mtype = 1;

    (void) fprintf(stderr,"msgget: msgget succeeded: msqid = %d\n", msqid);
    (void) strcpy(sbuf.mtext, "Did you get this?");
    (void) fprintf(stderr,"msgget: msgget succeeded: msqid = %d\n", msqid);
```

```
    buf_length = strlen(sbuf.mtext) + 1 ;

    /* Send a message.  */

    if (msgsnd(msqid, &sbuf, buf_length, IPC_NOWAIT) < 0) {
       printf ("%d, %d, %s, %d\n", msqid, sbuf.mtype, sbuf.mtext, buf_length);
        perror("msgsnd");
        exit(1);
    }

  else
     printf("Message: \"%s\" Sent\n", sbuf.mtext);
    exit(0);
}
```

The essential points to note here are:

- The Message queue is created with a basic key and message flag msgflg = IPC_CREAT | 0666 -- create queue and make it read and appendable by all.
- A message of type (sbuf.mtype) 1 is sent to the queue with the message "Did you get this?"

Receiving the preceding message as sent using message_send program is illustrated below. The full code listing for message_send.c's companion process, message_rec.c is as follows:

```
#include <sys/types.h>
#include <sys/ipc.h>
#include <sys/msg.h>
#include <stdio.h>

#define MSGSZ    128

/* Declare the message structure. */

typedef struct msgbuf {
    long    mtype;
    char    mtext[MSGSZ];
} message_buf;

main()
{
    int msqid;
    key_t key;
    message_buf  rbuf;

    /* Get the message queue id for the "name" 1234, which was created by the
server.  */

    key = 1234;

    if ((msqid = msgget(key, 0666)) < 0) {
        perror("msgget");
        exit(1);
    }
```

```
    /* Receive an answer of message type 1. */
    if (msgrcv(msqid, &rbuf, MSGSZ, 1, 0) < 0) {
        perror("msgrcv");
        exit(1);
    }
    /* Print the answer. */
    printf("%s\n", rbuf.mtext);
    exit(0);
}
```

The essential points to note here are:

- The Message queue is opened with msgget (message flag 0666) and the same key as message_send.c}.
- A message of the same type 1 is received from the queue with the message "Did you get this?" stored in rbuf.mtext.

3.5.6   Real-Time POSIX Signals

Signals are software representation of interrupts or exception occurrences. Signals asynchronously alter the control flow of a task. It is important to point out that no routine should be called from a signal handler that might cause the handler to block – it makes it impossible to predict which resources might be unavailable. Signals are used for many purposes:

- Exception handling
- Process notification of asynchronous event occurrence
- Process termination in abnormal situations
- Interprocess communication

However, there are several limitations of standard POSIX signals on their use in real-time applications. These include:

- Lack of signal queueing
- No signal delivery order
- Poor information content
- Asynchrony

POSIX real-time extensions (POSIX.4) improves the POSIX signals to applications. POSIX.4 defines a new set of application-defined real-time signals, and these signals are numbered from SIGRTMIN to SIGRTMAX. There must be RTSIG_MAX >8 signals in between these two limits.

The sigaction defines all the details that a process need to know when a signal arrives. As real-time signals can be queued, the queueing option for a real-time signal is chosen by setting bit SA_SIGINFO in the sa_flags field of the sigaction structure of the signal.

```
struct sigaction{
  void (*sa_handler)();
  sigset_t sa_mask;
  int sa_flags;//SA_NOCLDSTOP or SA_SIGINFO
  void (*sa_sigaction)(int, siginfo_t*, void*);
  /*used for real-time signals!! Also, ''SA_SIGINFO''
    is set in ''sa_flags.''
   */
};

int sigaction(int sig, const struct sigaction *reaction,
                  struct sigaction *oldreaction);
```

Real-time signals can carry extra data. SA_SIGINFO increases the amount of information delivered by each signal. If SA_SIGINFO is set, then the signal handlers have as an additional parameter a pointer to a data structure called a siginfo_t that contains the date value to be piggybacked.

The sigqueue() includes an application-specified value (of type sigval) that is sent as a part of the signal. It enables the queuing of multiple signals for any task. Real-time signals can be specified as offsets from SIGRTMIN. All signals delivered with sigqueue() are queued by numeric order, lowest numbered signals delivered first.

POSIX.4 provides a new and more responsive (or fast) synchronous signal-wait function called sigwaitinfo. Upon arrival of the signal, it does not call the signal handler (unlike sigsuspend), but unblocks the calling process.

3.5.7 Clocks and Timers

In developing real-time applications, clearly it is desirable to have facilities to set and get the time. For example, suppose a diagnostic program checks the health of the system periodically. Essentially, the program would execute one round of diagnostics and then wait for a notification to execute again, with the process repeating forever. This is accomplished by having a timer that is set to expire at a particular time interval. When the time interval expires, the program that set the timer is notified, usually through a signal delivery.

3.5.7.1 Time Management In order to generate a time reference, a timer circuit is programmed to interrupt the processor at a fixed rate. The internal system time is incremented at each timer interrupt. The interval of time with which the timer is programmed to interrupt defines the unit of time (also called "tick") in the system (time resolution). Typically, the system time is represented by a long integer (unsigned 32 bits) variable, whereas the value of the tick is stored in a float variable.

The values of the system lifetime (range) for some tick values (granularity) is shown in the Table 3.13. At any time, "sys_clock," a variable holding system time, contains the number of interrupts generated by the timer since the Epoch.

Table 3.13   System lifetime

Tick	Lifetime
1 microsecond	71.6 minutes
1 millisecond	50 days
10 millisecond	16 months
1 second	136 years

If k denotes system tick, and n is the value stored in sys_clock, then the actual time elapsed is kn.

3.5.7.2 POSIX Clock   POSIX allows many clocks to be supported by an implementation. Each clock has its own identifier of type clockid_t. The commonly supported "time-of-day clock" is the CLOCK_REALTIME clock, defined in the time.h header file. The CLOCK_REALTIME clock is a systemwide clock, visible to all processes running on the system. The CLOCK_REALTIME clock measures the amount of time that has elapsed since 00:00:00 January 1, 1970.

As mentioned, CLOCK_REALTIME is commonly used as the clock_id argument in all clock functions. Some of the common clock functions and their descriptions are given in the Table 3.14. The value returned by a clock function is stored in a data structure called timespec that has two fields of the long-integer type, namely tv_sec representing the value in number of seconds since the Epoch, and tv_nsec representing the value in nanoseconds.

Table 3.14   POSIX clock functions

Function	Description
clock_getres	Returns the resolution of the specified clock int clock_getres(clockid_t clock_id, struct timespec *res)
Clock_gettime	Returns the current value for the specified value int clock_gettime(clockid_t clock_id, struct timespec *tp)
Clock_settime	Sets the specified clock to the specified value int clock_settime(clockid_t clock_id, const struct timespec *tp)

3.5.7.3 Determining Clock Resolution The following example calls the clock_getres function to determine clock resolution:

```
#include <unistd.h>
#include <time.h>
main(){
struct timespec clock_resolution;
int stat;
   stat = clock_getres(CLOCK_REALTIME, &clock_resolution);
   printf(''Clock resolution is %d seconds, %ld nanoseconds\n'',
          clock_resolution.tv_sec, clock_resolution.tv_nsec); }
```

3.5.7.4 Retrieving System Time The clock_gettime function returns the value of the systemwide clock as the number of elapsed seconds since the Epoch. The timespec data structure (used for the clock_gettime function) also contains a member to hold the value of the number of elapsed nanoseconds not comprising a full second.

```
#include <unistd.h>
#include <time.h>
main(){
 struct timespec ts;
 clock_gettime(CLOCK_REALTIME, &ts);
 printf(''clock_gettime returns:\n'');
 printf(''%d seconds and %ld nanoseconds\n'', ts.tv_sec, ts.tv_nsec); }
```

3.5.7.5 System Clock Resolution The system clock resolution on DEC's Alpha system is 1/1024 seconds or 976 microseconds),[7] that is, the system maintains time by adding 976 microseconds at every clock interrupt. The actual time period between clock ticks is exactly 976.5625 microseconds. The missing 576 microseconds (1024 * 0.5625) are added at the end of the 1024th tick, that is the 1024th tick advances the system time by 1552 microseconds.

Note that if an application program requests a timer value that is not an exact multiple of the system clock resolution (an exact multiple of 976.5625 microseconds), the actual time period counted down by the system will be slightly larger than the requested time period. A program that asks for a periodic timer of 50 milliseconds will actually get a time period of 50.78 milliseconds.

3.5.7.6 Timer So far, mechanisms that allow setting and getting the time in POSIX have been discussed. Beyond this, it is desirable to time a process's execution so that it gets to run on the processor at a specific time interval. As discussed earlier, POSIX timers provide a mechanism to control the frequency of a program execution. In order to use a timer to time a process it is necessary to:

- Create the time object within the kernel.
- Generate a signal to get notification.
- Choose either relative or absolute timer.

[7] *Guide to Realtime Programming*, Digital Equipment Corp., March 1996.

3.5.7.7 Creating a Timer The first step is to create a timer for the application by using the `timer_create()` function.

```
#include<signal.h>
#include<time.h>
timer_t timer_create(clockid_t clock_id, struct sigevent *event,
                     timer_t *timer_id);
```

As per POSIX standard, different platforms can have multiple time bases, but every platform must support at least the CLOCK_REALTIME time base. A timer based upon the system clock called CLOCK_REALTIME can be created. The seconf argument event points to a structure that contains all the information needed concerning the signal to be generated. This is essentially used to inform the kernel about what kind of event the timer should deliver whenever it "fires." By setting it NULL, the system is forced to use default delivery, which is defined to be SIGALRM.

The return value from `timer_create()` is effectively a small integer that just acts as an index into the kernel's timer tables.

3.5.7.8 Type of Timers Having created the timer, it is necessary to decide what kind of timer functionality it will have – a one-shot timer or a repeating timer. A one-shot timer is armed with an initial expiration time, expires only once, and then is disarmed. A timer becomes a periodic or repeating timer with the addition of a repetition value. The timer expires, then loads the repetition interval, rearming the timer to expire after the repetition interval has elapsed. The function `timer_settime()` actually sets and starts the timer. The struct itimerspec simply incorporates two timespecs to form a high-resolution interval timer structure:

```
struct itimerspec{
      struct timespec it_value,
                      it_interval;
};
int timer_settime (timer_t timerid, int flag,
                   struct itimerspec *value,
                   struct itimerspec *oldvalue);
```

This function sets the next expiration time for the timer specified. If flag is set to Timer_ABSTIME, then the timer will expire when the clock reaches the absolute value specified by *value.it_value. If flag is not set to TIMER_ABSTIME, the timer will expire when the interval specified by value->it_value passes. If *value.it_interval is nonzero, then a periodic timer will go off every value->it_interval after value->it_value has expired. Any previous timer setting is returned in *oldvalue. For example, to specify a timer that executes only once, 10.5 seconds from now, specify the following values for the members of the itimerspec structure:

```
newtimer_setting.it_value.tv_sec = 10;
newtimer_setting.it_value.tv_nsec = 500000000;
```

```
newtimer_setting.it_interval.tv_sec = 0;
newtimer_setting.it_interval.tv_nsec = 0;
```

To arm a timer to execute 15 seconds from now and then at 0.25-second intervals, specify the following values:

```
newtimer_setting.it_value.tv_sec = 15;
newtimer_setting.it_value.tv_nsec = 0;
newtimer_setting.it_interval.tv_sec = 0;
newtimer_setting.it_interval.tv_nsec = 250000000;
```

3.5.8 Asynchronous Input and Output

I/O operation is a key component in any real-time application. The real-time program is usually responsible for tracking or controlling the external environment in some desired way. Some of the common I/O operations typically found in real-time systems include:

- Data gathering/output from/to devices.
- Data logging (e.g., for monitoring purposes).
- Multimedia applications (playback or recording).
- Operations on (real-time) databases, keyboards, mice, etc.
- I/O devices: joysticks, keyboards.

It is important to note that UNIX I/O is synchronous, that is, the execution of a program has to wait while the I/O takes place. For example, UNIX read calls blocks the calling process until the user buffer is filled up with data being requested, or an error occurs during the I/O operation. However, many real-time applications, and those applications requiring high-speed or high-volume data collection and/or low-priority journaling functions, need to perform I/O asynchronously, that is, the system performs the I/O in parallel with the application, which is free to perform other tasks while the data are read in or written. When the I/O completes, the application receives some kind of notification, usually by the delivery of a signal.

3.5.8.1 Associated Data Structures Asynchronous I/O (AIO) operations are submitted using a structure called the AIO control block, or aiocb. This control block contains asynchronous operation information, such as the initial point for the read operation, the number of bytes to be read, and the file descriptor on which the AIO operation will be performed. The aiocb structure contains the following members:

```
struct aiocb{
   int            aio_fildes;    // File descriptor
   off_t          aio_offset;    // File offset
   volatile void  *aio_buf;      // Pointer to buffer
   size_t         aio_nbytes;    // Number of bytes to transfer
```

```
    int             aio_reqprio;   // Request priority offset
    struct sigevent aio_sigevent;  // Signal structure
    int             aio_lio_opcode;// Specifies type of I/O operation
};
```

It is important to understand what actually happens when the aio_read/ aio_write(...) functions are called. In fact, the following code is performed:

```
    lseek(a.aio_fildes, ...); // Seek to position
    read(a.aio_fildes,...);   // Read data
    sigqueue(...);            // Queue a signal to a process
```

The AIO operation is depicted in Figure 3.23. Error handling for synchronous I/O is simplified by looking at errno. For AIO, the system maintains a return value, and an errno value, for each asynchronous operation separately. These two values from the system are obtained via functions aio_return and aio_error. Each function gives back the return value, or errno value, associated with the asynchronous operation at that moment.

Multiple I/O operations are permitted in the POSIX AIO specification. AIO allows a combination of several read and write operations into a single system call, lio_listio. The function lio_listio permits transfer of large amounts of I/O simultaneously. The function aio_suspend waits for particular AIO operations to complete. It does not say which I/O completed, it just returns 0 when it has determined that one of the asynchronous operations has finished. AIO can be canceled using aio_cancel, though it is not recommended.

3.5.9 POSIX Memory Locking

Virtual memory, which allows for mapping of virtual addresses to different physical locations, is useful in real-time systems. In addition to paging (and associated thrashing problems), the key disadvantage of page swapping in real-time systems is the lack of predictable execution time. It is not uncommon that an application demands responsiveness that may be measured in microseconds, and the program is waiting milliseconds or more while the operating system is involved in

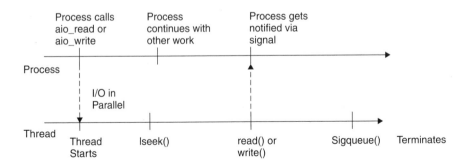

Figure 3.23 Asynchronous I/O operation.

disk access and in fetching the desired instructions in the memory. In a real-time system, it is often desirable to lock the important process's memory down so that the operating system does not page it, thereby making the execution times more predictable. In the case of many large processes, it is desirable to lock just the time-critical portions of these processes.

POSIX allows for a simple procedure to lock the entire process down.

```
#include <unistd.h>
#ifdef _POSIX_MEMLOCK
#include<sys/mman.h>
int mlockall(int flags);
int munlockall(void);
```

The function `mlockall` tries to lock all the memory down; this includes the program text, data, heap, and stack (Figure 3.24). Locking a process includes shared libraries that can be mapped in and other shared memory areas that the process may be using. Depending on the flags being specified, it will either lock all process's current mappings (`MCL_CURRENT`), or the process's current mapping and any future mappings that it may make (`MCL_CURRENT|MCL_FUTURE`). The function `munlock-all` unlocks all locked memory and notifies the system that it is okay to page this process's memory if the system must do so. Assuming that `mlockall` is called with the `MCL_FUTURE` flag being set, the rightmost column in Figure 3.24 illustrates the effect of memory locking upon execution of `malloc`. Instead of locking down the entire process, POSIX permits the user to lock down part of the process:

```
#include <unistd.h>
#ifdef _POSIX_MEMLOCK_RANGE
#include<sys/mman.h>
int mlock(void *address, size_t length);
int munlock(void *address, size_t length);
#endif
```

The function `mlock` locks down the address range being specified, and `munlock` unlocks a range (Figure 3.25). If `mlock` is called for a range of memory, then

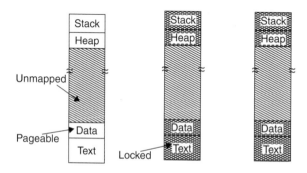

Before mlockall Call   After mlockall Call   After malloc Call

Figure 3.24 `mlockall` operation.

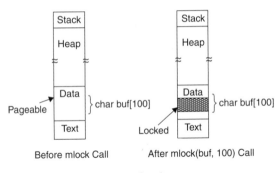

Figure 3.25 mlock operation.

calling munlockall unlocks the memory that has been locked with mlock. It is not possible to lock memory for all of a small section of code, then unlock it. Memory locking is a global technique that should not be performed by small, transient sections of code. In general, once a memory is locked, it should be locked down until the application is out of "real-time mode." The function mlock can cause a segmentation fault if an address is passed to it where there is not any memory for executable code.

As can be seen from this example, memory locking decreases execution times for the locked modules and, more importantly, can be used to guarantee execution time. At the same time, it makes fewer pages available for the application, encouraging contention.

3.6 EXERCISES

3.1 For the sample real-time systems described in Chapter 1, discuss which real-time architecture is most appropriate.
   (a)   Inertial measurement system
   (b)   Nuclear monitoring system
   (c)   Airline reservations system
   (d)   Pasta sauce bottling system
   (e)   Traffic light control
   Make whatever assumptions you like, but document them.

3.2 Should a task be allowed to interrupt itself? If it does, what does this mean?

3.3 What criteria are needed to determine the size of the run-time stack in a multiple-interrupt system? What safety precautions are necessary?

3.4 Identify some of the limitations of existing commercial real-time kernels for the development of different mission- and safety-critical applications.

3.5 What are the desirable features that a system should have to provide for predictability in time-critical applications?

3.6 Discuss the difference between static and dynamic, on-line and off-line, optimal, and heuristic scheduling algorithms.

3.7 Construct a cyclic executive with four procedures, A, B, C, D. Procedure A runs two times as frequently as B and C, and procedure A runs four times as frequently as D.

3.8 Discuss some of the advantages of EDF scheduling over RM scheduling and vice versa.

3.9 Show with an example that EDF is no longer an optimal scheduling policy if preemption is not allowed.

3.10 Explain what is meant by context-switching overhead, and how to account for it in the RM and EDF schedulability tests.

3.11 Exceptions can be used as a framework for error recovery. Define the following terms:

(a) A synchronous exception

(b) An asynchronous exception

(c) An application-detected error

(d) An environment-detected error

3.12 An operating system provides 256 fixed priorities to threads in the system, but only 32 priorities levels for their messages exchanged through message queues. Suppose that each sending thread chooses the priority of its messages by mapping the 256 thread priority levels to 32 message priority levels. Discuss some potential problems associated with this uniform mapping scheme. What kind of approach would you take?

3.13 Verify the schedulability under RM and construct the schedule of the following task set:

τ_i	e_i	p_i
τ_1	3	7
τ_2	5	16
τ_3	3	15

3.14 Verify the schedulability under EDF and construct the schedule of the following task set:

τ_i	e_i	p_i	D_i
τ_1	1	5	4
τ_2	2	8	6
τ_3	1	4	3

3.15 Give two different explanations why the following periodic tasks are schedulable by the RM algorithm.

τ_i	e_i	p_i
τ_1	0.8	2
τ_2	1.4	4
τ_3	2	8

3.16 The following system of periodic tasks is scheduled and executed according to a cyclic schedule. Draw an execution trace (timeline) showing two occurances of each task.

τ_i	e_i	p_i
τ_1	1	8
τ_2	4	15
τ_3	3	20
τ_4	8	22

3.17 Verify the schedulability under EDF, and construct the EDF schedule of the following task set in the interval (0,20).

τ_i	r_i	e_i	p_i	D_i
τ_1	0	2	8	6
τ_2	1	2	6	5
τ_3	0	4	12	10

3.18 Consider the following tasks with their resource requirements given as:

(a) $\tau_1 = (10, 4, 1; [A; 1])$, where the task executes for two time units, then requests the resource A.

(b) $\tau_2 = (7, 4, 2; [A; 1][B; 1])$, where the task executes for one time unit, then requests the resource A and then B.

(c) $\tau_3 = (4, 4, 3; [B; 1][C; 1])$, where the task executes for one time unit, then requests the resource B and then C.

(d) $\tau_4 = (0, 11, 4; [A; 5[B; 2]][C; 1])$, where the task executes for two time units, then requests the resource A and holds it for one time unit and makes a nested request for resource B, and then requests C.

Here, the notation $\tau_1 = (r_i, e_i, [R, t])$ represents that the task i is released at time r_i, has execution time e_i, priority π_i, and the critical section $[R; t]$ for the resource R and the execution time t. Also note that the representation $[R; t[S; w]]$ denotes the nested critical sections, that is, that the usage of resource R in turn includes the usage of resource S, and time t includes the time w of the critical section S.

Construct the schedule for the system using (a) the Priority Inheritance Protocol, (b) the Priority Ceiling Protocol.

3.19 What effect would size N of a ring buffer have on its performance? How would you determine the optimal size?

3.20 For a machine you are familiar with, discuss whether the counting semaphore implementation given in this chapter has any critical-region problems. That is, can the semaphore itself be interrupted in a harmful way?

3.21 Why is it not wise to disable interrupts before the while statement in the binary semaphore, P(S)?

3.22 Discuss the problems that can arise if the test and set in the P(S) operation are not atomic. What could happen if the simple assignment statement in the V(S) operation were not atomic?

3.23 Rewrite the save and restore routines assuming that eight general register (R0–R7) and the program counter are to be saved on a stack. Do this for

(a) 0-address machine

(b) 1-address machine

(c) 3-address machine

3.24 Rewrite the save and restore routines so that they save and restore to the head and tail of a ring buffer, respectively.

3.25 Rewrite the save and restore routines in 2-address code, assuming block move (BMOVE) and restore (BRESTORE) instructions are available. Make the necessary assumptions about the format of the instructions.

3.26 Rewrite the save and restore routines in the language of your choice so that they employ the push and pop procedures.

3.27 Write a pseudocode algorithm that allocates pages of memory on request. Assume that 100 pages of size 1 megabyte, 2 megabytes, and 4 megabyte are available. The algorithm should take the size of the page requested as an argument, and return a pointer to the desired page. The smallest available page should be used, but if the smallest size is unavailable, the next smallest should be used.

3.28 Write a pseudocode algorithm compacting 64 megabytes of memory that is divided into 1-megabyte pages. Use a pointer scheme.

3.29 A real-time system has a fixed number of resources of types A, B, and C. There are five tasks in the system, and the maximum amount of resources A, B, and C needed for each task is known. Implement a banker's algorithm scheme in the language of your choice.

3.30 Show how priority inheritance can cause deadlock and also multiple blocking. For example, consider the following sequence (with $\tau_1 \succ \tau_2$):

τ_1: Lock S_1; Lock S_2; Unlock S_2; Unlock S_1

τ_2: Lock S_2; Lock S_1; Unlock S_1; Unlock S_2

Here two tasks use two semaphores in a nested fashion but in reverse order.

3.31 Modify the write procedure for the ring buffer to handle the overflow condition.

3.32 Write a set of pseudocode routines to access (read from and write to) a 20-item ring buffer. The routines should use semaphores to allow more than one user to access the buffer.

3.33 Consider a binary semaphore, counting semaphore, queues, and mailboxes. Any three can be implemented with the fourth. It was shown how binary semaphores can be used to implement counting semaphores and vice versa, how mailboxes can be used to implement binary semaphores, and how mailboxes can be used to implement queues. For the remaining pairs, show how one can be used it implement the other.

3.34 The TANDS instruction can be used in a multiprocessing system to prevent simultaneous access to a global semaphore by two processors. The instruction is made indivisible by the CPU refusing to issue a DMA acknowledge (DMACK) signal in response to a DMA request (DMARQ) signal during execution of the instruction. The other processors sharing the bus are locked out of accessing memory. What are the real-time implications for a processor trying to access memory when another processor is executing a process that is looping for a semaphore using the following code?

```
getlock:  TANDS semaphore
          JNE getlock
```

If this busy wait must be used, is there a better way to test the semaphore in process 2 so that the bus is not tied up?

3.37 Obtain as much data as you can for as many of the existing commercial real-time systems as you can. Summarize your findings for each operating system, briefly, in narrative form.

3.38 Use your selection criteria and the information you have obtained to create a matrix of features by-products. In other words, present the findings you describe in step 2 more succinctly in tabular form.

3.39 For the following kinds of systems give your best recommendation as to the most likely commercial real-time operating system to use based on the selection criteria you developed

(a) A controller application for the fuel injection system of a passenger car.

(b) A hand-held game application.

(c) The F-16 navigation system.

(d) An animatronic robot used in the filming of a new science fiction movie.

(e) A medical device that reduces the time needed for an MRI scan.

Make whatever assumptions you like, but document them.

4

SOFTWARE REQUIREMENTS ENGINEERING

4.1 REQUIREMENTS-ENGINEERING PROCESS

Requirements engineering is the subdiscipline of software engineering that is concerned with determining the goals, functions, and constraints of software systems and the representation of these aspects in forms amenable to modeling and analysis. The goal is to create a requirements specification that is complete, correct, and understandable to both customers and developers. This last goal creates somewhat of a dilemma, as it indicates the duality of purpose of requirements documents: to provide insight for the customers to ensure the product under development meets their needs and expectations, and as a complete representation of the functions and constraints of the system as a basis for developers. In the real-time system domain this is further complicated by the need to represent timing and performance constraints as well as the more readily elicited functional requirements.

A generalized workflow for the requirements engineering phase is shown in Figure 4.1, where activities are represented as smoothed rectangles and the documents resulting from those activities are rectangles. The requirements engineering process begins with a preliminary study. This is an investigation into the motivation for the project and the nature of the problem. This investigation may consist of stakeholder perspectives and constraints, determination of project scope and feature priorities and, in real-time systems, some early analysis of the temporal

Some of this chapter has been adapted from Phillip A. Laplante, *Software Engineering for Image Processing*, CRC Press, Boca Raton, FL 2003.

Real-Time Systems Design and Analysis, By Phillip A. Laplante
ISBN 0-471-22855-9 © 2004 Institute of Electrical and Electronics Engineers

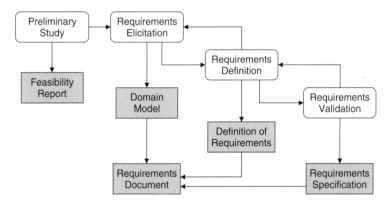

Figure 4.1 The requirements-engineering process depicting documentation in rectangles and activities in smoothed rectangles. Adapted from Sommerville [Sommerville00].

constraints imposed upon the system. One of the major deliverables of the preliminary study is a feasibility report that may advise discontinuing development of the software product. Most of the time this will not be the case, and the preliminary study will be followed by requirements elicitation.

Requirements elicitation involves gathering requirements through a variety of techniques that may include stakeholder interviews and questionnaires, focus groups, workshops, and prototyping. While requirements can be expressed in many forms ranging from standard text through mathematical formalisms, it is usual for requirements to be represented in the form of a domain model, that is, a model of the problem domain that may include such artifacts as use cases, entity-relationship diagrams, or context diagrams.

The next stage is requirements definition. It is important to define, as precisely as possible, each of the captured requirements so that they can be analyzed for completeness, noncontradiction, and correctness in the validation stage. The overall outcome of this process is a requirements document containing a software requirements specification (SRS), which is a description of the features, behaviors, and constraints of the final system. Precise software specifications provide the basis for analyzing the requirements, validating that they are the stakeholder's intentions, defining what the designers have to build, and verifying that they have done so correctly.

4.2 TYPES OF REQUIREMENTS

While there are a number of useful taxonomies of requirements, the most established being the simple functional versus nonfunctional classification, the standard scheme for real-time systems is defined by IEEE Standard [IEEE830]. Standard 830 [IEEE98] defines the following kinds of requirements:

1. Functional
2. External interfaces

3. Performance
4. Logical database
5. Design constraints
   - Standards compliance
   - Software systems attributes
6. Software system attributes
   - Reliability
   - Availability
   - Security
   - Maintainability
   - Portability

Requirements 2 through 6 are considered to be nonfunctional.

Functional requirements include a description of all system inputs and the sequence of operations associated with each input set. Either through case-by-case description or through some other general form of description (e.g., using universal quantification), the exact sequence of operations and responses (outputs) to normal and abnormal situations must be provided for every input possibility. Abnormal situations might include error handling and recovery, including failure to meet deadlines. In essence, functional requirements describe the complete deterministic behavior of the system. Generally, the functional requirements are allocated to software and hardware before requirements analysis begins, though constant trade-off analysis may cause these to shift further into the project life cycle.

External interface requirements are a description of all inputs and outputs to the system including

- Name of item
- Description of purpose
- Source of input or destination of output
- Valid range, accuracy, and/or tolerance
- Units of measure
- Timing
- Relationships to other inputs/outputs
- Screen formats/organization
- Window formats/organization
- Data formats
- Command formats

Performance requirements include the static and dynamic requirements placed on the software or on human interaction with the software as a whole. For a real-time system, static requirements might include the number of simultaneous users to be supported. The dynamic requirements might include the number of transactions and tasks and the amount of data to be processed within certain time periods for both normal and peak workload conditions.

Logical database requirements include the types of information used by various functions, such as frequency of use, accessing capabilities, data entities and their relationships, integrity constraints and data-retention requirements. Design constraint requirements are related to standards compliance and hardware limitations. Lastly, software system attribute requirements include reliability, availability, security, maintainability, and portability.

It is worth observing that the conventional nomenclature for functional versus nonfunctional requirements is unfortunate; however, because the terms functional/nonfunctional seem inappropriate for real-time systems. A more logical taxonomy would include a classification of behavior observable via execution and that which is not observable via execution (e.g., maintainability, portability).

4.3   REQUIREMENTS SPECIFICATION FOR REAL-TIME SYSTEMS

There appears to be no particularly dominant approach for specification of real-time applications [Bucci95]. In general, it seems that real-time systems engineers tend to use one or a combination of the following approaches:

- Top-down process decomposition or structured analysis.
- Object-oriented approaches.
- Program description languages (PDL) or pseudocode.
- High-level functional specifications that are not further decomposed.
- Ad hoc techniques, including simple natural language and mathematical description, and are always included in virtually every system specification.

There are three general classifications of specification techniques; informal, formal, and semiformal. Formal methods have a rigorous, mathematical basis. A small sampling of these approaches is discussed in the next section. A requirements specification technique is informal if it cannot be completely transliterated into a rigorous mathematical notation and associated rules. Informal specifications, for example, flowcharting, have little or no underlying mathematical structure, therefore, they cannot be properly analyzed. All that can be done with informal specifications is to find counterexamples of where the system fails to meet the requirements or where there are conflicts. This is simply not good enough for real-time systems where formal substantiation of performance characteristics of requirements is necessary. Approaches to requirements specification that defy classification as either formal or informal are sometimes called semiformal. Semiformal approaches, while not appearing to be mathematically based, might be. For example, some contend that Unified Modeling Language (UML) is semiformal because the Statechart is formal and other metamodeling techniques it employs have a pseudomathematical basis.

Others contend, however, that UML is not even semiformal, because it has serious holes and inconsistencies. There is a move to make UML more formal. The proposal for UML 2.0 contains formal components, and there are various attempts to formalize it further. The 2U consortium, for example, wants to define an unambiguous and clearly defined subset of UML via a UML-compliant series of modeling notations. In any case, UML largely enjoys the benefits of both informal and formal techniques and is widely used in real-time systems specifications and design.

4.4 FORMAL METHODS IN SOFTWARE SPECIFICATION

Formal methods attempt to improve requirements formulation and expression by the use and extension of existing mathematical approaches such as propositional logic, predicate calculus, and set theory. This approach is attractive because it offers a more scientific way to requirements specification. Writing formal requirements can often lead to error discovery in the earliest phases of the software life cycle, where they can be corrected quickly and at a low cost.

By their nature, specifications for most real-time systems usually contain some formality in the mathematical expression of the underlying interactions with the systems in which they are embedded. While this fact does not justify the claim that every real-time system specification is fully formalized, it does lead to some optimism that real-time systems can be made suitable for, at least, partial formalization.

Formal methods, however, are perceived to be difficult to use by even the most expertly trained and are sometimes error-prone. For these reasons, and because they are often believed to increase early life-cycle costs and delay projects, formal methods are frequently avoided.

Formal methods attempt to improve requirement formulation and expression by applying mathematics and logic. Formal methods employ some combination of predicate calculus (first-order logic), recursive function theory, Lambda calculus, programming language semantics, discrete mathematics, number theory, abstract algebra, and so on. One of the primary attractions of formal methods is that they offer a highly scientific approach to development. Formal requirements offer the possibility of discovering errors at the earliest phase of development, while the errors can be corrected quickly and at a low cost. Informal specifications might not achieve this goal, because while they can be used to refute a specific requirement by counterexample, the counterexamples might not be obvious because there is no rigorous approach to reasoning about them.

Formal methods are typically not intended to take on an all-encompassing role in system or software development. Instead, individual techniques are designed to optimize one or two parts of the development life cycle. There are three general uses for formal methods:

- *Consistency Checking* This is where system behavioral requirements are described using a mathematically based notation.

- *Model Checking*   State machines are used to verify whether a given property is satisfied under all conditions.
- *Theorem Proving*   Here, axioms of system behavior are used to derive a proof that a system will behave in a given way.

Formal methods also offer opportunities for reusing requirements. Embedded systems are often developed as families of similar products, or as incremental redesigns of existing products. For the first situation, formal methods can help identify a consistent set of core requirements and abstractions to reduce duplicate engineering effort. For redesigns, having formal specifications for the existing system provides a precise reference for baseline behavior, and provides a way to analyze proposed changes [Bowen95].

As a simple example of the use of formal methods, consider the following excerpt from the Software Requirements Specification for the nuclear monitoring system:

1.1 *If interrupt A arrives, then task B stops executing.*

1.2 *Task A begins executing upon arrival of interrupt A.*

1.3 *Either Task A is executing and Task B is not, or Task B is executing and Task A is not, or both are not executing.*

These requirements can be formalized by rewriting each in terms of their component propositions, namely:

p: interrupt A arrives

q: task B is executing

r: task A is executing

Then rewriting the requirements using these propositions and logical connectives yields:

1.1 $p \Rightarrow \neg q$

1.2 $p \Rightarrow r$

1.3 $(r \wedge \neg q) \vee (q \wedge \neg r) \vee (\neg q \wedge \neg r)$

Notice the difficulties in dealing with the precise articulation of temporal behavior. For example, in requirement 1.2, task A begins executing upon arrival of the interrupt, but does it continue executing? For how long?

In any case the consistency of these requirements can be checked by demonstrating that there is at least one set of truth values that makes each requirement true. This can be seen by writing the truth table, as seen in Table 4.1. Looking at the table, clearly in rows 2, 3, 4, and 8 and columns 6, 7, and 8, corresponding to requirements 1.1, 1.2, and 1.3, are all true, and hence this set of requirements is consistent.

Table 4.1 Truth table used to check the consistency of the set of three requirements

	1	2	3	4	5	6	7	8
	p	q	r	$\neg q$	$\neg r$	$p \Rightarrow q$	$p \Rightarrow r$	$(r \wedge \neg q) \vee (q \wedge \neg r) \vee (\neg q \wedge \neg r)$
1	T	T	T	F	F	T	T	F
2	T	T	F	F	T	T	T	T
3	T	F	T	T	F	T	T	T
4	T	F	F	T	T	T	T	T
5	F	T	T	F	F	F	F	F
6	F	T	F	F	T	F	T	T
7	F	F	T	T	F	T	F	T
8	F	F	F	T	T	T	T	T

Note that rows 2, 3, 4, and 8 in columns 6, 7, and 8 are all true, indicating that these requirements are consistent.

Consistency checking is particularly useful when there are large numbers of complex requirements. If automated tools were available to do the checking, it would seem that large specifications could be consistency checked this way. However, aside from the difficulties in formalizing the notation, finding a set of truth values that yield a composite truth value for the set of propositions is, in fact, the Boolean satisfiability problem, which is an NP complete problem. (see Chapter 7).

4.4.1 Limitations of Formal Methods

Formal methods have two limitations that are of special interest to real-time system developers. First, although formalism is often used in pursuit of absolute correctness and safety, it can guarantee neither. Second, formal techniques do not yet offer good ways to reason about alternative designs or architectures.

Correctness and safety are two of the original motivating factors driving adoption of formal methods. Nuclear, defense, and aerospace regulators in several countries now mandate or strongly suggests use of formal methods for safety-critical systems. Some researchers emphasize the "correctness" properties of particular mathematical approaches, without clarifying that mathematical correctness in the development process might not translate into real-world correctness in the finished system. After all, it is only the specification that must be produced and proven at this point, not the software product itself.

Formal software specifications must be converted to a design, and later, to a conventional implementation language at some point. This translation process is subject to all the potential pitfalls of any programming effort. For this reason, testing is just as important when using formal requirement methods as when using traditional ones, though the testing load can be reduced with formal methods. Formal verification is also subject to many of the same limitations as traditional testing, namely, that testing cannot prove the absence of bugs, only their presence.

Notation evolution is a slow, but ongoing process in the formal methods community. It can take many years from when a notation is created until it is adopted in industry. Possibly the biggest challenge in applying formal methods to real-time embedded systems is choosing an appropriate technique to match the problem. Still, to make formal models usable by a wide spectrum of people, requirement documents should use one or more nonmathematical notations, such as natural language, structured text, or diagrams.

4.4.2 Z

Z (pronounced *zed*),[1] introduced in 1982, is a formal specification language that is based on set theory and predicate calculus. As in other algebraic approaches, the final specification in Z is reached by a refinement process starting from the most abstract aspects of the systems. There is a mechanism for system decomposition known as the Schema Calculus. Using this calculus, the system specification is decomposed in smaller pieces called schemas where both static and dynamic aspects of system behavior are described.

The Z language does not have any support for defining timing constraints. Therefore, several extensions for time management have emerged. For example, Z has been integrated with real-time interval logic (RTIL), which provides for an algebraic representation of temporal behavior.

There are other extensions of Z to accommodate the object-oriented approach, which adds formalism for modularity and specification reuse. These extensions, define the system state space as a composition of the state spaces of the individual system objects. Most of these extensions also provide for information hiding, inheritance, polymorphism, and instantiation into the Z Schema Calculus. For example, one extension, Object-Z, includes all the aforementioned extensions and further integrates the concepts of temporal logic, making it suitable for real-time specification. In this language the object status is a sort of event history of object behavior, making the language more operational than the early version of Z.

4.4.3 Finite State Machines

The finite state automaton (FSA), finite state machine (FSM), or state transition diagram (STD) is a formal mathematical model used in the specification and

---

[1] Z is just one of many formal specification languages. Others, such as CSP and CCS, could have been used in this discussion instead. This is not a dismissal of these or other techniques – it is simply a sacrifice to brevity. The reader is encouraged to refer to other texts for a more thorough treatment, for example, [Burns90], [Shaw01].

design of a wide range of systems. Intuitively, finite state machines rely on the fact that many systems can be represented by a fixed number of unique states. The system may change state depending on time or the occurrence of specific events, a fact that is reflected in the automaton. The use of finite state machines in the design of real-time software is discussed in Chapter 5.

A finite state machine can be specified in diagrammatic, set-theoretic, and matrix representations. To illustrate a diagrammatic representation suppose that it is desired to model an avionics computer for a fighter aircraft. The inertial measurement unit described thus far resides within this avionics system. The avionics system computer can operate in one of five modes: takeoff (TAK), navigation (NAV), navigation/evasive (NAE), navigation/attack (NAA), and landing (LAN). The avionics system computer reacts to various signals received from other computers in the aircraft. These signals consist of: mission assignment (MA), enemy locked-on (LO), target detected (TD), mission complete (MC), enemy evaded (EE), enemy destroyed (ED). The initial state is TAK, and the only terminal state is LAN. The transition from state to state based on the appropriate signal is fairly clear from Figure 4.2.

Formally, the finite state machine in Figure 4.2 can also be represented mathematically by the five-tuple

$$M = \{S, i, T, \Sigma, \delta\} \tag{4.1}$$

where S is a finite, non-empty set of states; i is the initial state (i is a member of S); T is the set of terminal states ($T \subseteq S$); Σ is an alphabet of symbols or events used to mark transitions; δ is a transition function that describes the next state of the machine given the current state, and a symbol from the alphabet (an event). That is, $\delta : S \times \Sigma \rightarrow S$. In the avionics system example, the set of states, $S = \{TAK, NAV, NAE, NAA, LAN\}$, the initial state, $i = TAK$, the set of

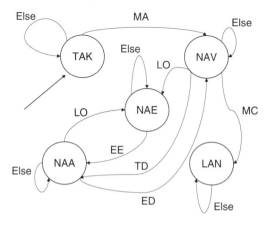

Figure 4.2 A graphical representation of a finite state machine for the avionics system in which the inertial measurement unit resides.

terminal states is {LAN}, and the alphabet, $\Sigma = \{$MA, LO, TD, MC, EE, ED$\}$. The transition function, δ, is embodied in the diagram, and can be described formally as a set of triples, but it is usually more convenient to represent the transition function with a transition table, as shown in Table 4.2.

As another example, consider three states of the process in a real-time operating system: suspended, ready, and executing. The events that cause transitions from one state to another can be clearly depicted by the partial finite state machine shown in Figure 4.3. A finite state machine that does not depict outputs during transition is called a Moore machine. However, outputs during transition can be depicted by a variation of the Moore machine called a Mealy machine. The Mealy machine can be described mathematically by a six-tuple,

$$M = \{S, i, T, \Sigma, \Gamma, \delta\} \tag{4.2}$$

Table 4.2 Transition table representation for the finite state machine in Figure 4.2

	MA	LO	TD	MC	EE	ED
TAK	NAV	TAK	TAK	TAK	TAK	TAK
NAV	NAV	NAE	NAA	LAN	NAV	NAV
NAE	NAE	NAE	NAE	NAE	NAA	NAE
NAA	NAA	NAE	NAA	NAA	NAA	NAV
LAN	LAN	LAN	LAN	LAN	LAN	LAN

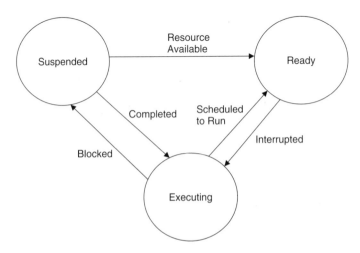

Figure 4.3 Partial finite state machine specification of a real-time operating system's behavior.

where the first five elements of the six-tuple are the same as for the Moore machine and a sixth parameter, Γ, that represents the set of outputs. The transition function is slightly different from before in that it describes the next state of the machine given the current state, and a symbol from the alphabet. The transition function is then $\delta : S \times \Lambda \to S \times \Gamma$. A general Mealy machine for a system with three states, three inputs, and three outputs is shown in Figure 4.4. The transition matrix for the FSM shown in Figure 4.4 is shown in Table 4.3.

Finite state machines are easy to develop, and code can be easily generated using tables to represent the transitions between states. They are also unambiguous, since they can be represented with a formal mathematical description. In addition, concurrency can be depicted by using multiple machines.

Because mathematical techniques for reducing the number of states exist, programs based on FSMs can be formally optimized. A rich theory surrounds finite state machines, and this can be exploited in the development of system specifications. On the other hand, the major disadvantage of FSMs is that the internal aspects, or "insideness" of modules cannot be depicted. That is, there is no way to indicate how functions can be broken down into subfunctions. In addition, inter-task communication for multiple FSMs is difficult to depict. Finally, depending on the system and alphabet used, the number of states can grow very large. Both of these problems, however, can be overcome through the use of statecharts.

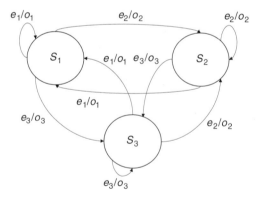

Figure 4.4 A generic Mealy machine for a three-state system with events e_1, e_2, e_3 and outputs o_1, o_2, o_3 [Laplante03c].

Table 4.3   Transition matrix for FSM in Figure 4.4

	S_1	S_2	S_3
e_1	S_1/S_1	S_1/S_1	S_1/S_1
e_2	S_2/O_2	S_2/O_2	S_2/O_2
e_3	S_3/O_3	S_3/O_3	S_3/O_3

4.4.4 Statecharts

Statecharts combine finite state machines with data flow diagrams and a feature called broadcast communication in a way that can depict synchronous and asynchronous operations. Statecharts can be described succinctly as statecharts = FSM + depth + orthogonality + broadcast communication (Figure 4.5). Here, FSM is a finite state machine, depth represents levels of detail, orthogonality represents the existence of separate tasks, and broadcast communication is a method for allowing different orthogonal processes to react to the same event. The statechart resembles a finite state machine where each state can contain its own FSM that describes its behavior. The various components of the statechart are depicted as follows:

1. The FSM is represented in the usual way, with capital letters or descriptive phrases used to label the states.
2. Depth is represented by the insideness of states.
3. Broadcast communications are represented by labeled arrows, in the same way as FSMs.
4. Orthogonality is represented by dashed lines separating states.
5. Symbols $a, b, \ldots, z$ represent events that trigger transitions, in the same way that transitions are represented in FSMs.
6. Small letters within parentheses represent conditions that must be true for the transitions to occur.

A significant feature of statecharts is the encouragement of top-down design of a module. For example, for any module (represented like a state in an FSM), increasing detail is depicted as states internal to it. In Figure 4.6 the system is composed of states A and B. Each of these in turn can be decomposed into states A1, A2, and B1 and B2, respectively, which might represent program modules. Those states can also be decomposed, and so forth. To the software designer, each nested substate within a state represents a procedure within a procedure.

Orthogonality depicts concurrency in the system for processes that run in isolation, called AND states. Orthogonality is represented by dividing the orthogonal components by dashed lines. For example, if state Y consists of AND components

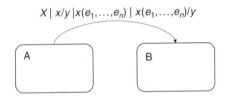

$$X \mid x/y \mid x(e_1, \ldots, e_n) \mid x(e_1, \ldots, e_n)/y$$

Figure 4.5 Statechart format where A and B are states, x is an event that causes the transition marked by the arrow, y is an optional event triggered by x, and $e_1, \ldots, e_n$ are conditions qualifying the event [Laplante03c].

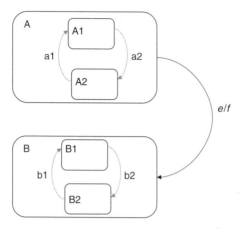

Figure 4.6 A statechart depicting insideness [Laplante03c].

A and D, Y is called the orthogonal product of A and D. If Y is entered from the outside (without any additional information), then the states A and D are entered simultaneously. Communication between the AND states can be achieved through global memory, whereas synchronization can be achieved through a capability of statecharts called broadcast communication.

Broadcast communication is depicted by the transition of orthogonal states based on the same event. For example, if the inertial measurement system switches from standby to ready mode, an event indicated by an interrupt can cause a state change in several processes. Another unique aspect of broadcast communication is the concept of the chain reaction; that is, events triggering other events. The implementation follows from the fact that statecharts can be viewed as an extension of Mealy machines, and output events can be attached to the triggering event. In contrast with the Mealy machine, however, the output is not seen by the outside world; instead it affects the behavior of an orthogonal component. For example, in Figure 4.7 suppose there exists a transition labeled e/f, and if event e occurs then event f is immediately activated. Event f could, in turn, trigger a transaction such as f/g. The length of a chain reaction is the number of transitions triggered by the first event. Chain reactions are assumed to occur instantaneously. In this system a chain reaction of length 2 will occur when the

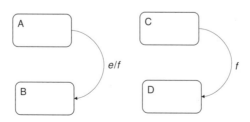

Figure 4.7 Statechart depicting a chain reaction [Laplante03c].

e/f transition occurs. As a further example, Figure 5 found in the Case Study in Section 4.10 illustrates a statechart corresponding to the traffic intersection control system.

Statecharts are excellent for representing embedded systems because they can easily depict concurrency while preserving modularity. In addition, the concept of broadcast communication allows for easy intertask.

In summary, the statechart combines the best of data flow diagrams and finite state machines. Finally, commercial products allow an engineer to graphically design a real-time system using statecharts, perform detailed simulation analysis, and generate Ada or C code. Furthermore Statecharts can be used in conjunction with both structured and object-oriented analysis.

4.4.5 Petri Nets

Petri nets are another formal method used to specify the operations to be performed in a multiprocessing or multitasking environment. While they have a rigorous foundation, they can also be described graphically. A series of circular bubbles called "places" are used to represent data stores or processes. Rectangular boxes are used to represent transitions or operations. The processes and transitions are labeled with a data count and transition function, respectively, and are connected by unidirectional arcs.

Table 4.4   Transition table for Petri net shown in Figure 4.8 [Laplante03c]

	P_1	P_2
Before firing	1	0
After firing	0	1

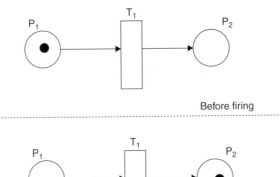

Before firing

After firing

Figure 4.8   Petri net firing rule [Laplante03c].

The initial graph is labeled with markings given by m_0, which represent the initial data count in the process. Net markings are the result of the firing of transitions. A transition, t, fires if it has as many inputs as required for output. In Petri nets, the graph topology does not change over time; only the "markings" or contents of the places do. The system advances as transitions "fire." To illustrate the notion of "firing," consider the Petri nets given in Figure 4.7 and the associated firing table given in Table 4.4.

As a somewhat more significant example, consider the Petri net in Figure 4.9. Reading from left to right and top to bottom indicates the stages of firings in the net. Table 4.5 depicts the firing table for the Petri net in Figure 4.8.

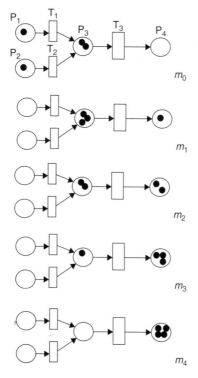

Figure 4.9 Behavior sequence of a slightly more complex Petri net [Laplante03c].

Table 4.5 Firing table for Petri net in Figure 4.9 [Laplante03c]

	P_1	P_2	P_3	P_4
m_0	1	1	2	0
m_1	0	0	3	1
m_2	0	0	2	2
m_3	0	0	1	3
m_4	0	0	0	4

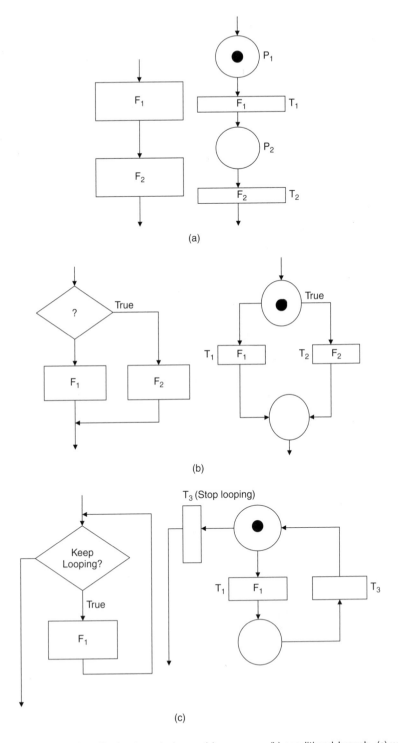

Figure 4.10 Flowchart-Petri net equivalence: (a) sequence, (b) conditional branch, (c) while loop [Laplante03c].

Petri nets can be used to model systems and to analyze timing constraints and race conditions. Certain Petri net subnetworks can model familiar flowchart constructs. Figure 4.10 illustrates these analogies.

Petri Nets are excellent for representing multiprocessing and multiprogramming systems, especially where the functions are simple. Because they are mathematical in nature, techniques for optimization and formal program proving can be employed. But Petri nets can be overkill if the system is too simple. Similarly, if the system is highly complex, timing can become obscured.

The model described herein is just one of a variety of available models. For example, there are timed Petri nets, which enable synchronization of firings, colored Petri nets, which allow for labeled data to propagate through the net, and even timed-colored Petri nets, which embody both features.

4.4.6 Requirements Analysis with Petri Nets

The Petri net is a powerful tool that can be used during analysis or design for deadlock and race-condition identification. For example, suppose a requirement contains a subnet that resembles Figure 4.11. Clearly, it is impossible to tell which of the transitions will fire, and in any case, only one of them will fire.

Petri nets can also be used to identify cycles that indicate a potential deadlock. For example, suppose a requirement can be modeled as in Figure 4.12, which is,

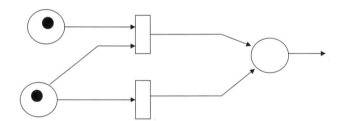

Figure 4.11 Race (or conflict) condition identification with Petri nets.

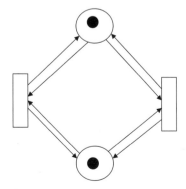

Figure 4.12 Deadlock in Figure 3.15 depicted using Petri nets.

in fact, a representation of Figure 3.15 involving two resources and two processes. Clearly, this scenario is one of deadlock. And while it is unlikely that such an obvious situation would be specified, Petri net analysis can be used to identify nonobvious cycles that appear as subgraphs in more complex diagrams.

4.5   STRUCTURED ANALYSIS AND DESIGN

Methods for structured analysis and structured design (SASD) have evolved over almost 30 years and are widely used in real-time applications, probably because the techniques are closely associated with the programming languages with which they co-evolved (Fortran and C) and in which many real-time applications are written. Structured methods appear in many forms, but the de facto standard is Yourdon's Modern Structured Analysis [Yourdon91].

Several extensions to structured analysis emerged in the 1980s to account for system dynamics and use if for the specification of real-time systems. These approaches include Gomaa's DARTS (design approach for real-time systems) [Gomaa00]. Ward and Mellor extended data flow diagrams by adding edges to represent control and state machines for representing behavior. Commercial tools that followed this approach, including Teamwork and Software Through Pictures®, emerged in the early 1990s.

Structured analysis for real-time systems is still based on the notion of the flow of data between successive transformations and provides little support to identify the concurrent processes that must be implemented in a given application. Depending upon the detail of the analysis, there is something arbitrary in identifying the system processes. This may result in the implementation of unnecessary processes and the possibility that a given process needs concurrency internally [Bucci95].

Perhaps the most commonly used form of structured analysis, Yourdon's Modern Structured Analysis uses three viewpoints to describe a system: an environmental model, a behavioral model, and an implementation model. The elements of each model are shown in Figure 4.13. The environmental model embodies the analysis aspect of SASD and consists of a context diagram and an event list. The purpose of the environmental model is to model the system at a high level of abstraction. The behavioral model embodies the design aspect of SASD as a series of data flow diagrams (DFDs), entity relationship diagrams (ERDs), process specifications, state transition diagrams, and a data dictionary. Using various combinations of these tools, the designer models the processes, functions, and flows of the system in detail. Finally, in the implementation model the developer uses a selection of structure charts, natural language, and pseudocode to describe the system to a level that can be readily translated to code.

Structured analysis (SA) is a way to try to overcome the problems of classic analysis using graphical tools and a top-down, functional decomposition method to define system requirements. SA deals only with aspects of analysis that can be structured: the functional specifications and the user interface. SA is used

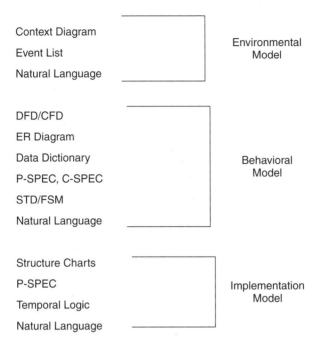

Context Diagram

Event List

Natural Language

Environmental
Model

DFD/CFD

ER Diagram

Data Dictionary

P-SPEC, C-SPEC

STD/FSM

Natural Language

Behavioral
Model

Structure Charts

P-SPEC

Temporal Logic

Natural Language

Implementation
Model

Figure 4.13   Elements of structured analysis and design [Laplante03c].

to model a system's context (where inputs come from and where outputs go), processes (what functions the system performs, how the functions interact, how inputs are transformed to outputs), and content (the data the system needs to perform its functions).

SA seeks to overcome the problems inherent in analysis through:

- Maintainability of the target document.
- Use of an effective method of partitioning.
- Use of graphics.
- Building a logical model of the system for the user before implementation.
- Reduction of ambiguity and redundancy.

The target document for SA is called the structured specification. It consists of a system context diagram, an integrated set of data flow diagrams showing the decomposition and interconnectivity of components, and an event list to represent the set of events that drive the system.

To illustrate the SA technique, consider the following simplified description of the inertial measurement system (IMU) (Figure 4.14). Some liberties have been taken with the notation, but this is not uncommon, as each organization tends to have its own "house style," that is, conventions that are dependent on tools being used or individual preferences. From the diagram, it is easy to see the four

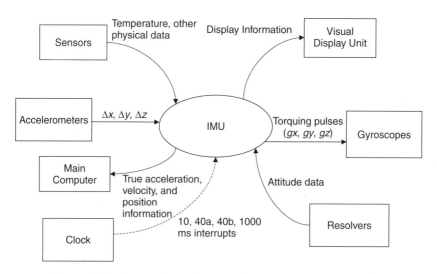

Figure 4.14   Context diagram for aircraft inertial measurement system.

interrupts in the system, each generated by an external device. These interrupts are identified as 10 ms (an interrupt generated by a clock every 10 ms), 40a ms (an interrupt generated by a clock every 40 ms), 40b ms (another 40 ms interrupt generated by a main computer requesting data from the IMU), and 1000 ms (an interrupt generated by a clock every second. It is clear from the diagram that the IMU provides display and diagnostic information to the main computer and to a display. It also outputs gyro torquing pulses in order to keep the gyroscopes properly oriented with respect to the inertial reference frame. The IMU takes information from temperature sensors, accelerometers, and position resolvers and computes true instantaneous acceleration, velocity, and position with respect to a starting point and orientation (the inertial reference frame). This information is provided to a main computer onboard the vehicle.

While the intent here is not to provide a complete system design, which means there are some omissions, a point to be made is that missing functionality is more easily identified during the requirements elicitation process if some form of graphical aid, such as the context diagram, is available. In the case of object-oriented analysis, a use case diagram will be helpful.

4.6 OBJECT-ORIENTED ANALYSIS AND THE UNIFIED MODELING LANGUAGE

As an alternative to the structured analysis approach to developing software requirements, consider using an object-oriented approach. There are various "flavors" of object-oriented analysis, each using its own toolsets. In the approach developed here, the system specification begins with the representation of externally accessible functionality as use cases.

In contrast to procedural programming, which employs algorithmic procedures, object-oriented programming uses a structure of collaborating objects, in which each part performs its specialized processing by reacting to inputs from its immediate neighbors.

The use of object-oriented approaches in real-time systems modeling provides several advantages including understandability and maintainability, modularity of design, concurrency, distributivity, improved management of complexity, enhanced reuse, increased extensibility, and excellent traceability. However, there are some disadvantages to an object-oriented approach to real-time systems, as will be discussed later.

4.6.1 Use Cases

Use cases are an essential artifact in object-oriented analysis and design (OOAD) and are described graphically using any of several techniques. The use-case diagram can be considered analogous to the context diagram in structured analysis in that it represents the interactions of the software system with its external environment. In the specification of an embedded system this is also where overall time constraints, sampling rates, and deadlines are specified.

Use cases are represented graphically as ellipses, with the actors involved represented by stick figures, as can be seen in Figure 4.15. The lines drawn

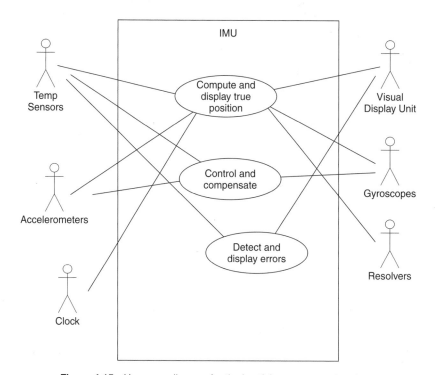

Figure 4.15 Use-case diagram for the inertial measurement system.

from the actor to the use case represent the communication between them. Each use case is, however, a document that describes scenarios of operation of the system under consideration as well as pre- and postconditions, and exceptions. In an iterative-development lifecycle these use cases will become increasingly refined and detailed as the analysis and design workflows progress. Interaction diagrams are then created to describe the behaviors defined by each use case. In the first iteration these diagrams depict the system as a black box, but once domain modeling has been completed, the black box is transformed into a collaboration of objects, as will be seen later. As a final example, Figure 3 in the Case Study in Section 4.10 illustrates the use case diagram for the traffic intersection control system.

4.6.2 Class Diagram

An analysis class diagram presents the static structure of the system, system abstractions, and their relationships. It contains classes that represent entities with common characteristics, including attributes and operations and associations that represent relationships between classes. The classes are represented by rectangles and the connection paths represent associations between classes. Classes require a name within the rectangle, whereas associations may not have an attached name.

The diamond attachment represents an aggregation relationship. If the diamond is filled, it is a dependent aggregation, otherwise it is independent, that is, the objects so aggregated can exist separately. Figure 4 in the Case Study in Section 4.10 illustrates an analysis class diagram for the traffic intersection control system.

4.6.3 Recommendations on Specification Approach for Real-Time Systems

The preceding discussions illustrate some of the challenges (in fact, one might consider them "habits") encountered by engineers specifying real-time systems:

- Mixing of operational and descriptive specifications.
- Combining low-level hardware functionality and high-level systems and software functionality in the same functional level.
- Omission of timing information.

It is risky to prescribe a preferred technique, because it is well known that there is no "silver bullet" when it comes to software specification and design, and each system should be considered on its own merits. Nevertheless, irrespective of the approach, real-time system modeling should incorporate the following best practices:

- Use consistent modeling approaches and techniques throughout the specification, for example, a top-down decomposition, structured, or object-oriented approaches.

- Separate operational specification from descriptive behavior.
- Use consistent levels of abstraction within models and conformance between levels of refinement across models.
- Model nonfunctional requirements as a part of the specification models, in particular, timing properties.
- Omit hardware and software assignment in the specification (another aspect of design rather than specification).

4.7 ORGANIZING THE REQUIREMENTS DOCUMENT

There are many ways to organize the SRS, but IEEE Standard 830-1998 is the IEEE's Recommended Practice for Software Requirements Specifications (SRS) [IEEE98], and provides a template of what an SRS should look like.

The SRS is described as a "binding contract among designers, programmers, customers, and testers," and it encompasses different design views or paradigms for system design. The recommended design views include some combination of decomposition, dependency, interface, and detail descriptions. Together with boilerplate front matter, these form a standard template for software requirements specifications, which is depicted in Figure 4.16. Sections 1 and 2 are self-evident; they provide front matter and introductory material for the SRS. The remainder

```
1. Introduction
        1.1 Purpose
        1.2 Scope
        1.3 Definitions and Acronyms
        1.4 References
        1.5 Overview
2. Overall Description
        2.1 Product Perspective
        2.2 Product Functions
        2.3 User Characteristics
        2.4 Constraints
        2.5 Assumptions and Dependencies
3. Specific Requirements
Appendices
Index
```

Figure 4.16 Recommended table of contents for an SRS from IEEE Standard 830-1998 [IEEE830].

of the SRS is devoted to the four description sections. The section headings can be broken down further using a technique such as structured analysis.

The IEEE 830 standard provides for several alternative means to represent the requirements specifications, aside from a function perspective. In particular, the software requirements can be organized by

- Functional mode (for example, "operational," "diagnostic," "calibration").
- User class (for example, "operator," "diagnostic").
- Object.
- Feature (what the system provides to the user).
- Stimulus (e.g., sensor 1, 2, and so forth).
- Functional hierarchy.
- Mixed (combining two or more of the preceding).

4.8 ORGANIZING AND WRITING REQUIREMENTS

The text structure of the SRS can be depicted by the number of statement identifiers at each hierarchical level. High-level requirements rarely have numbered statements below a depth of four (e.g., 3.2.1.5). Well-organized documents have a pyramidal structure to the requirements. Requirements with an hourglass structure mean too many administrative details. Diamond structured requirements indicate subjects introduced at higher levels were addressed at different levels of detail (Figure 4.17). Whatever approach is used in organizing the SRS, the IEEE 830 standard describes the characteristics of good requirements. That is, good requirements must be

1. *Correct* They must correctly describe the system behavior.
2. *Unambiguous* The requirements must be clear, not subject to different interpretations.
3. *Complete* There must be no missing requirements. Ordinarily, the note TBD [to be defined (later)] is unacceptable in a requirements document. IEEE 830 sets out some exceptions to this rule.

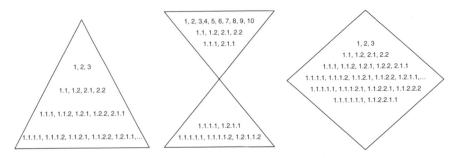

Figure 4.17 Triangle-, hourglass-, and diamond-shaped requirements configurations.

4. *Consistent* One requirement must not contradict another. Consistency can be checked using formal methods, as illustrated previously.

5. *Ranked for importance and/or stability* Not every requirement is as critical as another. By ranking the requirements, designers will find guidance in making trade-off decisions.

6. *Verifiable* A requirement that cannot be verified is a requirement that cannot be shown to have been met.

7. *Modifiable* The requirements need to be written in such a way so as to be easy to change. In many ways, this approach is similar to the information-hiding principle.

8. *Traceable* The benefits of traceability cannot be overstated. The requirements provide the starting point for the traceability chain. Numbering the requirements in hierarchical fashion can aid in this regard.

To meet these criteria and to write clear requirements documentation, there are several best practices that the requirements engineer can follow. These are as follows:

- Invent and use a standard format and use it for all requirements.
- Use language in a consistent way.
- Use "shall" for mandatory requirements.
- Use "should" for desirable requirements.
- Use text highlighting to identify key parts of the requirement.
- Avoid the use of technical language unless it is warranted.

To illustrate, consider the following bad requirements:

1. "The systems shall be completely reliable."
2. "The system shall be modular."
3. "The system shall be maintainable."
4. "The system will be fast."
5. "Errors shall be less than 99%."

These requirements are bad for a number of reasons. None are verifiable, for example, how is "reliability" supposed to be measured? Even requirement 5 is vague, that is, what does "less than 99% mean"?

Now consider the following requirements:

1. "Response times for all level one actions will be less than 100 ms."
2. "The cyclomatic complexity of each module shall be in the range or 10 to 40."
3. "95% of the transactions shall be processed in less than 1 s."
4. "An operator shall not have to wait for the transaction to complete."
5. "MTBF (mean time between failures) shall be 100 hours of continuous operation."

These requirements are better versions of the preceding ones. Each is measurable, because each makes some attempt to quantify the qualities that are desired. For example, cyclomatic complexity is a measure of modularity, MTBF is a measure of failures, and processing time is a measure of speed. Although these improved requirements could stand some refinement based on the context of requirements specification as a whole.

4.9 REQUIREMENTS VALIDATION AND REVIEW

Verification of the software product means ensuring that the software is conforming to the SRS. Verification is akin to asking the question "Am I building the software right?" in that it requires satisfaction of requirements.

Requirements validation, on the other hand, is tantamount to asking the question "Am I building the right software?" Too often, software engineers deliver a system that conforms to the SRS, only to discover that it is not what the customer really wanted.

Performing requirements validation involves checking the following:

1. Validity, that is, does the system provide the functions that best support the customer's needs?
2. Consistency, that is, are there any requirements conflicts?
3. Completeness, in other words, are all functions required by the customer included?
4. Realism, or, can the requirements be implemented given available budget and technology?
5. Verifiability, that is, can the requirements be checked?

There are a number of ways of checking the SRS for conformance to the IEEE 830 best practices and for validity. These approaches include:

1. Requirements reviews.
2. Systematic manual analysis of the requirements.
3. Prototyping.
4. Using an executable model of the system to check requirements.
5. Test-case generation.
6. Developing tests for requirements to check testability.
7. Automated consistency analysis.
8. Checking the consistency of a structured requirements description.

Of these approaches, automated checking is the most desirable and the least likely because of the context sensitivity of natural languages, and the impossibility of verifying such things as requirements completeness. However, simple tools can be developed to perform simple spelling and grammar checking (which is obviously undesirable, but also can indicate ambiguity and incompleteness), flagging of key words that may be ambiguous (e.g., "fast," "reliable"), identification of missing requirements (e.g., search for the phrase "to be determined") and overly complex sentences (which can indicate unclear requirements).

4.9.1 Requirements Validation Using Model Checking

Model checking is a formal method that can be used to perform analysis of requirements specifications, even partial ones. The aim, however, is to find errors, not prove correctness.

One methodology uses state machines to test for safety and liveness. The first step involves building a state model of the system, for example, using statecharts. Once this initial model is obtained, the state space size is estimated in order to assess the potential for automated validation. Next, the state space is reduced by identifying equivalence classes and by exploiting symmetries and subclasses. Finally, a symbolic representation of the main features of the requirements are derived. This representation represents a behavioral, temporal logic that emulates the coarse-grain behavior of the system. To check for fault-tolerance, faults are injected into the "live" model and this model is exercised to identify problems [Schneider98]. Model checking represents, in some ways, a very high-level prototype of the requirements.

4.9.2 Automated Checking of Requirements

Automated requirements checking is used to assess certain qualities of requirements specifications, not to assess the correctness of the SRS (for example, some of the challenges in automated consistency checking were discussed). One example of such a tool is NASA's Automated Requirements Measurement (ARM) tool [Wilson97].

Tools, like ARM use several requirements indicators at both a coarse-grain and fine-grain scale. Coarse-grain indicators include

- Size of requirements
- Text structure
- Specification depth
- Readability

Fine-grain measures look at the use of certain categories of words in the documents. Typical indicators are

- Imperatives
- Continuances
- Directives
- Options

Table 4.6   Imperatives found in requirements specifications and their purpose [Wilson97]

Imperative	Purpose
Shall	Dictates provision of fundamental capability
Must	Establishes performance requirements or constraints
Must not	Establishes performance requirements or constraints
Is required to	Used in specifications statements written in passive voice
Are applicable	Used to include, by reference, standards or other documentation as an addition to the requirements being specified
Responsible for	Used as an imperative for systems whose architectures are already defined
Will	Generally used to cite things that the operational or development environment are to provide to the capability being specified
Should	Not recommended for use

- Weak phrases
- Imperatives

Imperatives are given in Table 4.6.

Continuances follow an imperative and introduce the specification of requirements at a lower level. Continuances include:

- "Below"
- "As follows"
- "Following"
- "Listed"
- "In particular"
- "Support"

Directives are words and phrases that point to illustrative information:

- "Figure"
- "Table"
- "For example"

Options give the developer latitude in satisfying the specifications, and include:

- "Can"
- "May"
- "Optionally"

Weak phrases, which should be avoided in SRS, include:

- "Adequate"
- "As a minimum"

- "As applicable"
- "Be able to"
- "Be capable"
- "But not limited to"
- "Capability of"
- "Capability to"
- "Effective"
- "If practical"
- "Normal"
- "Provide for"
- "Timely"
- "TBD"

These fine-grained measures can, minimally, be used to measure certain size qualities of the SRS, such as:

- Lines of text
- Imperatives
- Subjects (unique words following imperatives)
- Paragraphs

Certain ratios can also be computed from these fine-grained measures, which can be used to judge the fitness of the specification of these ratios are shown in Table 4.7

Readability statistics, similar to those used to measure writing level, can be used as a quality measure for SRS. Readability statistics include:

- *Flesch Reading Ease Index*   Number of syllables/word and words/sentence.
- *Flesch-Kincaid Grade Level Index*   Flesch score converted to a grade level (standard writing is about seventh or eighth grade).

Table 4.7   Certain ratios derived from software requirements specifications and their purpose

Ratio	Purpose
Imperatives to subjects	Indicates level of detail
Lines of text to imperatives	Indicates conciseness
Number of imperatives found at each document levels	Counts the number of lower-level items that are introduced at a higher level by an imperative followed by a continuance
Specification depth to total lines of text	Indicates conciseness of the SRS

- *Coleman-Liau Grade Level Index*   Uses word length in characters and sentence length in words to determine grade level.
- *Bormuth Grade Level Index*   Same as Coleman-Liau.

Any of these requirements metrics can be incorporated into a metrics-management discipline, and if used consistently and intelligently, will improve the real-time system in the long run.

4.10 APPENDIX: CASE STUDY IN SOFTWARE REQUIREMENTS SPECIFICATION FOR FOUR-WAY TRAFFIC INTERSECTION TRAFFIC LIGHT CONTROLLER SYSTEM

The following is an excerpt from the SRS for the traffic intersection control system introduced in Chapter 1. It embodies many of the elements discussed in this chapter in more detail and provides a fully developed example of an object-oriented approach to requirements specification of a complex real-time system.

1 INTRODUCTION

Traffic controllers currently in use comprise simple timers that follow a fixed cycle to allow vehicle/pedestrian passage for a pre-determined amount of time regardless of demand, actuated traffic controllers that allow passage by means of vehicle/pedestrian detection, and adaptive traffic controllers that determine traffic conditions in real-time by means of vehicle/pedestrian detection and respond accordingly in order to maintain the highest reasonable level of efficiency under varying conditions. The traffic controller described in this specification is capable of operating in all three of these modes.

1.1 Purpose

This specification defines the software design requirements for an intersection control system for simple, four-way pedestrian/vehicular traffic intersections. The specification is intended for use by end users as well as software developers.

1.2 Scope

This software package is part of a control system for pedestrian/vehicular traffic intersections that allows for (1) a fixed cycle mode, (2) an actuated mode, (3) a fully adaptive automatic mode, (4) a locally controlled manual mode, (5) a remotely controlled manual mode and (6) an emergency preempt mode. In the fully adaptive automatic mode, a volume detection feature has been included so that the system is aware of changes in traffic patterns. Pushbutton fixtures are also included so the system can account for and respond to pedestrian traffic. The cycle is controlled by an adaptive algorithm that uses data from many inputs to achieve maximum throughput and acceptable wait-times for both pedestrians and motorists. A preempting feature allows emergency vehicles to pass through the intersection in a safe and timely manner by altering the state of the signals and the cycle time.

1.3 Definitions, Acronyms, Abbreviations

The following is a list of terms and their definitions as used in this document.

1.3.1 10-Base T
Physical connection formed by a twisted-pair as described in IEEE 802.3. Networking connection designed to transfer up to 10 megabits per second.

1.3.2 ADA
Americans With Disabilities Act.

1.3.3 API
Application Program Interface.

1.3.4 Approach
Any one of the routes allowing access to an intersection.

1.3.5 Arterial Road
A major traffic route or route used to gain access to a highway.

1.3.6 Aspect
The physical appearance of an illuminated traffic standard.

1.3.7 Attribute
Property of a class.

1.3.8 Cycle Time
The time required to complete an entire rotation (cycle) of traffic signals at any one intersection.

1.3.9 Direct Route
A route directly through the intersection that does not require the vehicle to turn.

1.3.10 DOT
Department of Transportation.

1.3.11 Downstream
The normal travel direction for vehicles.

1.3.12 Ethernet
The most commonly used local area networking method as described in IEEE 802.3.

1.3.13 Intersection
A system, including hardware and software, that regulates vehicle and pedestrian traffic where two or more major roads traverse. The class of intersection considered in this specification has only two roads.

1.3.14 Manual Override
A device located at and physically connected to each intersection control system that allows traffic regulatory personnel to control the intersection manually.

1.3.15 Method
Procedure within a class exhibiting an aspect of class behavior.

1.3.16 Message
An event thrown from one code unit and caught by another.

1.3.17 Occupancy Loop
A device used to detect the presence of vehicles in an approach or to count the passage of vehicles using an approach.

1.3.18 Offset
The time difference between cycle start times at adjacent intersections. Applies only to coordinated intersection control, which is not covered by this specification.

1.3.19 Orthogonal Route
A route through an intersection that requires a vehicle to turn.

1.3.20 Pedestrian Presence Detector
A button console located on the corner of an intersection which gives pedestrians who wish to cross a street the ability to alert the intersection control system to their presence.

1.3.21 Pedestrian Traffic Standard
Signals facing in the direction of pedestrian cross walks which have lighted indicators marked ''Walk'' and ''Don't Walk.''

1.3.22 Phase
The state of an intersection. A particular period of the regulatory traffic pattern.

1.3.23 Remote Override
A computer host that includes a software interface allowing a remote administrator to control the intersection remotely.

1.3.24 RTOS
Real-Time Operating System.

1.3.25 Secondary Road
A route that does not typically support high traffic volume or experiences less usage relative to another route.

1.3.26 SNMP (Simple Network Management Protocol)
The de facto standard for inter-network management, defined by RFC 1157.

1.3.27 Split
The duty cycle for a given phase, expressed as a decimal or percentage.

1.3.28 Vehicle Traffic Standard
A traditional traffic signal with red, yellow, and green indicators.

1.3.29 Upstream
Direction opposite to the normal direction of vehicle travel.

1.3.30 Vehicle Presence Detector
See Occupancy Loop.

1.3.31 WAN
Wide Area Network.

1.4 References

1. 10 base-T Ethernet (IEEE 802.3)
2. SNMP (RFC 1157)
3. "DEVELOPMENT OF AN ACTUATED TRAFFIC CONTROLPROCESS UTILIZING REAL-TIME ESTIMATED VOLUME FEEDBACK", September 2000

1.5 Overview

2 OVERALL DESCRIPTION

2.1 Intersection Overview

The intersection class to be controlled is illustrated in Figure 1.

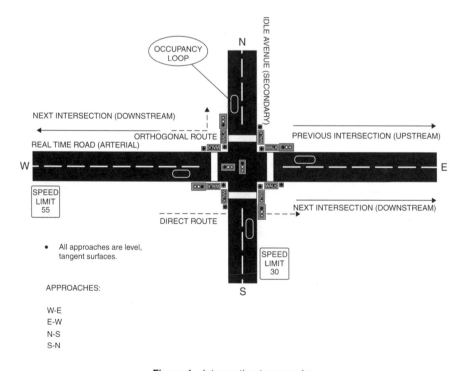

Figure 1 Intersection topography.

The target class of intersection has the following characteristics:

1. Four-way crossing.
2. Roadway gradients and curvatures are small enough to be neglected.
3. No right-turn or left-turn lanes or right-turn and left-turn signals (note, however, that the intersection is wide enough to allow vehicles passing directly through to pass to the right of vehicles turning left).
4. Intersecting roads of different priorities (e.g., one road may be an arterial while the other may be a secondary road) or of equal priority.
5. Two vehicle traffic standards per approach: one suspended by overhead cable, the other mounted on a pedestal.
6. One pedestrian crosswalk per approach.
7. Pedestrian traffic standards, pedestal mounted, on each side of each crosswalk.
8. Pedestrian presence detectors (pushbuttons) on each side of each crosswalk.
9. Stop-line vehicle presence detectors (loop detectors) in all approaches (one per approach) for detecting vehicle presence and for counting vehicles passing through the intersection.

2.2 Product Perspective

2.2.1 System Interfaces
These are described in detail in the sections below.

2.2.2 User Interfaces
2.2.2.1 Pedestrians
Pedestrian pushes button, generating service request to software and receives, in time, the "Walk" signal.

2.2.2.2 Motor Vehicles
In ACTUATED mode, vehicle enters the intersection, generating service request to software and receives, in time, the "Okay to Proceed" signal.

In ADAPTIVE mode, vehicle passes over the loop detector, increasing the vehicle count, which, in turn, causes an adjustment in intersection timings.

2.2.2.3 Emergency Vehicle
Emergency vehicle operator activates the "emergency vehicle override signal", generating priority service request to software and receives, in a preemptive time, the "Okay to proceed" signal.

2.2.2.4 Traffic Regulatory Personnel
Traffic regulatory personnel will remove the manual override device from the control box and press buttons to control the intersection manually.

2.2.2.5 Remote Operator
Remote operator uses a software control panel either to control the state of the intersection directly or to observe and manipulate the parameters and state of a specific intersection control system.

2.2.2.6 Maintainer
Maintainer accesses system through Ethernet port to perform maintenance.

2.2.3   *Hardware Interfaces*

The Intersection Control System hardware interfaces are summarized in Figure 2 on the following page.

2.2.3.1   *Major Hardware Components – Summary*

Table 1   Major intersection control system hardware components

Item	Description	Quantity
1	Intersection Controller Enclosure	1
1.1	Input Circuit Breaker	1
1.2	Input Transformer	1
1.3	Input Power Supply with UPS	1
1.4	Intersection Controller	1
1.5	Lamp Driver	20
1.6	Lamp Current Sensor	40
1.7	Green Signal Safety Relay	1
1.8	Manual Override Console	1
1.9	Vehicle Presence Detector Interface Unit (not shown in Figure 2)	4
1.10	Pedestrian Request Detector Interface Unit (not shown in Figure 2)	8
1.11	RJ-45 Ethernet Connector – DOT Network	1
1.12	RJ-45 Ethernet Connector – Maintenance	1
1.13	Enclosure Wiring	A/R
2	Vehicle Traffic Standard – Suspended	4
3	Vehicle Traffic Standard – Pole Mounted	4
4	Pedestrian Traffic Standard	8
5	Pedestrian Request Detector	8
6	Vehicle Presence Detector	4
7	Emergency Vehicle Transponder	1
10	Field Wiring	A/R

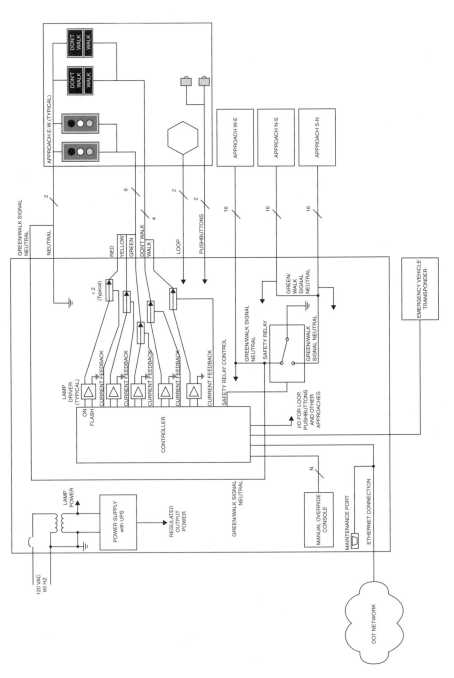

Figure 2 Intersection controller hardware (not all details and interconnects shown).

2.2.3.2   Wired Interfaces – Internal
Hard-wired connections between the intersection controller and the following hardware components within the intersection controller enclosure are provided:

1. Traffic Standard Lamp Drivers (20)
2. Traffic Standard Lamp Current Sensors (40)
3. Vehicle Presence Detector Interface Units (4)
4. Pedestrian Presence Detector Interface Units (4)
5. Green Signal Safety Relay (1)
6. Manual Override Console (1)
7. Maintenance Connector (2; 10-base T twisted pair)

2.2.3.3   Wired Interfaces – External
Hard-wired connections between the intersection control enclosure and the following external hardware components are provided:

1. Pedestrian Presence Detector
2. Pedestrian Traffic Standard
3. Vehicle Presence Detector
4. Vehicle Traffic Standard
5. Emergency Vehicle Transponder
6. DOT Wide-Area Network (WAN)

2.2.3.4   Emergency Vehicle Transponder
The emergency vehicle transponder is a radio frequency link between the intersection control system and the emergency vehicle override controller.

2.2.3.5   Ethernet Connection to DOT WAN
Interaction between the software system and the remote operator console is conducted over a standard 10 base-T local area network. Each intersection control system is identified with a unique, statically assigned IP address.

2.2.4   Software Interfaces
2.2.4.1   Operating System
The intersection controller interfaces to the RTOS via standard OS API calls.

2.2.4.2   Resource Managers
Interfaces to hardware are handled by resource managers not specified in this SRS. Resource managers are assumed to have direct access to the object model defined here.

2.2.4.3   Software Control Panel
The intersection control system must be able to interact with the software control panel to allow remote user access. This interface provides a remote user the ability to modify system parameters, perform maintenance functions, or assume manual control of the intersection. The standard protocol for this communication will be SNMP version 1.

2.2.5   Communications Interfaces

The system will utilize TCP/IP's SNMP interface for inter-system communication.

2.2.6   Memory Constraints

2.2.6.1   Flash Memory

Flash memory will be the memory media of choice for the system. The software will require no more than 32 MB of flash memory for RTOS, application program, and data.

2.2.6.2   RAM

RAM will be used for application execution. The system shall not require more than 32 MB of RAM. Upon boot, the RTOS, application program and static data needed for execution will be copied from flash into the RAM.

2.2.7   Operations

1. Automatic, unattended operation (normal operation)
2. Local manual operation (through override console)
3. Remote manual operation (through WAN port)
4. Local observed operation (through maintenance port)
5. Remote observed operation (through WAN port)
6. Remote coordinated operation (option; through WAN port)

2.2.8   Site Adaptation Requirements

This is summarized in Section 2.1, above.

2.3   Product Functions

The Intersection Control System provides the following functions:

1. Control of the intersection vehicle traffic standards.
2. Control of the intersection pedestrian traffic standards.
3. Collection and processing of traffic history from all approaches.
4. Adaptive control of intersection timings in response to traffic flow.
5. Actuated control of intersection in response to vehicle presence.
6. Timed control of intersection in response to a fixed scheme.
7. Handling of pedestrian crossing requests.
8. Handling of emergency vehicle pre-emption.
9. Intersection control in response to manual override commands.
10. Intersection control in response to remote override commands.
11. Management of traffic history and incident log databases.
12. Handling of maintenance access requests from the maintenance port.
13. Handling of maintenance access requests from the DOT WAN.

2.4   User Characteristics

2.4.1   Pedestrians

General population, including persons with disabilities.

2.4.2   Motor Vehicle
Automobiles and trucks, depending on roadway use limitations.

2.4.3   Traffic Regulatory Personnel
Authorized DOT, police, or other personnel trained in use of the Manual Override console. Must have key to the system enclosure.

2.4.4   System Administrators
Authorized DOT personnel with training in the use of this system.

2.5   Constraints

System Constraints include the following:

1. Regulatory policy (e.g., ADA).
2. DOT specifications.
3. Local ordinances.
4. Hardware limitations.
5. Minimum time for pedestrian to cross.
6. Minimum stopping distance for vehicles.
7. Momentary power droops/outages.
8. Interfaces to other applications.
9. Audit functions.
10. Higher-order language requirements (OO language supported by RTOS required).
11. Network protocols (e.g., SNMP).
12. Reliability requirements.
13. Criticality of the application.
14. Security considerations.
15. Safety considerations.

2.6   Assumptions and Dependencies

1. SI units are used for all physical quantities.
2. Commercially available RTOS is used.
3. Hardware interfaces have resource managers (drivers) already developed and available for integration with the software system specified here.
4. DOT WAN will use SNMP to communicate with intersection control system.
5. Watchdog circuitry forces safe default intersection state through hardware.

3   SPECIFIC REQUIREMENTS

This section describes the basic functional elements of the intersection control system. In particular, the software object model is described in detail, with attributes and methods enumerated. External interfaces to users, hardware, and other software elements are described, and background on the adaptive algorithm to be used is provided.

3.1   External Interface Requirements

3.1.1   User Interfaces

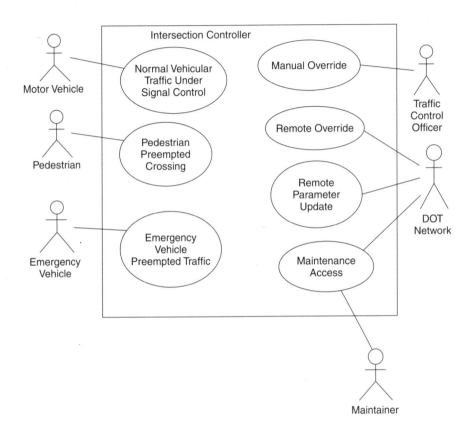

Figure 3   Top-level use-case diagram.

1. Vehicle Presence Detector – User: **Motor Vehicle**
2. Pedestrian Presence Detector – User: **Pedestrian**
3. Emergency Vehicle Override – User: **Emergency Vehicle**
4. Manual Override – User: **Traffic Control Officer**
5. Remote Override – User: **DOT Officer**
6. Maintenance Interface – User: **Maintainer**

3.1.2   Hardware Interfaces

1. Vehicle
2. Pedestrian crossing pushbutton
3. Traffic standard
4. Walk signal

5. Hardware watchdog
6. Uninterruptible power supply

3.1.3   Software Interfaces

1. RTOS API calls.
2. Hardware resource manager interfaces.

3.1.4   Communications Interfaces

1. Interface to RTOS TCP/IP stack.

3.2   Classes/Objects

Figure 4 depicts the classes constituting the intersection control system software application.

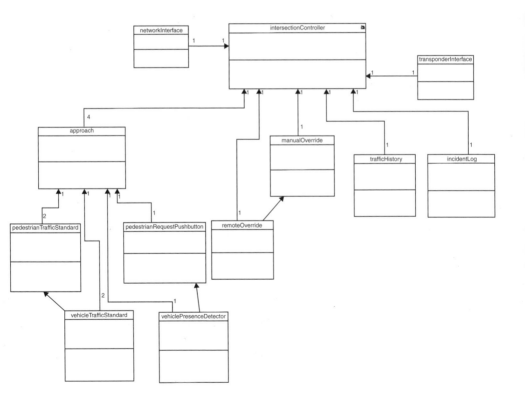

Figure 4   Preliminary intersection controller class diagram.

3.2.1 Intersection Controller

The Intersection Controller is responsible for managing the following functions:

1. Initialization.
2. Instantiation of contained objects.
3. Control of the intersection vehicle traffic standards.
4. Control of the intersection pedestrian traffic standards.
5. Collection and processing of traffic history from all approaches.
6. Adaptive control of intersection timings in response to traffic flow.
7. Actuated control of intersection in response to vehicle presence.
8. Timed control of intersection in response to a fixed scheme.
9. Handling of pedestrian crossing requests.
10. Handling of emergency vehicle pre-emption.
11. Intersection control in response to manual override commands.
12. Intersection control in response to remote override commands.
13. Management of traffic history and incident log databases.
14. Handling of maintenance access requests from the maintenance port.
15. Handling of maintenance access requests from the DOT WAN (wide area network).

Table 2 below illustrates the attributes, methods, and events of the Interface Controller class and Figure 5 illustrates the controller functional sequencing.

Table 2 Intersection controller class

Intersection Controller		
	Name	Description
Attributes	Approaches	Array of Approach objects.
	Manual Override	Represents the Manual Override console.
	Remote Override	Represents the Remote Software console.
	Traffic History	Contains the traffic history for up to at least seven (7) days.
	Incident Log	Contains the incident log for up to at least seven (7) days.
	Network Interface	Object that provides an interface from the Network resource manager (driver) to the Intersection Controller object.
	Emergency Vehicle Interface	Object that provides an interface between the Emergency Vehicle transponder and the Intersection Controller object.
	Mode	Current operating mode of the Intersection Controller.

Table 2   (*continued*)

Intersection Controller		
	Name	Description
	Priority	Relative priority of the approaches.
	Cycle Time	Time to complete a full traversal of all intersection phases.
	Splits	Array of numbers defining the fraction of the cycle time allocated to each phase.
	Current Phase	Current intersection phase.
	Phase Time Remaining	Time remaining until the intersection moves to the next phase in the sequence.
	Commanded Green Signal Safety Relay State	Based on the Current Phase, this attribute holds the value required for the Green Signal Safety Relay resource manager, which is responsible for driving the relay.
	Detected Green Signal Safety Relay State	This holds the actual state of the Green Signal Safety Relay.
Methods	Initialize	
	Advance Phase	Advance the intersection phase to the next phase in the sequence.
	Calculate Cycle Parameters	Calculates the cycle time and splits for the next cycle based on traffic data.
Events	Phase Time Remaining Value Reaches 0	Fires when the Phase Time Remaining timer reaches 0.
	Override Activated	Fires when either the Manual Override or Remote override is activated.
	Override Canceled	Fires when Overrides are deactivated.
	Watchdog Timeout	Fires on a watchdog trip.
	Error	Fires when an error occurs. Takes the Error code is a parameter.

The corresponding traffic standard aspects are shown in Figure 6 below.

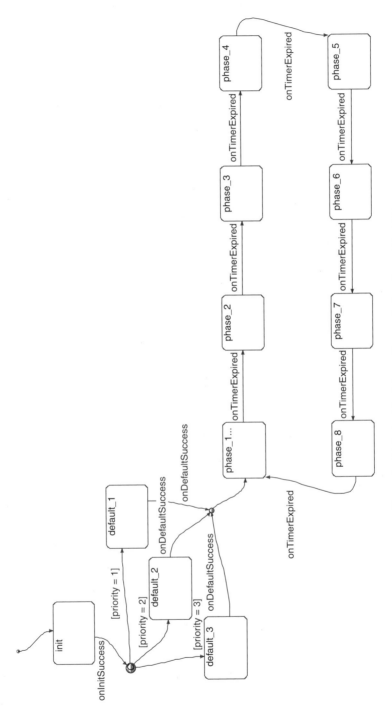

Figure 5 Statechart for intersection controller phase sequence.

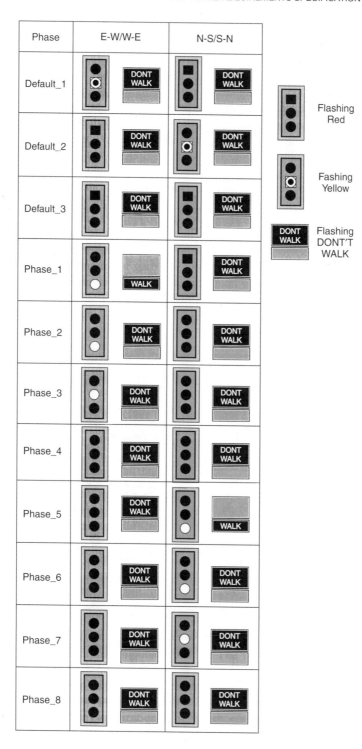

Figure 6 Traffic standard aspects for each phase.

3.2.2 Approach
This is the programmatic representation of an intersection approach.
The Approach object is responsible for managing the following functions:

1. Instantiation of contained objects.
2. Control of the traffic standards associated with the approach.
3. Handling of pedestrian crossing events.
4. Handling of loop detector entry and exit events.
5. Tracking the vehicle count.

Table 3 below illustrates the attributes, methods, and events of the Approach class.

Table 3 Approach class

Approach			
		Name	Description
Attributes		Pedestrian Traffic Standard	Object representing the two pedestrian traffic standards associated with the approach.
		Vehicle Traffic Standards	Object representing the two vehicle traffic standards associated with the approach.
		Pedestrian Service Button	Object representing the two pedestrian service pushbuttons associated with the approach.
		Vehicle Presence Detector	Object representing the proximity detection loop, located at the stop line, associated with the approach.
		Vehicle Count	Count of vehicles passing through the approach.
		Indication	Array used to store the indications actually being displayed on all associated traffic standards.
		Current Aspect	Current commanded aspect corresponding to the Intersection Controller phase.
		Speed Limit	Value (in km/h) of the speed limit associated with the approach.
Methods		Set Aspect	Set the displayed aspect to the Commanded Aspect.
		Get Aspect	Get the actual displayed aspect based on signals from the current sensor hardware resource manager.
		Increment Count	Increase the vehicle count by 1.
		Reset Count	Reset the vehicle count to 0.

Table 3 *(continued)*

Approach		
	Name	Description
Events	Pedestrian Request	Fires when a pedestrian request has been made.
	Vehicle Entry	Fires when the loop detector detects vehicle entry.
	Vehicle Exit	Fires when the loop detector detects vehicle exit.

3.2.3   Pedestrian Traffic Standard

This is the programmatic representation of a pedestrian crossing signal.

The Pedestrian Traffic Standard object is responsible for managing the following functions:

1. Displaying the commanded indication aspect from the Approach.
2. Determining the indication actually displayed.

Table 4 below illustrates the attributes, methods, and events of the Pedestrian Service Button class.

Table 4   Pedestrian traffic standard class

Pedestrian Traffic Standard		
	Name	Description
Attributes	Commanded Aspect	Commanded aspect from the Intersection Controller.
Methods	Set Indication	Set the displayed indication to the Commanded Indication.
	Get Indication	Get the actual displayed indication based on signals from the current sensor hardware resource manager.

3.2.4   Vehicle Traffic Standard

This is the programmatic representation of a vehicle traffic signal.

The Vehicle Traffic Standard object is responsible for managing the following functions:

1. Displaying the commanded aspect from the Intersection Controller.
2. Determining the aspect actually displayed.

Table 5 below illustrates the attributes, methods, and events of the Vehicle Traffic Standard class.

Table 5   Vehicle traffic standard class

Vehicle Traffic Standard		
	Name	Description
Attributes	Commanded Aspect	Commanded aspect from the Intersection Controller.
Methods	Set Indication	Set the displayed indication to the Commanded Indication.
	Get Indication	Get the actual displayed indication based on signals from the current sensor hardware resource manager.

3.2.5   Pedestrian Service Button

This is an object representing the set of pushbutton consoles located on opposite sides of the crosswalk associated with an approach.

The Pedestrian Service Button object is responsible for managing the following functions:

1. Filtering of pushbutton service requests.
2. Generation of Pedestrian Service Request event.

Table 6 below illustrates the attributes, methods, and events of the Pedestrian Service Button class.

Table 6   Pedestrian service button class

Pedestrian Service Button		
	Name	Description
Attributes	Request Masked	Indicates whether pedestrian service pushbutton signals should be ignored or processed.
	Request State	Indicates whether or not a pedestrian service request is active.
Methods	Set Request State	In response to a signal from the pushbutton hardware resource manager, determine whether or not to modify the Request State and raise an event.
	Reset Request State	Clear the Request State.

Table 6 (*continued*)

Pedestrian Service Button		
	Name	Description
	Ignore Request State	Masks subsequent pedestrian button operations.
	Listen Request State	Respond to subsequent pedestrian button operations.
Events	Pedestrian Service Request	Indicates that a valid pedestrian service request is active.

3.2.6 Vehicle Presence Detector

This is an object representing the proximity detection loop located near the stop line associated with an approach. The object class is based on the Pedestrian Service Button class.

The Vehicle Presence Detector object is responsible for managing the following functions:

1. Filtering of vehicle service requests (ACTUATED mode).
2. Generation of Vehicle Service Request event (ACTUATED mode).
3. Maintenance of the vehicle count statistic (FIXED, ACTUATED, and ADAPTIVE mode).

Table 7 below illustrates the attributes, methods, and events of the Pedestrian Service Button class.

Table 7 Vehicle presence detector class

Vehicle Presence Detector		
	Name	Description
Attributes	Request State	Indicates whether or not a vehicle service request is active (ACTUATED mode).
Methods	Set Request State	Set the Request State.
	Reset Request State	Clear the Request State.
Events	Vehicle Entry	Indicates that the detector loop is occupied.
	Vehicle Exit	Indicates that the detector loop is no longer occupied.

3.2.7 Manual Override

This is an object representing the set of pushbuttons on the manual override console.

The Manual Override object is responsible for managing the following functions:

1. Triggering the appropriate mode change.
2. Generation and handling of events required to control intersection phase.

Table 8 below illustrates the attributes, methods, and events of the Manual Override class.

Table 8   Manual override class

Manual Override		
	Name	Description
Attributes	None	None
Methods	None	None
Events	Override Activated	Fires when the override is activated.
	Override Canceled	Fires when the override is deactivated.
	Advance Phase	Fires in response to the ADVANCE button on the override console being pressed.

3.2.8   Remote Override

This is an object representing the commands available on the Remote Software console. Additionally, the object provides an interface for remote access to and update of intersection traffic data and cycle parameters for coordinated intersection control (option).

The Remote Override object is responsible for managing the following functions:

1. Triggering the appropriate mode change.
2. Generation and handling of events required to control intersection phase.

Table 9 below illustrates the attributes, methods, and events of the Remote Override class.

Table 9   Remote override class

Remote Override		
	Name	Description
Attributes	None	None
Methods	Process Command	Processes the events generated by the object, modifying the appropriate attribute or calling the appropriate method of the Intersection Controller object.

Table 9 (*continued*)

Remote Override		
	Name	Description
	Get Status	Retrieves the all parameter and other status data used as inputs to the Calculate Cycle Parameters adaptive control algorithm.
	Set Parameters	Sets the cycle timing parameters as calculated by the remote host.
Events	Override Activated	Fires when the override is activated.
	Override Canceled	Fires when the override is deactivated.
	Advance Phase	Fires in response to the ADVANCE command from the Remote Software console.

3.2.9 Emergency Vehicle Interface

This is an object that manages the wireless transponder interface to authorized emergency vehicles and accesses the Intersection Control object in order to display the correct traffic signals, allowing the emergency vehicle priority access to the intersection.

The Emergency Vehicle Interface object is responsible for managing the following functions:

1. Triggering the appropriate mode change.
2. Reception of emergency vehicle preemption requests.
3. Decryption and validation of emergency vehicle preemption requests.
4. Generation and handling of events required to control intersection phase.

Table 10 below illustrates the attributes, methods, and events of the Emergency Vehicle Interface class.

Table 10 Emergency vehicle interface class

Emergency Vehicle Interface		
	Name	Description
Attributes	None	None
Methods	None	None

(*continued*)

Table 10 *(continued)*

Emergency Vehicle Interface		
	Name	Description
Events	Preempt Activated	Fires when preemption is activated.
	Preempt Canceled	Fires when preemption is deactivated.
	Preempt Timeout	Fires when the preempt cancellation timeout interval expires.

3.2.10 Network Interface

This is an object that manages communication via the Ethernet port.

The Network Interface object is responsible for managing the following functions:

1. Routing control messages to the appropriate objects.
2. Transferring traffic history and incident log data.
3. Management of maintenance operations.

Table 11 below illustrates the attributes, methods, and events of the Network Interface class.

Table 11 Network interface class

Emergency Vehicle Interface		
	Name	Description
Attributes	None	None
Methods	Process Message	Analyzes and routes network messages.
	Receive Message	Receives network messages.
	Send Message	Sends network messages.
Events	None	None

3.2.11 Traffic History

This is an object that manages the stored traffic history.

The Traffic History object is responsible for managing the following functions:

1. Storage and retrieval of traffic history database records.
2. Clearing of traffic history in response to a command from a remote host.

Table 12 below illustrates the attributes, methods, and events of the Traffic History class.

Table 12   Traffic history class

Traffic History			
		Name	Description
Attributes	Record		An array of structures, each of which holds a single traffic history record.
	First Record		Index of the first active record.
	Last Record		Index of the record most recently added.
	Record Pointer		Index used to sequence through the Traffic History records.
Methods	Write Record		Writes a database record at the current position or at a specified position.
	Read Record		Reads a database record at the current position or at a specified position.
	Move Record Pointer		Moves record pointer as specified.
	Clear Database		Returns the database to an empty state.
Events	EOF		Fires when the last record is reached.
	Database Full		Fires when all allocated space for the database is used. Since the database is a FIFO structure, records will begin to be overwritten.

3.2.12   Incident Log
This is an object that manages the stored incident log.
   The Incident Log object is responsible for managing the following functions:

1. Storage and retrieval of incident log database records.
2. Clearing of incident in response to a command from a remote host.

Incidents are generated by the following events:

1. Error conditions.
2. Traffic History database full.
3. System resets.

4. Mode changes, including emergency vehicle preempts.

5. Maintenance actions, as updated by maintenance personnel through portable test equipment (laptop).

Table 13 below illustrates the attributes, methods, and events of the Traffic History class.

Table 13   Incident log class

Incident Log		
	Name	Description
Attributes	Record	An array of structures, each of which holds a single traffic history record.
	First Record	Index of the first active record.
	Last Record	Index of the record most recently added.
	Record Pointer	Index used to sequence through the Traffic History records.
Methods	Write Record	Writes a database record at the current position or at a specified position.
	Read Record	Reads a database record at the current position or at a specified position.
	Move Record Pointer	Moves record pointer as specified.
	Clear Database	Returns the database to an empty state.
Events	EOF	Fires when the last record is reached.
	Database Full	Fires when all allocated space for the database is used. Since the database is a FIFO structure, records will begin to be overwritten.

3.3   Performance Requirements

3.3.1   Timing Requirements

3.3.1.1   Summary

Table 14 below provides a summary of all timing requirements.

Table 14 Software timing requirements

ID	Designation	Applies to Mode(s)	Object/ From Event	Object/ To Response	Min Time (ms)	Max Time (ms)
1.	Initialization	All	Hardware/Reset Signal	Intersection Controller/Initialization Complete	–	4900
2.	Set Default Phase	All	Intersection Controller/Initialization Complete	All Traffic Standards/Display of commanded phase	–	100
3.	Start Normal Operation	ACTUATED ADAPTIVE TIMED	Intersection Controller/Initialization Complete	All Traffic Standards/Display of Phase 1	–	500
4.	Advance Phase – Normal	ACTUATED ADAPTIVE TIMED	Intersection Controller/Phase Time Remaining Reaches 0	All Traffic Standards/Display of Commanded Phase	–	100
5.	Advance Phase – Local	LOCAL_MANUAL	Manual Override/Receipt of Advance Phase signal from Manual Override Panel	All Traffic Standards/Display of Commanded Phase	–	100

(continued)

215

Table 14 (*continued*)

ID	Designation	Applies to Mode(s)	Object/ From Event	Object/ To Response	Min Time (ms)	Max Time (ms)
6.	Advance Phase – Remote	REMOTE_MANUAL	Remote Override/Receipt of Advance Phase signal from Network Interface	All Traffic Standards/Display of Commanded Phase	–	100
7.	Calculate Cycle Parameters – Actuated	ACTUATED	Pedestrian Detector or Vehicle Detector/ Pedestrian Request signal - or - Vehicle Request signal	Intersection Controller/Cycle Time and Splits updated	–	100
8.	Calculate Cycle Parameters – Adaptive	ADAPTIVE	Intersection Controller/Start of last phase in cycle	Intersection Controller/Cycle Time and Splits updated	–	250
9.	Critical Error – Display Defaults	All	Any/Critical Error	All Traffic Standards/Display of Default State	–	50
10.	Critical Error – Alarm	All	Any/Critical Error	Network Interface/Initiation of Alarm Transmission	–	1000
11.	Critical Error – Reset	All	Any/Critical Error	Intersection Controller/System reset	4500	5000
12.	Write Error Log	All	Any/Any Error	Incident Log/Write completed	–	500

#	Name	Modes				Value
13.	Set Phase	ACTUATED ADAPTIVE MANUAL REMOTE	Intersection controller/Advance phase	All Traffic Standards/display commanded phase	–	100
14.	Get Phase	ACTUATED ADAPTIVE MANUAL REMOTE	Intersection controller/Get phase	Intersection Controller/Displayed phase determined	–	150
15.	Check Phase	ACTUATED ADAPTIVE MANUAL REMOTE	Intersection controller/Check phase	Intersection Controller/Phase check status returned	–	10
16.	Pedestrian Request Latching	ACTUATED ADAPTIVE TIMED	Resource Manager/Pedestrian Request signal	Pedestrian Detector/Latching of Pedestrian DetectorPending state	–	10
17.	Pedestrian Request Reset	ACTUATED ADAPTIVE TIMED	Intersection Controller/Completion of phase(s) during which pedestrian requests may be accepted	Pedestrian Detector Detector/Clearing of Pedestrian DetectorPending state	–	100
18.	Pedestrian Request Processing	ACTUATED ADAPTIVE TIMED	Pedestrian Detector/Latching of Pedestrian DetectorPending state	Intersection Controller/Updating of cycle time for next two (2) cycles; Updating of all splits for next two (2) cycles.	–	100

Table 14 *(continued)*

ID	Designation	Applies to Mode(s)	Object/ From Event	Object/ To Response	Min Time (ms)	Max Time (ms)
19.	Vehicle Entrance	FIXED ACTUATED ADAPTIVE	Resource manager/Vehicle Entry signal	Vehicle Detector/Vehicle Entry state set	–	10
20.	Vehicle Exit	FIXED ACTUATED ADAPTIVE	Resource manager/Vehicle Exit signal	Vehicle Detector/Vehicle Entry state cleared	–	10
21.	Vehicle Request Processing	FIXED ACTUATED ADAPTIVE	Vehicle Detector/Entry State set	Intersection controller/process vehicle request	–	100
22.	Vehicle Reset Request State	ACTUATED ADAPTIVE	Intersection controller/reset vehicle request state	Vehicle Detector/Clear vehicle entry state	–	100
23.	Vehicle Count Update	FIXED ACTUATED ADAPTIVE	Vehicle Detector/Entry State cleared	Approach/Bump count	–	50
24.	Vehicle Count Fetch	FIXED ACTUATED ADAPTIVE	Intersection controller/Get Count	Intersection Controller/Count returned	–	100
25.	Vehicle Count Reset	FIXED ACTUATED ADAPTIVE	Intersection controller/Phase change	Approach/Reset count	–	100
26.	Get Cycle Parameters	REMOTE	Remote Override/Parameter request	Network interface/Packet ready to send	–	100

27.	Update Cycle Parameters	REMOTE	Remote Override/Parameter update	Intersection controller/Parameters updated	–	100
28.	Process Message	EMERGENCY PREEMPT	Emergency Vehicle Interface/Activate	Intersection controller/Mode changed	–	200
29.	Process Command	EMERGENCY PREEMPT	Emergency Vehicle Interface	Intersection controller/emergency vehicle operations	–	100
30.	Process Message	REMOTE	Remote operations	Intersection controller/Network Interface	–	200
31.	Fetch Database	REMOTE	Remote operations	Intersection controller/Network Interface	–	1000
32.	Add Record	FIXED ACTUATED ADAPTIVE	Intersection controller	Traffic History	–	200
33.	Clear Database	FIXED ACTUATED ADAPTIVE	Intersection controller	Traffic History	–	200

(continued)

ID	Designation	Applies to Mode(s)	Object/ From Event	Object/ To Response	Min Time (ms)	Max Time (ms)
34.	Add Record	FIXED ACTUATED ADAPTIVE	Intersection controller	Incident Log	–	200
35.	Clear Database	FIXED ACTUATED ADAPTIVE	Intersection controller	Incident Log	–	200

Notes:

[1] The timing requirements for vehicle detection are based on the following considerations:

Minimum vehicle length = **8 ft**
Minimum following distance in motion = **4 ft**
Loop width = **4 ft**
Loop detects entrance with leading overlap = **2 ft**
Loop detects exit with trailing overlap = **1 ft**
Maximum vehicle speed = **65 mph(= 95.3 ft/s)**
Vehicle speed for minimum gap time (see below) = **10 mph (= 14.67 ft/s)**
Minimum presence pulse width = 9 ft/95.3 ft/s = **94.4 ms**
Minimum gap time (time between exit and next vehicle entrance) = 3 ft/14.67 ft/s = **204.5 ms**

This is illustrated in Figure 7 and Figure 8 below.

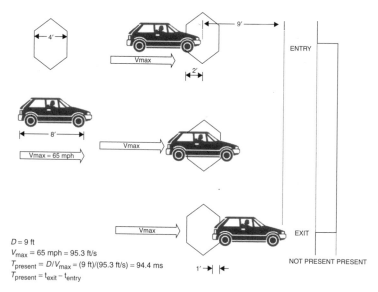

$D = 9$ ft
$V_{max} = 65$ mph = 95.3 ft/s
$T_{present} = D/V_{max} = (9$ ft$)/(95.3$ ft/s$) = 94.4$ ms
$T_{present} = t_{exit} - t_{entry}$

Figure 7   Minimum presence pulse width.

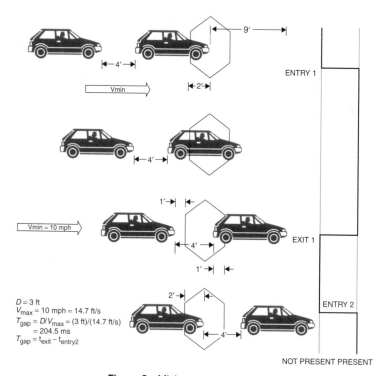

$D = 3$ ft
$V_{max} = 10$ mph = 14.7 ft/s
$T_{gap} = D/V_{max} = (3$ ft$)/(14.7$ ft/s$)$
      $= 204.5$ ms
$T_{gap} = t_{exit} - t_{entry2}$

Figure 8   Minimum gap pulse width.

It is necessary to determine the speed at which the minimum gap time between vehicles occurs in order to determine that time. The distance a following vehicle must cover in order to trigger the loop detector after the leading vehicle has exited is given by

$$D_{gap}(v) = \begin{cases} 4 \text{ ft, } v < 10 \text{ mph} \\ \dfrac{v^2}{2a} - \dfrac{\left(10 \text{ mph} \cdot \dfrac{88}{60}\right)^2}{2a} + 3 \text{ ft, } v \geq 10 \text{ mph} \end{cases}$$

Given this, the gap time is given by

$$T_{gap}(v) = \frac{D_{gap}(v)}{v}, \qquad v > 0$$

It can be shown that the minimum gap time occurs for $v = 10$ mph.

4.11   EXERCISES

4.1   What are the problems and ramifications of translating a requirement from one modeling technique (e.g., finite state machines) to another (e.g., Petri nets)?

4.2   Who should write, analyze, and critique software requirements specifications?

4.3   Under what circumstances and when should software requirements specifications be changed?

4.4   Use statecharts instead of finite state machines to represent the visual inspection system as it is described in the examples. Do the same using Petri nets instead.

4.5   Redraw the use-case diagram for the IMU in Figure 4.15 to include calibration and diagnostic modes.

4.6   For a system with which you are familiar, find three good requirements and three bad requirements in the software requirements specification. Rewrite the bad requirements so that they comply with IEEE 830.

4.7   Using the structured analysis draw a context diagram for a credit card system defined below. You should further refine the context diagram to depict more details of the functionality of the system. You are free to make assumptions as needed, but make sure that you have stated them clearly.

The credit card system under consideration handles transactions for retail stores. For example, a transaction might consist of buying a music CD from your favorite music store (say, HMV). The diagram should include functions for retrieving and checking a credit card record for a customer, approving and recording each transaction, and maintaining a log of transactions for each retail store. The system should maintain files of credit card holders, current transactions, and accounts payable (approved transactions) for each store.

4.8   Consider a hospital's patient monitoring system. Each patient is connected to machines monitoring blood pressure, heart rate, and EKG. These machines issue a Boolean signal indicating a FAIL or WORKING condition. The results of each of these machined are ORed together to form a signal called ALARM. The ALARM signals for each of the rooms (one patient per room) are then ORed together and sent to the nurse's station. If any machine on any patient indicates a failure, the emergency alarm is sounded and the nurse is directed to the appropriate patient and machine. Draw a data flow diagram for such a system.

4.9 Discuss the advantages of built-in-test software in enhancing fault-tolerance capabilities of real-time systems.

4.10 Using a data flow diagram, design a process controller for an environment monitoring system that collects data from a set of air-quality sensors situated around a city. There are 100 sensors located in 5 neighborhoods. Each sensor must be polled four times per second. When more than 40% of the sensors in a particular neighborhood indicate that the air quality is below an acceptable level, local warning lights are activated. All sensors return the readings to a central processing unit that generates reports every 15 minutes on the air quality.

4.11 Draw a state machine model of the control software for a simple VCR. State clearly all the assumptions regarding specific features of the chosen VCR.

4.12 For the Petri net shown in Figure 4.18, identify all possible transitions that would cause the Petri net to be in a dead state, that is, no further transitions are possible.

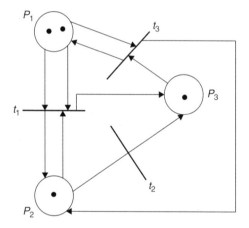

Figure 4.18 Petri net for Exercise 4.12.

4.13 What would be the appropriate combination of techniques to write the software specifications for the:

(a) Inertial measurement unit

(b) Nuclear monitoring system

(c) Airline reservations system

(d) Pasta sauce bottling system

Use a Web search to select any appropriate commercial processor and support tool set.

5

SOFTWARE SYSTEM DESIGN

Software design is the process concerned with translating the models of the problem domain generated during analysis into the models of the solution that are needed for implementation. It is during this phase that decisions are made concerning responsibility fulfillment and assignment, system architecture and deployment, separation of concerns, and layering and modularization. Successful design is achieved when these decisions result in a model that exhibits a number of properties.

5.1 PROPERTIES OF SOFTWARE

Software can be characterized by any of a number of qualities. External qualities are those that are visible to the user, such as usability and reliability, and are of concern to the end user. Internal qualities are those that may not be necessarily visible to the user, but help the developers to achieve improvement in external qualities. For example, good requirements and design documentation might not be seen by the typical user, but these are necessary to achieve improvement in most of the external qualities. A specific distinction between whether a particular quality is external or internal is not often made because they are so closely tied. Moreover, the distinction is largely a function of the software itself and the kind of user involved.

Some of this chapter has been adapted from Phillip A. Laplante, *Software Engineering for Image Processing*, CRC Press, Boca Raton, FL, 2003.

Real-Time Systems Design and Analysis, By Phillip A. Laplante
ISBN 0-471-22855-9 © 2004 Institute of Electrical and Electronics Engineers

While it is helpful to describe these qualities, it is equally desirable to quantify them. Quantification of these characteristics of software is essential in enabling users and designers to talk succinctly about the product and for software process control and project management. More importantly, however, it is these properties or qualities that will be embodied in the real-time design.

5.1.1 Reliability

Reliability is a measure of whether a user can depend on the software. This notion can be informally defined in a number of ways. For example, one definition might be "a system that a user can depend on." Other loose characterization of a reliable software system includes:

- The system "stands the test of time."
- There is an absence of known catastrophic errors; that is, errors that render the system useless.
- The system recovers "gracefully" from errors.
- The software is robust.

For real-time systems, other informal characterizations of reliability might include:

- Downtime is below a certain threshold.
- The accuracy of the system is within a certain tolerance.
- Real-time performance requirements are met consistently.

While all of these informal characteristics are desirable in real-time systems, they are difficult to measure. Moreover, they are not truly measures of reliability, but of other attributes of the software.

There is specialized literature on software reliability that defines this quality in terms of statistical behavior, that is, the probability that the software will operate as expected over a specified time interval. These characterizations generally take the following approach. Let S be a software system, and let T be the time of system failure. Then the reliability of S at time t, denoted $r(t)$, is the probability that T is greater than t; that is,

$$r(t) = P(T > t) \tag{5.1}$$

This is the probability that a software system will operate without failure for a specified period of time.

Thus, a system with reliability function $r(t) = 1$ will never fail. However, it is unrealistic to have such expectations. Instead, some reasonable goal should be set, for example, in the nuclear monitoring system that the failure probability be no more than 10^{-9} per hour. This represents a reliability function of $r(t) = (0.99999999)^t$, with t in hours. Note that as $t \to \infty, r(t) \to 0$.

Another way to characterize software reliability is in terms of a real-valued failure function. One failure function uses an exponential distribution where the abscissa is time and the ordinate represents the expected failure intensity at that time (Equation 5.2).

$$f(t) = \lambda e^{-\lambda t} \qquad t \geq 0 \tag{5.2}$$

Here the failure intensity is initially high, as would be expected in new software, as faults are detected during testing. However, the number of failures would be expected to decrease with time, presumably as failures are uncovered and repaired (Figure 5.1). The factor λ is a system-dependent parameter.

A second failure model is given by the "bathtub curve" shown in Figure 5.2. Brooks notes that while this curve is often used to describe the failure function of hardware components, it might also be useful in describing the number of errors found in a certain release of a software product [Brooks95].

The interpretation of this failure function is clear for hardware; a certain number of product units will fail early due to manufacturing defects. Later, the failure intensity will increase as the hardware ages and wears out. But software does not wear out. If systems seem to fail according to the bathtub curve, then there has to be some plausible explanation.

It is clear that a large number of errors will be found early in a particular software product, just as in the exponential model of failure. But why would the failure intensity increase much later? There are at least three possible explanations. The first is that the errors are due to the effects of patching the software for

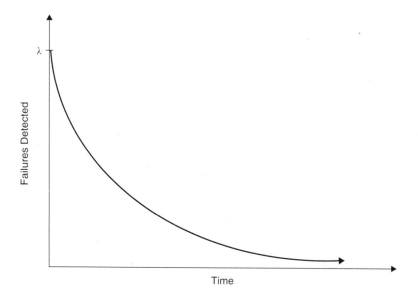

Figure 5.1 An exponential model of failure represented by the failure function $f(t) = \lambda e^{-\lambda t}, t \geq 0$. Here λ is a system-dependent parameter [Laplante03c].

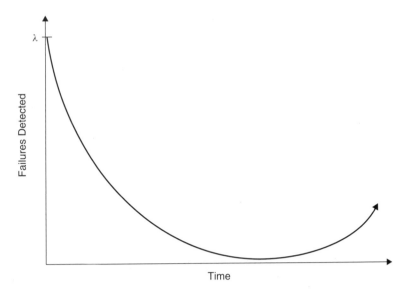

Figure 5.2   A software failure function represented by the bathtub curve [Laplante03c].

various reasons. The second reason is that late software failures are really due to failure of the underlying hardware. Finally, additional failures could appear because of the increased stress on the software by expert users. That is, as users master the software and begin to expose and strain advanced features, it is possible that certain poorly tested functionality of the software is beginning to be used.

Often the traditional quality measures of mean time to first failure (MTFF) or mean time between failures (MTBF) are used to stipulate reliability in the software requirements specification. This approach to failure definition places great importance on the effective elicitation (gathering) and specification of functional requirements, because the requirements define the software failure.

5.1.2   Correctness

Software correctness is closely related to reliability, and the terms are often used interchangeably. The main difference is that minor deviation from the requirements is strictly considered a failure and hence means the software is incorrect. However, a system may still be deemed reliable if only minor deviations from the requirements are experienced. In real-time systems correctness incorporates correctness of outputs and deadline satisfaction. This definition is closer to the formal methods definition of correctness in terms of correctness of outputs if the software halts and establishing that the software halts.

5.1.3   Performance

Performance is a measure of some required behavior. One method of measuring performance is based on mathematical or algorithmic complexity, either in terms of O, o, Ω, or θ notations. Another approach involves directly timing the behavior of the completed system with logic analyzers and similar tools. Finally, a simulation of the finished system might be built with the specific purpose of estimating performance.

5.1.4 Usability

Often referred to as ease of use, or user friendliness, usability is a measure of how easy the software is for humans to use. This quantity is an elusive one. Properties that make an application user friendly to novice users are often different from those desired by expert users or the software designers. Use of prototyping can increase the usability of a software system because, for example, interfaces can be built and tested by the user.

Usability is difficult to quantify. However, informal feedback can be used, as can user feedback from surveys, and problem reports can be used in most cases.

5.1.5 Interoperability

This quality refers to the ability of the software system to coexist and cooperate with other systems. For example, in real-time systems the software must be able to communicate with various devices using standard bus structures and protocols.

A concept related to interoperability is that of an open system. An open system is an extensible collection of independently written applications that cooperate to function as an integrated system. Open systems differ from open source code, which is source code that is made available to the user community for moderate improvement and correction. An open system allows the addition of new functionality by independent organizations through the use of interfaces whose characteristics are published. Any applications developer can then take advantage of these interfaces, and thereby create software that can communicate using the interface. Open systems allow different applications written by different organizations to interoperate. Interoperability can be measured in terms of compliance with open system standards.

5.1.6 Maintainability

Anticipation of change is a general principle that should guide the software engineer. A software system in which changes are relatively easy to make has a high level of maintainability. In the long run, design for change will significantly lower software life cycle costs and lead to an enhanced reputation for the software engineer, the software product, and the company.

Maintainability can be broken down into two contributing properties: evolvability and repairability. Evolvability is a measure of how easily the system can be changed to accommodate new features or modification of existing features. Software is repairable if it allows for the fixing of defects.

Measuring these qualities of software is not always easy, and often is based on anecdotal observation only. This means that changes and the cost of making them are tracked over time. Collecting this data has a twofold purpose. First, the costs of maintenance can be compared to other similar systems for benchmarking and project management purposes. Second, the information can provide experiential learning that will help to improve the overall software production process and the skills of the software engineers.

5.1.7 Portability

Software is portable if it can easily run in different environments. The term environment refers to the hardware on which the system runs, operating system, or other software with which the system is expected to interact. Because of the specialized hardware with which they interact, special care must be taken in making real-time systems portable.

Portability is achieved through a deliberate design strategy in which hardware-dependent code is confined to the fewest code units as possible. This strategy can be achieved using either object-oriented or procedural programming languages and through object-oriented or structured approaches. Both of these will be discussed throughout the text.

Portability is difficult to measure, other than through anecdotal observation. Person months required to move the software are the standard measure of this property.

5.1.8 Verifiability

A software system is verifiable if its properties, including all of those previously introduced, can be verified. In real-time systems, verifiability of deadline satisfaction is of the utmost importance. This topic is discussed further in Chapter 7.

Table 5.1   Some software properties and the means for measuring them

Software Quality	Possible Measurement Approach
Correctness	Probabilistic measures, MTBF, MTFF
Interoperability	Compliance with open standards
Maintainability	Anecdotal observation of resources spent
Performance	Algorithmic complexity analysis, direct measurement, simulation
Portability	Anecdotal observation
Reliability	Probabilistic measures, MTBF, MTFF, heuristic measures
Usability	User feedback from surveys and problem reports
Verifiability	Software monitors

One common technique for increasing verifiability is through the insertion of software code that is intended to monitor various qualities, such as performance or correctness. Modular design, rigorous software engineering practices, and the effective use of an appropriate programming language can also contribute to verifiability.

5.1.9 Summary of Software Properties and Associated Metrics

So far it has been emphasized that measurement of the software properties is essential throughout the software life cycle. A summary of the software qualities just discussed and possible ways to measure them are given in Table 5.1.

5.2 BASIC SOFTWARE ENGINEERING PRINCIPLES

Software engineering has been criticized for not having the same kind of underlying rigor as other engineering disciplines. While it may be true that there are few formulaic principles, there are many fundamental rules that form the basis of sound software engineering practice. The following sections describe the most general and prevalent of these.

5.2.1 Rigor and Formality

Because software development is a creative activity, there is an inherent tendency toward informal ad hoc techniques in software specification, design, and coding. But the informal approach is contrary to good software engineering practice.

Rigor in software engineering requires the use of mathematical techniques. Formality is a higher form of rigor in which precise engineering approaches are used. In the case of the real-time system, formality further requires that there be an underlying algorithmic approach to the specification, design, coding, and documentation of the software.

5.2.2 Separation of Concerns

Separation of concerns is a divide-and-conquer strategy that software engineers use. There are various ways in which separation of concerns can be achieved. In terms of software design and coding it is found in modularization of code and in object-oriented design. There may be separation in time, for example, developing a schedule for a collection of periodic computing tasks with different periods.

Yet another way of separating concerns is in dealing with qualities. For example, it may be helpful to address the fault-tolerance of a system while ignoring other qualities. However, it must be remembered that many of the qualities of software are interrelated, and it is generally impossible to affect one without affecting the other, possible adversely.

5.2.3 Modularity

Some separation of concerns can be achieved in software through modular design. Modular design involves the decomposition of software behavior in encapsulated software units, and can be achieved in either object-oriented or procedurally oriented programming languages.

Modularity is achieved by grouping together logically related elements, such as statements, procedures, variable declarations, and object attributes, in an increasingly fine-grained level of detail (Figure 5.3). The main benefit of modularity is high cohesion and low coupling. With respect to the code units, cohesion represents intramodule connectivity and coupling represents intermodule connectivity. Coupling and cohesion can be illustrated informally as in Figure 5.4, which shows software structures with high cohesion and low coupling (a) and low cohesion and high coupling (b). The inside squares represent statements or data, arcs indicate functional dependency. Cohesion relates to the relationship of the elements of a module. High cohesion implies that each module represents a single part of the problem solution. Therefore, if the system ever needs modification, then that part that needs to be modified exists in a single place, making it easier to change.

Constantine and Yourdon identified seven levels of cohesion in order of strength [Pressman00]:

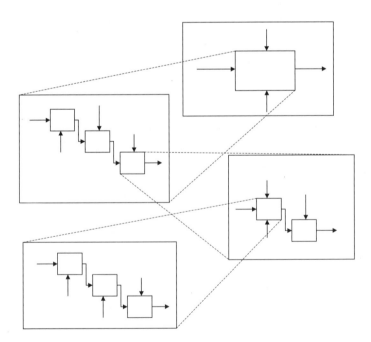

Figure 5.3 Modular decomposition of code units. The arrows represent inputs and outputs in the procedural paradigm. In the object-oriented paradigm they represent associations. The boxes represent encapsulated data and procedures in the procedural paradigm. In the object-oriented paradigm they represent classes [Laplante03c].

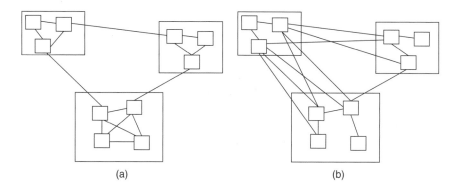

Figure 5.4 Software structures with (a) high cohesion and low coupling and (b) low cohesion and high coupling. The inside squares represent statements or data; arcs indicate functional dependency.

- *Coincidental* – Parts of module are not related, but simply bundled into a single module.
- *Logical* – Parts that perform similar tasks are put together in a module
- *Temporal* – Tasks that execute within the same time span are brought together.
- *Procedural* – The elements of a module make up a single control sequence.
- *Communicational* – All elements of a module act on the same area of a data structure.
- *Sequential* – The output of one part in a module serves as input for some other part.
- *Functional* – Each part of the module is necessary for the execution of a single function.

Coupling relates to the relationships between the modules themselves. There is a great benefit in reducing coupling so that changes made to one code unit do not propagate to others, that is, they are hidden. This principle of "information hiding," also known as Parnas Partitioning, is the cornerstone of all software design [Parnas79]. Low coupling limits the effects of errors in a module (lower "ripple effect") and reduces the likelihood of data-integrity problems. In some cases, however, high coupling due to control structures may be necessary. For example, in most graphical user interfaces control coupling is unavoidable, and indeed desirable.

Coupling has also been characterized in increasing levels as:

1. *No Direct Coupling* – All modules are completely unrelated.
2. *Data* – When all arguments are homogeneous data items; that is, every argument is either a simple argument or data structure in which all elements are used by the called module.

3. *Stamp* – When a data structure is passed from one module to another, but that module operates on only some of the data elements of the structure.

4. *Control* – One module passes an element of control to another; that is, one module explicitly controls the logic of the other.

5. *Common* – If two modules both have access to the same global data.

6. *Content* – One module directly references the contents of another.

To further illustrate both coupling and cohesion, consider the class-structure diagram shown in Figure 5.5 The figure illustrates two points. The first is the straightforward difference between the same system embodying low coupling and high cohesion versus high coupling and low cohesion. The second point is that the proper use of graphical design techniques can positively influence the eventual design.

5.2.4 Anticipation of Change

As has been mentioned, software products are subject to frequent change either to support new hardware or software requirements or to repair defects. A high maintainability level of the software product is one of the hallmarks of outstanding commercial software.

Real-time engineers know that their systems are frequently subject to changes in hardware, algorithms, and even application. Therefore these systems must be designed in such a way as to facilitate changes without degrading the other

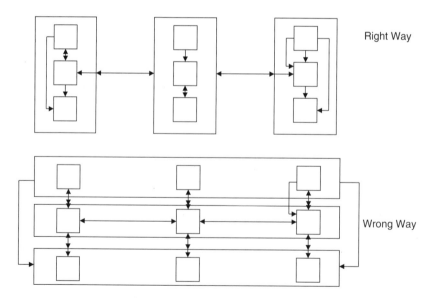

Figure 5.5 Coupling and cohesion. The right way: low coupling and high cohesion. The wrong way: high coupling and low cohesion.

desirable properties of the software. Anticipation of change can be achieved in the software design through appropriate techniques, through the adoption of an appropriate software life cycle model and associated methodologies, and through appropriate management practices.

5.2.5 Generality

In solving a problem, the principle of generality can be stated as the intent to look for the more general problem that may be hidden behind it. An obvious example, designing the inertial measurement system for a specific application, is less general than designing it to be adaptable to a wide range of applications.

Generality can be achieved through a number of approaches associated with procedural and object-oriented paradigms. For example, Parnas' information hiding can be used in procedural languages. Although generalized solutions may be more costly in terms of the problem at hand, in the long run, the costs of a generalized solution may be worthwhile.

5.2.6 Incrementality

Incrementality involves a software approach in which progressively larger increments of the desired product are developed. Each increment provides additional functionality, which brings the product closer to the final one. Each increment also offers an opportunity for demonstration of the product to the customer for the purposes of gathering requirements and refining the look and feel of the product.

5.2.7 Traceability

Traceability is concerned with the relationships between requirements, their sources, and the system design. Regardless of the process model, documentation and code traceability are paramount. A high level of traceability ensures that the software requirements flow down through the design and code and then can be traced back up at every stage of the process. This would ensure, for example, that a coding decision can be traced back to a design decision to satisfy a corresponding requirement.

Traceability is particularly important in real-time systems because often design and coding decisions are made to satisfy hardware constraints that may not be easily associated with a requirement. Failure to provide a traceable path from such decisions through the requirements can lead to difficulties in extending and maintaining the system.

Generally, traceability can be obtained by providing links between all documentation and the software code. In particular, there should be links:

- From requirements to stakeholders who proposed these requirements.
- Between dependent requirements.
- From the requirements to the design.
- From the design to the relevant code segments.

- From requirements to the test plan.
- From the test plan to test cases.

One way to achieve these links is through the use of an appropriate numbering system throughout the documentation. For example, a requirement numbered 3.2.2.1 would be linked to a design element with a similar number (the numbers do not have to be the same as long as the annotation in the document provides traceability). In practice, a traceability matrix is constructed to help cross reference documentation and code elements (Table 5.2). The matrix is constructed by listing the relevant software documents and the code unit as columns, and then each software requirement in the rows.

Constructing the matrix in a spreadsheet software package allows for providing multiple matrices sorted and cross referenced by each column as needed. For example, a traceability matrix sorted by test case number would be an appropriate appendix to the text plan. The traceability matrices are updated at each step in the software life cycle. For example, the column for the code unit names (e.g., procedure names, object class) would not be added until after the code is developed. Finally, a way to foster traceability between code units is through the use of data dictionaries, which are described later.

5.3 THE DESIGN ACTIVITY

The design activity is involved in identifying the components of the software design and their interfaces from the Software Requirements Specification. The principal artifact of this activity is the Software Design Description (SDD).

During the design period, in particular, the real-time systems engineer must design the software architecture, which involves the following tasks:

- Performing hardware/software trade-off analysis.
- Designing interfaces to external components (hardware, software, and user interfaces).

Table 5.2   A traceability matrix sorted by requirement number

Requirement Number	Software Design Document Reference Number(s)	Test Plan Reference Number(s)	Code Unit Name(s)	Test Case Number(s)
3.1.1.1	3.1.1.1 3.2.4	3.1.1.1 3.2.4.1 3.2.4.3	Process_A	3.1.1.A 3.1.1.B
3.1.1.2	3.1.1.2	3.1.1.2	Process_B	3.1.1.A 3.1.1.B
3.1.1.3	3.1.1.3	3.1.1.3	Process_C	3.1.1.A 3.1.1.B 3.1.1.C

- Designing interfaces between components.
- Making the determination between centralized or distributed processing schemes.
- Determining concurrency of execution.
- Designing control strategies.
- Determining data storage, maintenance, and allocation strategy.
- Designing database structures and handling routines.
- Designing the start-up and shutdown processing.
- Designing algorithms and functional processing.
- Designing error processing and error message handling.
- Conducting performance analyses.
- Specifying physical location of components and data.
- Designing any test software identified in test planning.
- Creating documentation for the system including (if applicable):
  Computer System Operator's Manual
  Software User's Manual
  Software Programmer's Manual
- Conducting internal reviews.
- Developing the detailed design for the components identified in the software architecture.
- Developing the test cases and procedures to be used in the formal acceptance testing.
- Documenting the software architecture in the form of the SDD.
- Presenting the design detail information at a formal design review.

This is an intimidating set of tasks that is further complicated by the fact that many of them must occur in parallel or be iterated several times iteratively. There is no algorithm, per se, for conducting these tasks. Instead, it takes many years of practice, experience, learning from the experience of others, and good judgment to guide the software engineer through this maze of tasks.

Two methodologies, process- or procedural-oriented and object-oriented design (OOD), which are related to structured analysis and object-oriented analysis, respectively, can be used to begin to perform the design activities from the Software Requirements Specification produced by either structured analysis or structured design. Each methodology seeks to arrive at a model containing small, detailed components.

5.4 PROCEDURAL-ORIENTED DESIGN

Procedural-oriented design methodologies, such as structured design (SD), involve top-down or bottom-up approaches centered on procedural languages such as C and Fortran. The most common of these approaches utilize design decomposition via Parnas Partitioning.

5.4.1 Parnas Partitioning

Software partitioning into software units with low coupling and high cohesion can be achieved through the principle of information hiding. In this technique, a list of difficult design decisions or things that are likely to change is prepared. Modules are then designated to "hide" the eventual implementation of each design decision or feature from the rest of the system. Thus, only the function of the module is visible to other modules, not the method of implementation. Changes in these modules are therefore not likely to affect the rest of the system.

This form of functional decomposition is based on the notion that some aspects of a system are fundamental, whereas others are arbitrary and likely to change. Moreover, it is those arbitrary things, which are likely to change, that contain "information." Arbitrary facts are hard to remember and usually require lengthier descriptions; therefore, they are the sources of complexity.

The following steps can be used to implement a design that embodies information hiding.

1. Begin by characterizing the likely changes.
2. Estimate the probabilities of each type of change.
3. Organize the software to confine likely changes to a small amount of code.
4. Provide an "abstract interface" that abstracts from the differences.
5. Implement "objects," that is, abstract data types and modules that hide changeable data structures.

These steps reduce coupling and increase module cohesion. Parnas also indicated that although module design is easy to describe in textbooks, it is difficult to achieve. He suggested that extensive practice and examples are needed to illustrate the point correctly [Parnas79].

As an example, consider a portion of the display function of a graphics system associated with the nuclear monitoring system and shown in hierarchical form in Figure 5.6. It consists of graphics that must be displayed (e.g., a representation of the reactor core or sensor data) and are essentially composed from circles and boxes. Different objects can also reside in different display windows. The implementation of circles and boxes is based on the composition of line-drawing calls. Thus, line drawing is the most basic hardware-dependent function. Whether the hardware is based on pixel, vector, turtle, or other type of graphics does not matter; only the line-drawing routine needs to be changed. Hence, the hardware dependencies have been isolated to a single code unit.

Parnas partitioning "hides" the implementation details of software features, design decisions, low-level drivers, and so on, in order to limit the scope of impact of future changes or corrections. By partitioning things likely to change, only that module needs to be touched when a change is required, without the

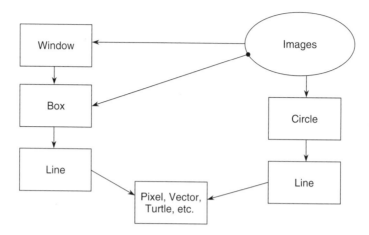

Figure 5.6 Parnas Partitioning of graphics rendering software [Laplante03c].

need to modify unaffected code. This technique is particularly applicable and useful in embedded systems; since they are so directly tied to hardware, it is important to partition and localize each implementation detail with a particular hardware interface. This allows easier future modification due to hardware interface changes and reduces the amount of code affected.

If in designing the software modules, increasing levels of detail are deferred until later (subordinate code units), then the software the approach is top-down. If, instead, the design detail is dealt with first and then increasing levels of abstraction are used to encapsulate those details, then the approach is bottom-up.

For example in Figure 5.6, it would be possible to design the software by first describing the characteristics of the various components of the system and the functions that are to be performed on them, such as opening, closing, and sizing windows. Then the window functionality could be broken down into its constituent parts, such as boxes and text. These could be subdivided even further, that is, all boxes consist of lines, and so on. The top-down refinement continues until the lowest level of detail needed for code development has been reached.

Alternatively, it is possible to begin by encapsulating the details of the most volatile part of the system, the hardware implementation of a single line or pixel, into a single code unit. Then working upward, increasing levels of abstraction are created until the system requirements are satisfied. This is a bottom-up approach to design.

5.4.2 Structured Design

Structured design (SD) is the companion methodology to structured analysis. Structured design is a systematic approach concerned with the specification of the software architecture and involves a number of techniques, strategies, and tools. SD provides a step-by-step design process that is intended to improve software quality and reduce risk of failure, to increase reliability, flexibility, maintainability, and effectiveness. The data flow diagrams (DFD) partition system functions and document that partitioning inside the specification.

5.4.2.1 *Transitioning from Structured Analysis to Structured Design*

Structured analysis (SA) is related to SD in the same way that a requirements representation is related to the software architecture, that is, the former is functional and flat and the latter is modular and hierarchical. In this regard data structure diagrams are used to give information about logical relationships in complex data structures.

The transition mechanisms from SA to SD are manual and involve significant analysis and trade-offs of alternative approaches. Normally, SD proceeds from SA in the following manner. Once the context diagram is drawn, a set of DFDs is developed. The first DFD, the level 0 diagram, shows the highest level of system abstraction. Subdividing processes to lower and lower levels until they are ready for detailed design renders new DFDs with successive levels of increasing detail. This decomposition process is called leveling.

In a typical DFD boxes represent terminators that are labeled with a noun phrase that describes the system, agent, or device from which data enters or to which data exits. Each process depicted by a circle is labeled as a verb phrase describing the operation to be performed on the data, although it may be labeled with the name of a system or operation that manipulates the data. Solid arcs are used to connect terminators to processes and between processes to indicate the flow of data through the system. Each arc is labeled with a noun phrase that describes the data. Dashed arcs are discussed later. Parallel lines indicate data stores, which are labeled by a noun phrase naming the file, database, or repository where the system stores data.

Each DFD should typically have between five and nine processes [DeMarco78]. The descriptions for the lowest level, or primitive, processes are called process specifications, or P-SPECs, and are expressed in either structured English, pseudocode, decision tables, or decision trees, and are used to describe the logic and policy of the program (Figure 5.7). For example, consider the inertial measurement system and refer to its context diagram shown in Figure 4.26. The associated level 0 DFD is shown in Figure 5.8.

Process 1 in the level 0 DFD can be further decomposed, as shown into the level 1 DFD shown in Figure 5.9. In this case, short of the equations, DFD 1 for gyro torquing can be used to produce P-SPECs which would include the necessary equations to convert the attitude adjustment to the appropriate number of torquing pulses. These are shown in Figure 5.10.

Continuing with the example, DFD 2: processor accelerometer data is sown in Figure 5.11. To illustrate a further decomposition, DFD 2.3 for compensate accelerometer data is shown in Figure 5.12. Finally, an associated P-SPEC for DFD 2.3 is given in Figure 5.13. In addition to the DFDs, SD uses a data dictionary to document and control interfaces. Entity relationship diagrams are frequently used to define the relationship between the components of the system, much as in the object-oriented paradigm. The data dictionary documents each

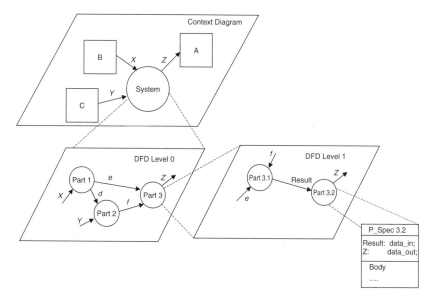

Figure 5.7 Context diagram evolution from context diagram to level 0 DFD to level 1 DFD, and finally, to a P-SPEC, which is suitable for coding [Laplante03b].

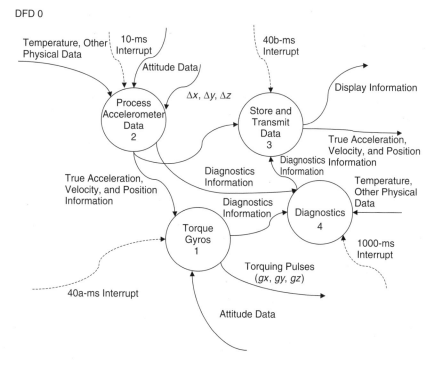

Figure 5.8 Level 0 DFD for the inertial measurement system. The dashed arcs represent control flow in the system.

DFD 1: Torque Gyros

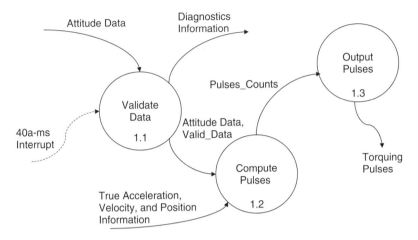

Figure 5.9   DFD 1: Gyro torquing.

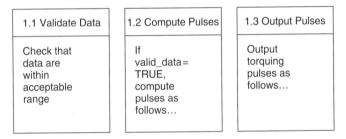

Figure 5.10   P-SPECs for DFD 1: Torque gyros.

interface flow in the DFD. Data structure diagrams are also used to describe information about logical relationships in complex data structures.

5.4.2.2 Data Dictionaries A data dictionary is a repository of data about data that describes every data entity in the system. This dictionary is an essential component of the structured design, and includes entries for data flows, control flows, data stores, data elements, control elements. Each entry is identified by name, range, rate, units, and so forth. The dictionary is organized alphabetically for ease of use. Other than that there is no standard format, but every design element should have an entry in it. Most CASE tools provide the data dictionary feature. For example, each entry might be organized to contain the following information:

Entry type (data flow, data store, terminator, process)
Name

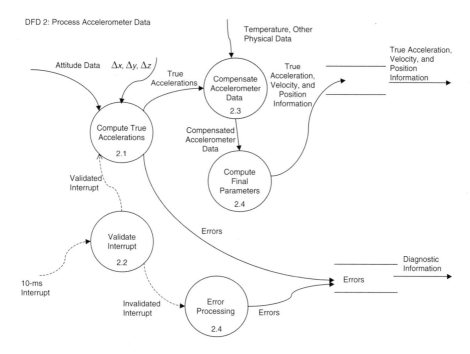

Figure 5.11   Level 1 DFD 2: Compensate accelerometer data.

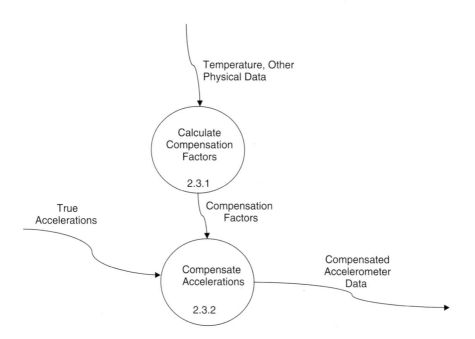

Figure 5.12   DFD 2.3 compensate accelerometer data.

```
┌─────────────────────────────────────┐
│  2.3.2 Compensate Accelerations      │
├─────────────────────────────────────┤
│  begin                               │
│                                      │
│        cax = f(ax,...)               │
│        cay = f(ay,...)               │
│        caz = f(az,...)               │
│  end                                 │
│                                      │
│                                      │
└─────────────────────────────────────┘
```

Figure 5.13 P-SPEC corresponding to functionality 2.3.2.

Alias

Description

Found in

For the inertial measurement system in particular one entry might appear as follows:

Type:	*Data flow*
Name:	*Compensation factors*
Alias:	*Comp. factors*
Description:	Accelerometer compensation factors used for temperature and mass unbalance corrections.
Found in:	2.3.1

The missing information for modules "Found in" will be added as the code is developed. In this way, data dictionaries help to provide substantial traceability between code elements.

5.4.2.3 *Problems with Structured Analysis and Structured Design in Real-Time Applications*   There are several apparent problems in using structured analysis and structured design (SASD) to model the real-time systems, including difficulty in modeling time and events. For example, concurrency is not easily depicted in this form of SASD.

Another problem arises in the context diagram. Control flows are not easily translated directly into code because they are hardware dependent. In addition, the control flow does not really make sense since there is no connectivity between portions of it, a condition known as "floating."

Details of the underlying hardware also need to be known for further modeling of Process 1. What happens if the hardware changes? What if the algorithm or even the sensitivity levels change because of the installation of new hardware? In this case the changes would need to propagate into the level 1 DFD for each process, any subsequent levels and, ultimately, into the code.

Clearly making and tracking changes in structured design is fraught with danger. Moreover, any change means that significant amounts of code would need to be rewritten, recompiled, and properly linked with the unchanged code to make the system work. None of these problems arise using the object-oriented paradigm.

5.4.2.4 Real-Time Extensions of Structured Analysis and Structured Design
It is well known that the standard SASD methodology is not well equipped for dealing with time, as it is a data-oriented and not a control-oriented approach. In order to address this shortcoming, Hatley and Pirbhai extended the SASD method by allowing for the addition of control flow analysis. To do this the following artifacts were added to the standard approach: arcs made of dashed lines to indicate the flow of control information, and solid bars indicating "stored" control commands (control stores), which are left unlabeled [Hatley87].

Additional tools, such as Mealy finite state machines, are used to represent the encapsulated behavior and process activations. The addition of the new control flows and control stores allow for the creation of a diagram containing only those elements called a control flow diagram (CFD). These CFDs can be decomposed into C-SPECs (control specifications), which can then be described by a finite state machine. The relationship between the control and process models is shown in Figure 5.14. Although the Hatley-Pirbhai extensions suggest that the CFD and C-SPECs stand alone, the CFD by itself makes little sense. Hence, the CFD and DFD are generally combined as shown in Figure 5.11.

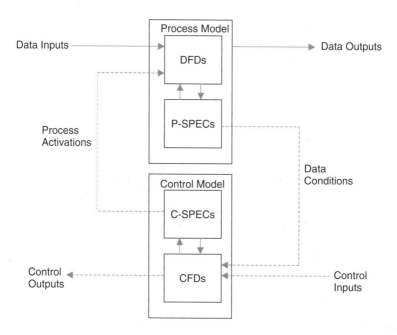

Figure 5.14 The relationship between the control and process model [Laplante03c].

5.4.3   Design in Procedural Form Using Finite State Machines

One of the advantages of using finite state machines in the software requirements specification and later in the software design is that they are easily converted to code and test cases. For example, consider the inertial measurement system. The tabular representation of the state transition function (Table 4.2), which describes the system's high-level behavior, can be easily transformed into a design using the pseudocode shown in Figure 5.15. Each procedure associated with the operational modes (TAK, NAV, NAE, NAA, LAN) will be structured code that can be viewed as executing in one of any number of process states at an instant in time. This functionality can be described by the pseudocode shown in Figure 5.15.

The pseudocode shown in Figures 5.15 and 5.16 can be coded easily in any procedural language, or even an object-oriented one. Alternatively, the system behavior can be described with a `case` statement or nested `if then` statements such that, given the current state and receipt of a signal, a new state is assigned. This is illustrated below:

```
TAK:
        If MA then NAV
        Else TAK

NAV:
        If  TD then NAA
        If MC then LAN
        If LO then NAE
        Else NAV
```

```
typedef states: (state 1,...,state n);        {n is# of states}
        alphabet: (input 1,...,input n);
        table_row: array [1..n] of states;
procedure move_forward;        {advances FSM one state}
var
        state: states;
        input: alphabet;
        table: array [1..m] of table_row;  {m is the size of the alphabet}
begin
        repeat
          get(input); {read one token from input stream}
          state:=table[ord(input)] [state];        {next state}
          execute_process (state);
          until input = EOF;
end;
```

Figure 5.15 Pseudocode that can implement the behavior of the finite state machine shown in Figure 4.1 [Laplante03c].

```
Procedure execute_process (state: states);
begin
        case state of
        state 1: process 1;              {execute process 1}
        state 2: process 2;              {execute process 2}

  ...

        state n: process n;              {execute process n}
end
```

Figure 5.16 Finite state machine code for executing a single operational process in the avionics system. Each process can exist in multiple states, allowing for partitioning of the code into appropriate modules [Laplante03c].

```
NAE:
        If EE then NAA
        Else NAE

NAA:
        If LO then NAE
        If ED then NAV
        Else NAA

LAN:
        Else LAN
```

The advantage of finite state machine design over the case statement, of course, is that the former is more flexible and compact.

5.5 OBJECT-ORIENTED DESIGN

Object-oriented programming languages are those characterized by data abstraction, inheritance, polymorphism and messaging. Data abstraction was defined earlier. Inheritance allows the software engineer to define new objects in terms of previously defined objects so that the new objects "inherit" properties. Function polymorphism allows the programmer to define operations that behave differently, depending on the type of object involved. Messaging allows objects to communicate and invoke the methods that they support.

Object-oriented languages provide a natural environment for information hiding through encapsulation. The state, data, and behavior of objects are encapsulated and accessed only via a published interface or private methods. For example, in the inertial measurement system it would be appropriate to design a class called accelerometer with attributes describing its physical implementation and methods describing its output, compensation, and so forth.

Object-oriented design is an approach to systems design that views the system components as objects and data processes, control processes, and data stores that are encapsulated within objects. Early forays into object-oriented design were led by attempts to reuse some of the better features of conventional methodologies, such as the DFDs and entity relationship models by reinterpreting them in the context of object-oriented languages. This can be seen in the Unified Modeling Language (UML). Over the last several years the object-oriented framework has gained significant acceptance into the software engineering community.

5.5.1 Benefits of Object Orientation

The benefits of object orientation in combining data and behavior into an encapsulated entity have been discussed already. The real advantages of applying object-oriented paradigms are the future extensibility and reuse that can be attained, and the relative ease of future changes.

Software systems are subject to near-continuous change: requirements change, merge, emerge, and mutate; target languages, platforms, and architectures change; and most significantly the way the software is employed in practice changes. This flexibility places a considerable burden on the software design: How can systems that must support such widespread change be built without compromising quality? There are four basic principles of object-oriented engineering that can answer this question and they have been recognized as supporting reuse.

5.5.1.1 *Open–Closed Principle*   First recorded by Meyer [Meyer00], the open–closed principle (OCP) states that classes should be open to extension, but closed to modification. That is, it should be possible to extend the behavior of a class in response to new or changing requirements, but modification to the source code is not allowed. While these expectations may seem at odds, the key is abstraction. In object-oriented systems a superclass can be created that is fixed, but can represent unbounded variation by subclassing. This aspect is clearly superior to structured approaches and top-down design in, for example, changes in accelerometer compensation algorithms, which would require new function parameter lists and wholesale recompilation of any modules calling that code in the structured design.

5.5.1.2 *Once and Only Once*   While certainly not a new idea, Beck [Beck99] put a name to the principle that any aspect of a software system – be it an algorithm, a set of constants, documentation, or logic – should exist in only one place. This isolates future changes, makes the system easier to comprehend and maintain, and through the low coupling and high cohesion that the principle instills, the reuse potential of these aspects increases. The encapsulation of state and behavior in objects, and the ability to inherit properties between classes allows for the rigorous application of these ideas in an object-oriented system, but is difficult to implement in structured techniques. More importantly, in structured techniques, once and only once (OAOO) needs to be breeched frequently for reasons of performance, reliability, availability, and often, for security.

5.5.1.3 *Dependency Inversion Principle*   The dependency inversion principle (DIP) states that high-level modules should not depend upon low-level modules. Both should depend upon abstractions. This can be restated as: Abstractions should not depend upon details, details should depend upon abstractions. Martin introduced this idea as an extension to OCP with reference to the proliferation of dependencies that exist between high- and low-level modules [Martin96]. For example, in a structured decomposition approach, the high-level procedures reference the lower-level procedures, but changes often occur at the lowest levels. This infers that high-level modules or procedures that should be unaffected by such detailed modifications may be affected due to these dependencies. Again, consider the case where the accelerometer characteristics change and even though perhaps only one routine needs to be rewritten, the calling module(s) need to be modified and recompiled as well. A preferable situation is to reverse these dependencies, as is evident in the Liskov substitution principle (LSP). The intent here is to allow dynamic changes in the preprocessing scheme, which is achieved by ensuring that all the accelerometer objects conform to the same interface, and are therefore interchangeable.

5.5.1.4   *Liskov Substitution Principle*   Liskov expressed the principle of the substitutivity of subclasses for their base classes as:

> *What is wanted here is something like the following substitution property: If for each object o_1 of type S there is an object o_2 of type T such that for all programs P defined in terms of T, the behavior of P is unchanged when o_1 is substituted for o_2 then S is a subtype of T. [Liskov88]*

This principle has led to the concept of type inheritance and is the basis of polymorphism in object-oriented systems, where instances of derived classes can be substituted for each other, provided they fulfill the obligations of a common superclass.

5.5.2   Design Patterns

Developing software is hard and developing reusable software is even harder. Designs should be specific to the current problem, but general enough to address future problems and requirements. Experienced designers know not to solve every problem from first principles, but to reuse solutions encountered previously, that is, they find recurring patterns and use them as a basis for new designs. This is simply an embodiment of the principle of generality.

While object-oriented systems can be designed to be as rigid and resistant to extension and modification as in any other paradigm, object-orientation has the ability to include distinct design elements that can cater to future changes and extensions. These "design patterns" were first introduced to the mainstream of software engineering practice by Gamma, Helm, Johnson, and Vlissides, and are commonly referred to as the "Gang of Four (GoF)" patterns [Gamma94].

The formal definition of a pattern is not consistent in the literature. Simply, a pattern is a named problem–solution pair that can be applied in new contexts, with advice on how to apply it in novel situations. This text is concerned with three pattern types: architectural patterns, design patterns, and idioms. An architectural pattern occurs across subsystems; a design pattern occurs within a subsystem, but is independent of the language; an idiom is a low-level pattern that is language specific.

In general, a pattern consists of four essential elements: a name, such as "strategy," "bridge," "façade"; the problem to be solved; the solution to the problem; and the consequences of the solution. More specifically, the problem describes when to apply the pattern in terms of specific design problems, such as how to represent algorithms as objects. The problem may describe class structures that are symptomatic of an inflexible design. Finally, the problem section might include conditions that must be met before it makes sense to apply the pattern.

The solution describes the elements that make up the design, though it does not describe a particular concrete design or implementation. Rather, the solution provides how a general arrangement of objects and classes solves the problem. Consider, for example, the previously mentioned GoF patterns. They describe 23 patterns, each organized by either creational, behavioral, or structural in its intent (Table 5.3). Table 5.3 is provided for illustration only, and it is not the intention to describe any of these patterns in detail. Other patterns have evolved, particularly for real-time systems, that provide various approaches to addressing the real-time communication and synchronization problem (e.g., [Douglass03; Schmidt00]).

5.5.3 Object-Oriented Design Using the Unified Modeling Language

The UML is widely accepted as the de facto standard language for the specification and design of software-intensive systems using an object-oriented approach. By bringing together the "best-of-breed" in specification techniques, the UML

Table 5.3 The set of design patterns popularized by the Gang of Four [Gamma94]

Creational	Behavioral	Structural
Abstract factory	Chain of responsibility	Adapter
Builder	Command	Bridge
Factory method	Interpreter	Composite
Prototype	Iterator	Decorator
Singleton	Mediator	Facade
	Memento	Flyweight
	Observer	Proxy
	State	
	Strategy	
	Template method	
	Visitor	

has become a family of languages (diagram types), and users can choose which members of the family are suitable for their domain.

The UML is a graphical language based upon the premise that any system can be composed of communities of interacting entities and that various aspects of those entities, and their communication can be described using the set of nine diagrams: use case, sequence, collaboration, statechart, activity, class, object, component, and deployment. Of these, five depict behavioral views (use case, sequence, collaboration, statechart, and activity), while the remaining are concerned with architectural or static aspects.

With respect to real-time systems it is these behavioral models that are of interest. The use case diagrams document the dialog between external actors and the system under development; sequence and collaboration diagrams describe interactions between objects; activity diagrams illustrate the flow of control between objects; and statecharts represent the internal dynamics of active objects. The principle artifacts generated when using the UML and their relationships are shown in Figure 5.15.

While not aimed specifically at embedded system design, some notion of time has been included in the UML through the use of sequence diagrams. Other modeling tools are needed, however. Statecharts and use case diagrams have already been discussed, and the rest are introduced below. While discussed only briefly below, many of these diagrams are illustrated in the extensive design case study at the end of this chapter.

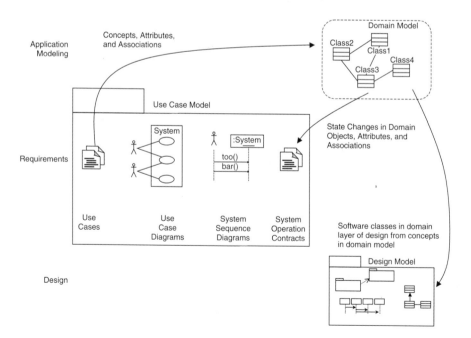

Figure 5.15 The UML and its role in specification and design. (Adapted from [Larman02]).

5.5.3.1 *Activity*   Activity diagrams are closely related to the flow chart and are used for the same purpose, that is, to show flow of control. Typically, they are used to model dynamic aspects of a system. However, unlike flowcharts, they can model concurrent computational steps and the flow of objects as they move from state to state at different points in the flow of control.

5.5.3.2 *Class Diagram*   During system design the class diagram defines the actual class attributes and methods implemented in an object-oriented programming language. Design pattern architectures are explored and physical requirements assessed during design. Design patterns provide guidance on how the defined class attributes, methods, and responsibilities should be assigned to objects. Physical requirements require the programmer to revisit the analysis class diagram where new classes for the system requirements are defined. Figure 10 in the Appendix at the end of this chapter is a design class diagram for the traffic intersection controller.

5.5.3.3 *Collaboration Diagram*   Collaboration diagrams show the messages passed between objects through the basic associations between classes. In essence, they depict the behavior on class diagrams. Collaboration diagrams are the most emphasized of UML interaction diagrams because of their clarity and expression of more information. The collaboration diagram contains classes, associations, and message flows between classes. Figures 4 through 9 in the Appendix at the end of the chapter are collaboration diagrams for the traffic intersection controller.

5.5.3.4 *Component*   These diagrams are made up of components, interfaces, and relationships. Components represent preexiting entities. Interfaces represent the functionality of components that are directly available to the user. Relationships represent conceptual relationships between components [Holt01].

5.5.3.5 *Deployment*   Deployment diagrams consist of nodes, representing real-world aspects of a system and links that show relationships between nodes.

5.5.3.6 *Object Diagram*   Object diagrams realize part of the static model of a system and are closely related to class diagrams. They show the insides of things in the class diagrams and their relationships. Moreover, they are a model or snapshot of the run-time system at a given point in time.

5.5.3.7 *Sequence Diagram*   Sequence diagrams are composed of three basic elements: objects, links, and messages, which are exactly the same as for the collaboration diagram. However, the objects shown in a sequence diagram have a lifeline associated with them, which shows a logical time line. The time line is present whenever the object is active, and is represented graphically as a vertical line with logical time traveling down the line. The objects for the sequence diagram are shown going horizontally across the page and are shown staggered down the diagram depending on when they are created [Holt01]. Figure 13 in the Appendix at the end of the chapter illustrates the sequence diagram for the traffic control system.

5.5.3.8 Modeling Time Explicitly

It is clear from the previous description that the UML in its current form does not provide sufficient facilities for the specification and analysis of real-time systems. It is also stated, however, that the UML is a family of languages, and there is no compelling reason for not adding to the family if a suitable language is found. Unfortunately, the majority of appropriate candidates are formal methods – specification languages with a sound mathematical basis – and these are traditionally shunned by the user community.

As stated earlier, the domain model is created based upon the use cases and, through further exploration of system behavior via the interaction diagrams, the domain model evolves systematically into the design class diagram. The construction of the domain model is, therefore, analogous to the analysis stage in SASD described earlier. In domain modeling the central objective is to represent the real-world entities involved in the domain as concepts in the domain model. This is a key aspect of object-oriented systems and is seen as a significant advantage of the paradigm, since the resultant model is "closer" to reality than in alternative modeling approaches, including SASD. Part of the design class diagram that results from evolution of the domain model is shown in Figure 5.15.

Most development in object-oriented design has been done with little or no provision for real-time requirements. Two methodologies that do make such provisions, real-time object-oriented modeling (ROOM) [Selic94] and HRT-HOOD [Burns90] have gained some penetration into industry. But only HRT-HOOD provides a method for guaranteeing deterministic timing behavior in hard real-time systems. This is achieved by eliminating inheritance, introducing a small number of class stereotypes (active, passive, protected, cyclic, and sporadic), and restricting the synchronization mechanism for class operations [de la Puente00]. However, the Unified Process Model (UPM) with UML has been used successfully, with a number of extensions. And, as of this writing, UML 2.0, with significant extensions for real-time applications, is about to be released.

5.5.3.9 Modeling Time in Object-Oriented Systems

Behavioral aspects of the design can be represented by a number of different diagrams in the UML. Perhaps the most popular choice is to use sequence diagrams. Other techniques for modeling time outside of the UML include the use of Q models [Motus94] and various temporal logics.

5.5.3.10 Object-Oriented or Structured Design

The preceding observations beg the question of whether object-oriented design is more suitable then SD for the embedded real-time systems and the inertial measure unit in particular. SD and object-oriented design are often compared and contrasted, and, indeed, they are similar in some ways. This should be no surprise, since both have their roots in the work of Parnas and his predecessors [Parnas79], [Parnas72]. Table 5.4 provides a side-by-side comparison of the methodologies.

Both structured and object-oriented analysis (OOA) are full life-cycle methodologies and use some similar tools and techniques. However, there are major

Table 5.4   A comparison of structured analysis and object-oriented analysis

	SA	OOA
System components	Functions	Objects
Data processes Control processes Data stores	Separated through internal decomposition	Encapsulated within objects
Characteristics	Hierarchical Structure Classifications of functions Encapsulation of knowledge within functions	Inheritance Classification of objects Encapsulation of knowledge within objects

differences. SA describes the system from a functional perspective and separates data flows from the functions that transform them, while OOA describes the system from the perspective of encapsulated entities that possess both function and form.

Additionally, object-oriented models include inheritance, while structured design does not. Although SD has a definite hierarchical structure, this is a hierarchy of composition rather than heredity. This shortcoming leads to difficulties in maintaining and extending both the specification and design.

The purpose of this discussion is not to dismiss SA, or even to conclude that it is better than OOA in all cases. An overriding indicator of suitability of OOA versus SA to real-time systems is the nature of the application. To see this, consider the vertices of the triangle in Figure 5.16 representing three distinct viewpoints of a system: data, actions, and events.

Events represent stimuli and responses such as measurements in process control systems, as in the case study. Actions are rules that are followed in complex algorithms, such as "compensate," "torque," and "calibrate" in the case of the inertial measurement system. The majority of early computer systems were focused on

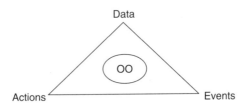

Figure 5.16   A project's applicability to either object-oriented or structured analysis according to system focus.

one, or at most two, of these vertices. For example, early, non-real-time image-processing systems were data and action intensive, but did not encounter much in the way of stimuli and response.

Real-time systems are usually data intensive and would seem well suited to structured analysis. But real-time systems also include control information, which is not well suited to SD. It is likely that a particular real-time system is as much event or activity based as it is data based, which makes it quite suitable for object-oriented techniques.

5.6 APPENDIX: CASE STUDY IN SOFTWARE REQUIREMENTS SPECIFICATION FOR FOUR-WAY TRAFFIC INTERSECTION TRAFFIC LIGHT CONTROLLER SYSTEM

To further illustrate the concepts of design, the Software Requirements Specification given in the case study of Chapter 4 is used to provide a corresponding object-oriented design for the traffic control system. Many of these figures have been referenced in the previous sections. The Appendix serves to further explicate the object-oriented design process, many of its artifacts, and provides a useful example of an object-oriented design document.

1 INTRODUCTION

Traffic controllers currently in use comprise simple timers that follow a fixed cycle to allow vehicle/pedestrian passage for a predetermined amount of time regardless of demand, actuated traffic controllers that allow passage by means of vehicle/pedestrian detection, and adaptive traffic controllers that determine traffic conditions in real-time by means of vehicle/pedestrian detection and respond accordingly in order to maintain the highest reasonable level of efficiency under varying conditions. The traffic controller described in this design document is capable of operating in all three of these modes.

1.1 Purpose

The purpose of this document is to provide a comprehensive set of software design guidelines to be used in the development phase of the application. This specification is intended for use by software developers.

1.2 Scope

This software package is part of a control system for pedestrian/vehicular traffic intersections that allows for (1) a fixed cycle mode, (2) an actuated mode, (3) a fully adaptive automatic mode, (4) a locally controlled manual mode, (5) a remotely controlled manual mode, and (6) an emergency preempt mode. In the fully adaptive automatic mode, a volume detection feature has been included so that the system is aware of changes in traffic patterns. Pushbutton fixtures are also included so the system can account for and respond to pedestrian traffic. The cycle is controlled by an adaptive algorithm that uses data from many inputs to achieve maximum throughput and acceptable wait-times for both pedestrians and motorists. A preempting feature allows emergency vehicles to pass through the intersection in a safe and timely manner by altering the state of the signals and the cycle time.

This document follows the structure provided in the object-oriented software requirements specification (SRS) template found in IEEE 830-1999 ([2]) and adopted in [1] rather than that defined in IEEE 1016-1998 ([3]) due to the fact that, as acknowledged in the IEEE standard itself, IEEE 1016 is not suitable as a basis for representing object-oriented designs.

1.3   Definitions and Acronyms

In addition to those given in [1], the following terms are defined here.

1.3.1   Accessor   A method used to access a private attribute of an object.

1.3.2   Active Object   An object that owns a thread and can initiate control activity. An instance of active class.

1.3.3   Collaboration   A group of objects and messages between them that interact to perform a specific function.

As defined in [5], a collaboration is "The specification of how an operation or classifier, such as a use case, is realized by a set of classifiers and associations playing specific roles used in a specific way. The collaboration defines an interaction."

1.3.4   Mutator   A method used to modify a private attribute of an object.

1.4   References

[1] T2-SRS Rev. A, "Software Requirements Specification (SRS) for an Intersection Traffic Control System."

[2] IEEE 830-1999.

[3] IEEE 1016-1998.

[4] Real Time UML, Second Edition- Developing Efficient Objects for Embedded Systems, Bruce Powel Douglas, Addison-Wesley, New York, 1999.

[5] OMG Unified Modeling Language Specification, Version 1.4, September 2001.

2 OVERALL DESCRIPTION

2.1   Intersection Overview

The intersection class to be controlled is illustrated in Figure 1. This figure has been repeated from [1].

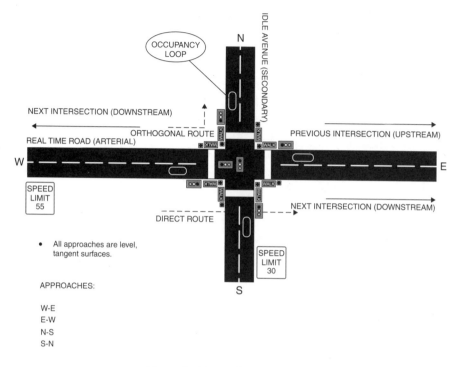

Figure 1   Intersection topography.

The target class of intersection is described in detail in [1].

2.2   Intersection Software Architecture

The intersection controller software architecture consists of the major components shown in Figure 2.

Name: IntersectionController- System Architecture
Author: Team 2
Version: 1.0
Created: 09-Dec-2002 10:02:10
Updated: 09-Dec-2002 10:14:34

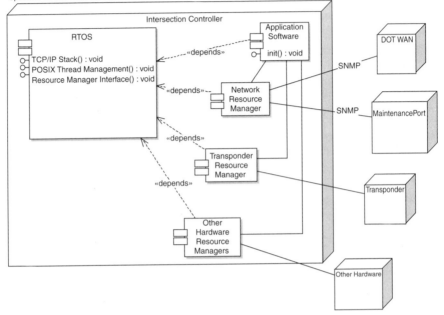

Figure 2   Intersection controller software architecture.

2.2.1 *Real-Time Operating System (RTOS)*   The RTOS selected for the intersection controller is QNX Neutrino 6.2 for the iX86 family of processors.

2.2.2 *Application Software*   Application software is written in C++ and is compiled using QNX Photon tools and the GNU gcc 2.95 compiler.

2.2.3 *Resource Managers*   Resource managers are written in C++ using the QNX Driver Development Kit. Note that these have been developed by another team and so have not been covered in detail in this document.

3   DESIGN DECOMPOSITION

This section provides a detailed object-oriented decomposition of the intersection controller software design. The decomposition is based on the use cases and preliminary class model described in [1].

The decomposition makes use of the Unified Modeling Language (UML), supplemented by text descriptions, to define the details of the design. This representation provides the design views described in IEEE 1016 ([3]) within the framework of object-oriented design, as shown in Table 1.

Table 1   **IEEE 1016 design views**

Design View	Represented By	SDD Reference
Decomposition view	Classes in class diagram	Figure 10
Interrelationship view	Associations in class diagram	Figure 10
Interface view	Collaboration diagrams	Figure 4 through Figure 9
Detailed view	Attribute and method details; behavioral diagrams	For each class

3.1   Major Software Functions (Collaborations)

Based on the use case diagram provided in [1], the major functions of the intersection controller have been grouped into UML collaborations (represented by dashed ovals) as shown in Figure 3. Collaboration details are described in the following paragraphs.

Name: IntersectionController- Top-Level Users
Author: Team 2
Version: 1.0
Created: 30-Nov-2002 09:01:19
Updated: 03-Dec-2002 07:35:44

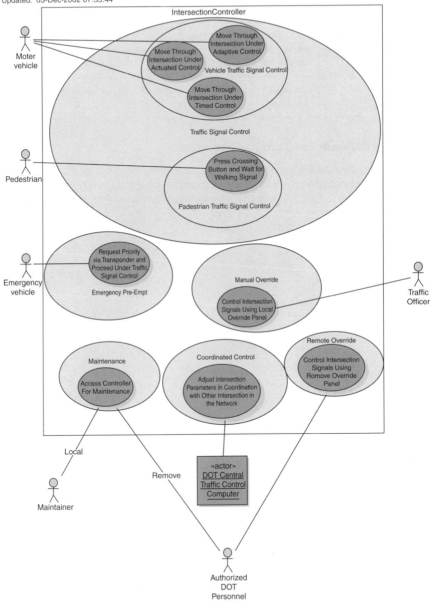

Figure 3   Intersection controller collaborations.

3.1.1   Collaboration Messages   The tables below provide a listing of the messages (method calls and events) passed between objects in each collaboration defined above. Messages with an ''on. . .'' prefix correspond to events.

3.1.1.1   Traffic Signal Control

Table 2   IntersectionController – traffic signal control collaboration messages

ID	Message	From Object	To Object
1	setAspect(Aspect)	m_Intersection Controller	m_Approach[0]
1.1	getAspect()	m_Intersection Controller	m_Approach[0]
1.2	getCount()	m_Intersection Controller	m_Approach[0]
2	ignoreState()	m_Approach[0]	m_PedestrianDetector[0]
2.1	watchState()	m_Approach[0]	m_PedestrianDetector[0]
2.2	resetState()	m_Approach[0]	m_PedestrianDetector[0]
3	onEntryStateSet(void)	m_VehicleDetector	m_Approach[0]
3.1	onEntryStateCleared(void)	m_VehicleDetector	m_Approach[0]
4	onPedestrianRequest()	m_PedestrianDetector[0]	m_Approach[0]
5	onPedestrianRequest()	m_Approach[0]	m_Intersection Controller
5.1	onVehicleEntry(int)	m_Approach[0]	m_Intersection Controller
6	setIndication(Indication)	m_Approach[0]	m_PedestrianTrafficStandard[0]
6.1	getIndication()	m_Approach[0]	m_PedestrianTrafficStandard[0]
7	setIndication(Indication)	m_Approach[0]	m_VehicleTrafficStandard[0]
7.1	getIndication()	m_Approach[0]	m_VehicleTrafficStandard[0]

3.1.1.2   Emergency Preempt

Table 3   IntersectionController – emergency preempt collaboration messages

ID	Message	From Object	To Object
1	onActivate()	Emergency Vehicle Transponder	m_EmergencyPreempt
1.1	onDeactivate()	Emergency Vehicle Transponder	m_EmergencyPreempt
2	onPreemptRequest()	m_EmergencyPreempt	m_Intersection Controller

(continued)

Table 3 (*continued*)

ID	Message	From Object	To Object
2.1	onPreemptCleared()	m_EmergencyPreempt	m_Intersection Controller

3.1.1.3 Manual Override

Table 4 IntersectionController – manual override collaboration messages

ID	Message	From Object	To Object
1	onActivate(OverrideType)	Manual Control Panel	m_ManualOverride
1.1	onDeactivate()	Manual Control Panel	m_ManualOverride
2	onSetPhase()	Manual Control Panel	m_ManualOverride
3	onOverrideActivated(OverrideType)	m_ManualOverride	m_Intersection Controller
3.1	onOverrideDeactivated(OverrideType)	m_ManualOverride	m_Intersection Controller
4	setPhase()	m_ManualOverride	m_Intersection Controller

3.1.1.4 Remote Override

Table 5 IntersectionController – remote override collaboration messages

ID	Message	From Object	To Object
1	onActivate(OverrideType)	m_Network	m_RemoteOverride
1.1	onDeactivate(OverrideType)	m_Network	m_RemoteOverride
2	onSetPhase()	m_Network	m_RemoteOverride
3	onOverrideActivated(OverrideType)	m_RemoteOverride	m_Intersection Controller
3.1	onOverrideDeactivated(OverrideType)	m_RemoteOverride	m_Intersection Controller
4	setPhase()	m_RemoteOverride	m_Intersection Controller
5	sendPacket(void*)	m_RemoteOverride	m_Network

3.1.1.5   Coordinated Control

Table 6   **IntersectionController – coordinated control collaboration messages**

ID	Message	From Object	To Object
1	setMode(Mode)	m_RemoteOverride	m_Intersection Controller
2	setParameters()	m_RemoteOverride	m_Intersection Controller
3	getStatus()	m_RemoteOverride	m_Intersection Controller
4	onSetParameters(Parameters*)	m_Network	m_RemoteOverride
5	onGetStatus()	m_Network	m_RemoteOverride
6	sendPacket(void*)	m_RemoteOverride	m_Network

3.1.1.6   Maintenance

Table 7   **IntersectionController – maintenance collaboration messages**

ID	Message	From Object	To Object
1	getStatus()	m_Maintenance	m_Intersection Controller
2	goFirst()	m_Maintenance	m_IncidentLog
2.1	read()	m_Maintenance	m_IncidentLog
2.2	goNext()	m_Maintenance	m_IncidentLog
2.3	isEOF()	m_Maintenance	m_IncidentLog
3	flush()	m_Maintenance	m_IncidentLog
4	getStatus()	m_Network	m_Maintenance
5	readDatabase(int)	m_Network	m_Maintenance
6	sendPacket(void*)	m_Maintenance	m_Network

3.1.2   Collaboration Diagrams   The collaborations described above are depicted in Figure 4 through Figure 9.

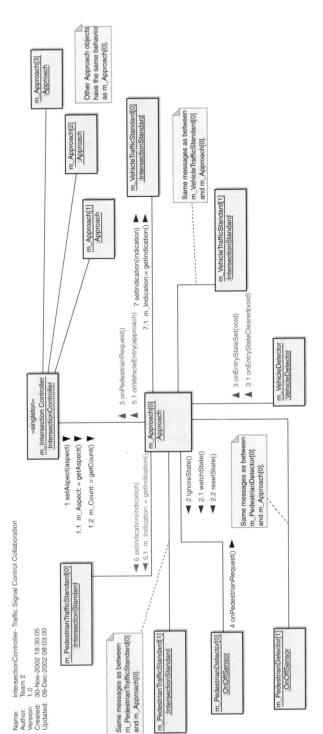

Name: IntersectionController- Traffic Signal Control Collaboration
Author: Team 2
Version: 1.0
Created: 30-Nov-2002 18:30:05
Updated: 09-Dec-2002 08:03:00

Figure 4   Traffic signal control.

Name: IntersectionController- Emergency Preempt Collaboration
Author: Team 2
Version: 1.0
Created: 03-Dec-2002 14:56:09
Updated: 03-Dec-2002 15:08:11

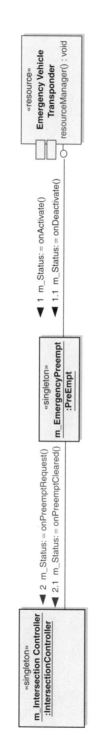

Figure 5 Emergency preempt.

Name: IntersectionController- Manual Override Collaboration
Author : Team 2
Version : 1.0
Created : 03-Dec-2002 08:50:43
Updated : 05-Dec-2002 18:38:06

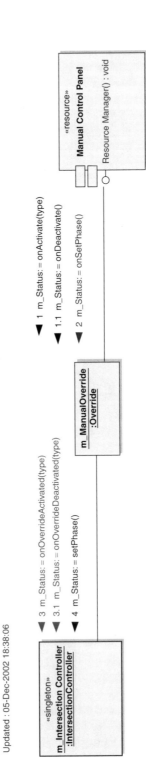

Figure 6   Manual override.

Name: IntersectionController- Remote Override Collaboration
Author: Team 2
Version: 1.0
Created: 03-Dec-2002 11:55:19
Updated: 05-Dec-2002 18:46:02

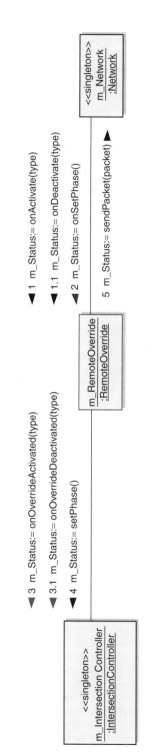

Figure 7   Remote override.

Name: IntersectionController- Coordinated Control Collaboration
Author: Team 2
Version: 1.0
Created: 03-Dec-2002 15:11:01
Updated: 03-Dec-2002 15:27:15

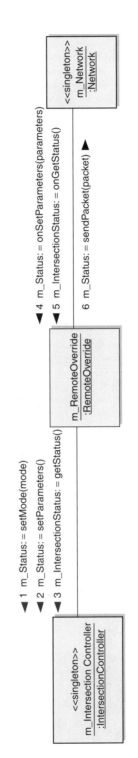

Figure 8 Coordinated control.

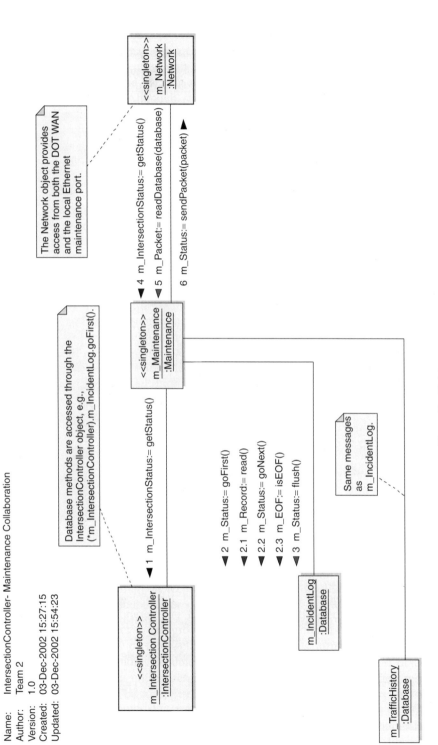

Name: IntersectionController- Maintenance Collaboration
Author: Team 2
Version: 1.0
Created: 03-Dec-2002 15:27:15
Updated: 03-Dec-2002 15:54:23

The Network object provides access from both the DOT WAN and the local Ethernet maintenance port.

Database methods are accessed through the IntersectionController object, e.g., (*m_IntersectionController).m_IncidentLog.goFirst().

<<singleton>>
m_Network
:Network

4 m_IntersectionStatus:= getStatus()
5 m_Packet:= readDatabase(database)
6 m_Status:= sendPacket(packet)

<<singleton>>
m_Maintenance
:Maintenance

1 m_IntersectionStatus:= getStatus()

<<singleton>>
m_Intersection Controller
:IntersectionController

2 m_Status:= goFirst()
2.1 m_Record:= read()
2.2 m_Status:= goNext()
2.3 m_EOF:= isEOF()
3 m_Status:= flush()

m_IncidentLog
:Database

Same messages as m_IncidentLog.

m_TrafficHistory
:Database

Figure 9 Maintenance.

269

3.2   Class Model

Figure 10 depicts the classes constituting the intersection control system software application. The diagram reflects the preliminary class structure defined in [1], but with additional detail and, in some cases, addition of classes and reallocation of responsibilities.

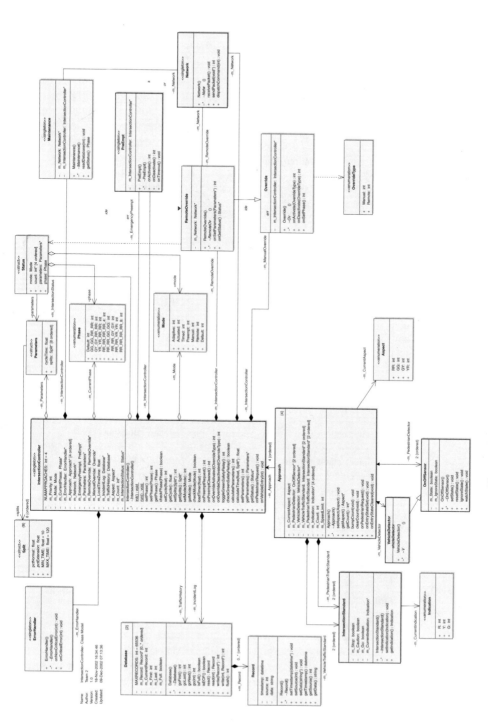

Figure 10 IntersectionController class diagram.

271

Classes corresponding to active objects (i.e., objects with their own thread of control) are shown in Figure 10 with **bold** outlines. The active object instances are summarized in Table 8.

Table 8 Active objects

Level	Object Name
1.	m_IntersectionController
1.1.	m_IntersectionController::m_Approach[0]
1.1.1.	m_IntersectionController::m_Approach[0]::m_VehicleTrafficStandard[0]
1.1.2.	m_IntersectionController::m_Approach[0]::m_VehicleTrafficStandard[1]
1.1.3.	m_IntersectionController::m_Approach[0]::m_VehicleTrafficStandard[0]
1.1.4.	m_IntersectionController::m_Approach[0]::m_PedestrianTrafficStandard[1]
1.1.5.	m_IntersectionController::m_Approach[0]::m_PedestrianDetector[0]
1.1.6.	m_IntersectionController::m_Approach[0]::m_PedestrianDetector[1]
1.1.7.	m_IntersectionController::m_Approach[0]::m_VehicleDetector
1.2.	m_IntersectionController::m_Approach[1]
1.2.1.	m_IntersectionController::m_Approach[1]::m_VehicleTrafficStandard[0]
1.2.2.	m_IntersectionController::m_Approach[1]::m_VehicleTrafficStandard[1]
1.2.3.	m_IntersectionController::m_Approach[1]::m_VehicleTrafficStandard[0]
1.2.4.	m_IntersectionController::m_Approach[1]::m_PedestrianTrafficStandard[1]
1.2.5.	m_IntersectionController::m_Approach[1]::m_PedestrianDetector[0]
1.2.6.	m_IntersectionController::m_Approach[1]::m_PedestrianDetector[1]
1.2.7.	m_IntersectionController::m_Approach[1]::m_VehicleDetector
1.3.	m_IntersectionController::m_Approach[2]
1.3.1.	m_IntersectionController::m_Approach[2]::m_VehicleTrafficStandard[0]
1.3.2.	m_IntersectionController::m_Approach[2]::m_VehicleTrafficStandard[1]
1.3.3.	m_IntersectionController::m_Approach[2]::m_VehicleTrafficStandard[0]
1.3.4.	m_IntersectionController::m_Approach[2]::m_PedestrianTrafficStandard[1]

Table 8 (*continued*)

Level	Object Name
1.3.5.	m_IntersectionController::m_Approach[2]::m_PedestrianDetector[0]
1.3.6.	m_IntersectionController::m_Approach[2]::m_PedestrianDetector[1]
1.3.7.	m_IntersectionController::m_Approach[2]::m_VehicleDetector
1.4.	m_IntersectionController::m_Approach[3]
1.4.1.	m_IntersectionController::m_Approach[3]::m_VehicleTrafficStandard[0]
1.4.2.	m_IntersectionController::m_Approach[3]::m_VehicleTrafficStandard[1]
1.4.3.	m_IntersectionController::m_Approach[3]::m_VehicleTrafficStandard[0]
1.4.4.	m_IntersectionController::m_Approach[3]::m_PedestrianTrafficStandard[1]
1.4.5.	m_IntersectionController::m_Approach[3]::m_PedestrianDetector[0]
1.4.6.	m_IntersectionController::m_Approach[3]::m_PedestrianDetector[1]
1.4.7.	m_IntersectionController::m_Approach[3]::m_VehicleDetector
2.	m_IntersectionController::m_Network
3.	m_IntersectionController::m_EmergencyPreempt

3.3   Class Details

3.3.1   *IntersectionController*   The IntersectionController class is responsible for managing the following functions:

1. Initialization.
2. Instantiation of contained objects.
3. Overall control of the intersection vehicle traffic standards.
4. Overall control of the intersection pedestrian traffic standards.
5. Collection and processing of traffic history from all approaches.
6. Adaptive control of intersection timings in response to traffic flow.
7. Actuated control of intersection in response to vehicle presence.
8. Timed control of intersection in response to a fixed scheme.
9. Overall handling of pedestrian crossing requests.
10. Handling of emergency vehicle preemption.
11. Intersection control in response to manual override commands.
12. Intersection control in response to remote override commands.
13. Management of traffic history and incident log databases.

14. Handling of maintenance access requests from the maintenance port.
15. Handling of maintenance access requests from the DOT WAN.

Figure 11 illustrates the attributes, methods, and events of the IntersectionController class.

```
┌─────────────────────────────────────────────────┐
│                  <<singleton>>                    │
│              IntersectionController               │
├─────────────────────────────────────────────────┤
│  −   NUMAPPROACHES: int = 4                        │
│  −   m_Priority: int                              │
│  −   m_Mode: Mode                                 │
│  −   m_CurrentPhase: Phase*                        │
│  −   m_ErrorHandler: ErrorHandler*                 │
│  −   m_Approach: Approach* [4 ordered]             │
│  −   m_Network: Network*                           │
│  −   m_EmergencyPreempt: PreEmpt*                  │
│  −   m_Parameters: Parameters*                     │
│  −   m_RemoteOverride: RemoteOverride*             │
│  −   m_ManualOverride: Override*                   │
│  −   m_LocalTimeZone: float                        │
│  −   m_IncidentLog: Database*                      │
│  −   m_TrafficHistory: Database*                   │
│  −   m_Aspect: Aspect*                             │
│  −   m_Count: int*                                 │
│  −   m_IntersectionStatus: Status*                 │
├─────────────────────────────────────────────────┤
│  +   IntersectionController()                      │
│  −*  ~IntersectionController()                     │
│  −   init() : void                                │
│  +   run() : void                                 │
│  +   setPhase() : int                             │
│  +   setPhase(Phase) : int                        │
│  +   getPhase() : Phase                           │
│  +   checkPhase(Phase) : boolean                  │
│  +   setCycle(float) : int                        │
│  +   getCycle() : float                           │
│  +   setSplits(Split*) : int                      │
│  +   getSplits() : Split*                         │
│  +   setMode(Mode) : int                          │
│  +   getMode() : Mode                             │
│  +   checkMode(Mode) : boolean                    │
│  +   loadTimer(float) : int                       │
│  +   onPreemptRequest() : int                     │
│  +   onPreemptCleared() : int                     │
│  +   onOverrideActivated(OverrideType) : int      │
│  +   onOverrideDeactivated(OverrideType) : int    │
│  +   toggleGreenSafetyRelay() : int               │
│  +   checkGreenSafetyRelay() : boolean            │
│  +   calculateParameters() : int                  │
│  +   calculateTime(float, Split*) : float         │
│  +   setParameters() : int                        │
│  +   getParameters() : Parameters*                │
│  +   getStatus() : Status*                        │
│  +   onPedestrianRequest() : void                 │
│  +   onVehicleEntry(int) : void                   │
└─────────────────────────────────────────────────┘
```

Figure 11 IntersectionController class.

3.3.1.1 *IntersectionController Relationships*

- Association link from class *Status*
- Association link to class *PreEmpt*
- Association link to class *Network*
- Association link from class *PreEmpt*
- Association link to class *Database*
- Association link from class *Override*
- Association link to class *Mode*
- Association link from class *Maintenance*
- Association link to class *Database*
- Association link to class *Parameters*
- Association link to class *RemoteOverride*
- Association link to class *Phase*
- Association link to class *ErrorHandler*
- Association link to class *Approach*

3.3.1.2 *IntersectionController Attributes*

Table 9   IntersectionController class-attributes

Attribute	Type	Notes
NUMAPPROACHES	private: *int*	Constant defining the number of approaches in the intersection.
m_Priority	private: *int*	Indicates the relative priority of the approaches. Values are as follows: 1. E-W/W-E approach pair has priority = 1. 2. N-S/S-N approach pair has priority = 2. 3. Both approach pairs have equal priority = 3. This attribute is used to determine which of the three default states should be set when the intersection initializes or is set to operate in Default mode either by an override command or by an error condition.
m_Mode	private: *Mode*	The object m_Mode, an instance of the Mode enumeration class, indicates the method currently being used to control the intersection. Valid values for this attribute are shown in the class diagram.

Table 9 (*continued*)

Attribute	Type	Notes
		The setPhase() method checks for changes in this value at the beginning of each cycle and changes the control scheme if required. Changes to Preempt, Manual or Remote modes are handled by specific events; these events cause the control scheme to change immediately rather than at the beginning of the next cycle.
m_CurrentPhase	private: *Phase*	This is an enumeration of class Phase that also serves as an index into the m_Split array (since C++ automatically casts enumerated types as arrays where required) denoting which portion of the cycle is currently active.
		The Default phase is used during initialization and in response to override commands and critical system faults. Phases GG_GG_RR_RR (1) to RR_RR_RR_RR_8 (8) are used in normal operation.
m_ErrorHandler	private: *ErrorHandler*	Pointer to the m_ErrorHandler object.
m_Approach	private: *Approach*	This is an array of type Approach and a length of NUMAPPROACHES. This array represents each of the four entrances to an intersection. See the Approach class for more details. The m_Approach array is declared as follows: Approach m_Approach[NUMAPPROACHES] Where NUMAPPROACHES is a compile-time constant.
m_Network	private: *Network*	This object is the instance of the Network class that provides an abstraction layer between the network resource manager and the m_IntersectionController object.
m_EmergencyPreempt	private: *PreEmpt*	This is a pointer to the instance of the PreEmpt class that provides an abstraction layer between the emergency vehicle transponder resource manager and the m_IntersectionController object.

(*continued*)

Table 9   (*continued*)

Attribute	Type	Notes
m_Parameters	private: *Parameters*	Structure holding the intersection parameters, which are the cycle time and the splits array.
m_RemoteOverride	private: *RemoteOverride*	This is the instance of the RemoteOverride class representing the Remote Software console. This object abstracts requests made from the off-site software control panel from the main application.
m_ManualOverride	private: *Override*	This is the instance of the Override class representing the Manual Override console. The object serves as a broker, abstracting the main application from any requests made from the Manual Override console, which is located at the site of the traffic control system.
m_LocalTimeZone	private: *float*	Given as an offset in hours to UTC (GMT).
m_TrafficHistory	private: *Database*	This is the instance of the Database class that is used to log statistical data regarding traffic levels at the intersection being controlled. The data are stored in the system's flash memory store. See Section 2.2.6.1 in [1] for more information about the flash memory included in the system.
m_IncidentLog	private: *Database*	This object, which is another instance of the Database class, logs abnormal events observed by the system on the site of the intersection. Data recorded by this object will be stored in the system's flash memory store. See Section 2.2.6.1 in [1] for more information about the flash memory included in the system.
m_Aspect	private: *Aspect*	Detected Aspect from each m_Approach object; Aspect[4].
m_Count	private: *int*	Vehicle count from each m_Approach object; int[4].
m_IntersectionStatus	private: *Status*	

3.3.1.3   IntersectionController Methods

Table 10   **IntersectionController class-methods**

Method	Type	Notes
IntersectionController ()	public:	Constructor.
~IntersectionController ()	private abstract:	Destructor.
init ()	private static: *void*	This is the first code unit executed when the equipment becomes active. This function performs the following basic tasks: 1. Test memory and hardware. 2. Gather all environmental information (initial mode, priority, approach parameters). 3. Set all the components of the intersection to their default states. 4. Start the first cycle in normal mode.
run ()	public static: *void*	
setPhase ()	public: *int*	Moves the intersection to the next phase in the cycle. This method is invoked in response to the following events: 1. Phase timer reaches 0 (in Actuated, Fixed and Adaptive modes). 2. Remote Override *onSetPhase(void)* event fired (in Remote mode). 3. Manual Override *onSetPhase(void)* event fired (in Manual mode). The following tasks are performed by this method: 1. Change the *m_CurrentPhase* attribute according to the assignment operation *m_CurrentPhase = (m_CurrentPhase++)* mod 9. 2. Change the state of the Green Signal Safety Relay as required by the new value of *m_CurrentPhase*. 3. Check the state of the Green Signal Safety Relay and raise an error if there is a discrepancy. 4. Manipulate the attributes of the *m_Approach* objects as required by the new Current Phase.

(*continued*)

Table 10   (*continued*)

Method	Type	Notes
		5. Calculate the phase time as *calculateTime(m_Cycle, m_Splits[m_CurrentPhase])*. 6. Load the Phase Time Remaining timer with the calculated phase time by invoking *loadTimer(calculateTime(m_Cycle, m_Splits[m_CurrentPhase]))*. 7. Check that the phase setting is displayed properly by the approaches and raise an error if there is a discrepancy.
setPhase (Phase)	public: *int*	param: phase [Phase - in] Moves the intersection to the specified phase.
getPhase ()	public: *Phase*	Determine the displayed intersection phase by querying all Aspect objects and determining their aspects. Used by the checkPhase method
checkPhase (Phase)	public: *boolean*	param: phase [Phase - in] Returns True if the displayed phase agrees with the commanded phase (passed as a parameter), False otherwise.
setCycle (float)	public: *int*	param: time [float - in] Mutator for the cycle time attribute.
getCycle ()	public: *float*	Accessor for the cycle time attribute.
setSplits (Split*)	public: *int*	param: splits [Split* - inout] Mutator for the splits attribute.
getSplits ()	public: *Split**	Accessor for the splits attribute.
setMode (Mode)	public: *int*	param: mode [Mode - in] Mutator for the attribute m_Mode.
getMode ()	public: *Mode*	Accessor for the attribute m_Mode.
checkMode (Mode)	public: *boolean*	param: mode [Mode - in]

Table 10 (*continued*)

Method	Type	Notes
loadTimer (float)	public: *int*	param: time [float - in] Loads the phase timer (utilizing OS timer services) with the phase time, specified as a parameter.
onPreemptRequest ()	public: *int*	Emergency preempt request event from the m_EmergencyPreempt object. This method performs the following tasks: 1. Save the current value of *m_Mode*. 2. Set the mode to Preempt. 3. Set the intersection phase to allow the emergency vehicle to pass safely under traffic signal control.
onPreemptCleared ()	public: *int*	Event that terminates preempted operation and returns the intersection to normal operating mode. This method performs the following tasks: 1. Restores the previous mode. 2. Sets the intersection to the default state. 3. Returns the intersection to normal operation.
onOverrideActivated (*OverrideType*)	public: *int*	param: type [OverrideType - in] Override activation event from either the *m_ManualOverride* or *m_RemoteOverride* object. The parameter type indicates which override is involved. This method performs the following tasks: 1. Save the current value of *m_Mode*. 2. Set the mode to Manual or Remote, depending on the value of parameter type. 3. Set the intersection to the Default phase.
onOverrideDeactivated (*OverrideType*)	public: *int*	param: type [OverrideType - in]

(*continued*)

Table 10   (*continued*)

Method	Type	Notes
		Override cancellation event from either the *m_ManualOverride* or *m_RemoteOverride* object. The parameter type indicates which override is involved. This method performs the following tasks: 1. Restore the previous value of *m_Mode*. 2. Set the intersection to the Default phase. 3. Returns the intersection to normal operation.
toggleGreenSafetyRelay ()	public: *int*	Toggles the state of the Green Safety Relay.
checkGreenSafetyRelay ()	public: *boolean*	Checks that the Green Safety Relay is in the proper state for the active intersection phase.
calculateParameters ()	public: *int*	Adaptive algorithm for determining intersection timing parameters for the next cycle.
calculateTime (*float, Split**)	public: *float*	param: cycle [float - in] param: split [Split* - in] Used to calculate the actual phase time from the values of *m_Parameters.cycleTime* and *m_Parameters.splits*.
setParameters ()	public: *int*	Mutator for the intersection timing parameters.
getParameters ()	public: *Parameters**	Accessor for the intersection timing parameters.
getStatus ()	public: *Status**	Method used to access the overall status of the intersection.
onPedestrianRequest ()	public: *void*	Event triggered by a valid pedestrian crossing request.
onVehicleEntry (*int*)	public: *void*	param: approach [int - in] Event triggered by a vehicle entering the vehicle detection loop.

3.3.1.4 *IntersectionController Behavioral Details*

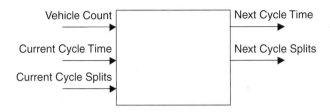

Figure 12 Black box representation of adaptive algorithm.

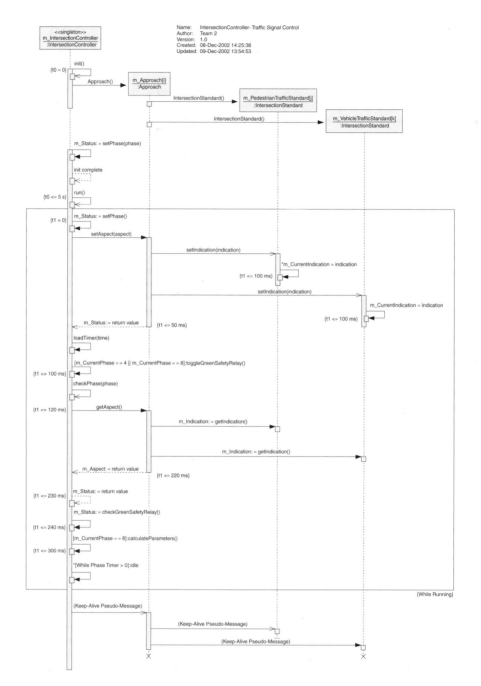

Figure 13   Traffic signal control sequence diagram.

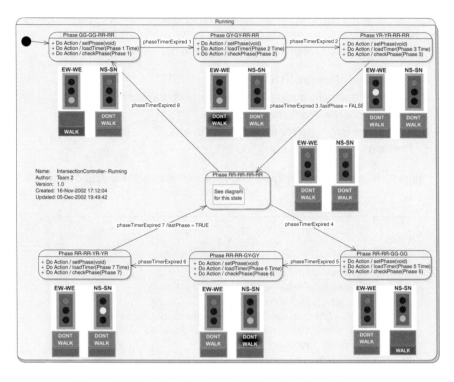

Figure 14   Statechart for intersectionController phase sequence.

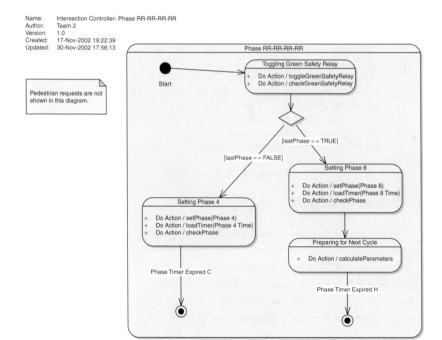

Figure 15   Statechart for Phases 4 and 8.

3.3.2   Approach   This is the programmatic representation of an individual entrance into the intersection.

The Approach class is responsible for managing the following functions:

1. Instantiation of contained objects.
2. Control of the traffic standards associated with the approach.
3. Handling of pedestrian crossing events.
4. Handling of loop detector entry and exit events.
5. Tracking the vehicle count.

Figure 16 illustrates the attributes, methods, and events of the Approach class.

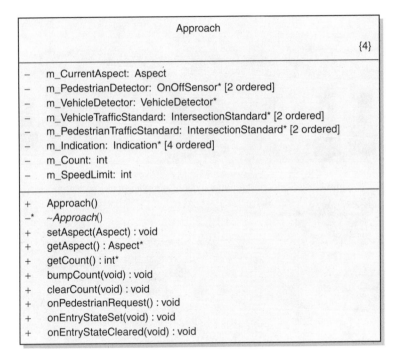

Figure 16   Approach class.

3.3.2.1   Aspect Relationships

- Association link to class *IntersectionStandard*
- Association link to class *Aspect*
- Association link to class *IntersectionStandard*
- Association link to class *VehicleDetector*
- Association link to class *OnOffSensor*
- Association link from class *IntersectionController*

3.3.2.2   Approach Attributes

Table 11   Approach class-attributes

Attribute	Type	Notes
m_CurrentAspect	private: *Aspect*	Current Approach aspect corresponding to the current intersection phase.
m_PedestrianDetector	private: *OnOffSensor**	Pointer to an array of objects of the *OnOffSensor* class, which provide an abstraction layer for the pedestrian crossing request pushbuttons.
m_VehicleDetector	private: *VehicleDetector**	Pointer to an object of class *VehicleDetector* (superclass of *OnOffSensor*), providing an abstraction layer for the vehicle detection loop.
m_VehicleTrafficStandard	private: *IntersectionStandard*	Pointer to an array of *IntersectionStandard* objects representing the vehicle traffic standards associated with the approach.
m_PedestrianTrafficStandard	private: *IntersectionStandard**	Pointer to an array of *IntersectionStandard* objects representing the pedestrian traffic standards associated with the approach.
m_Indication	private: *Indication**	Pointer to an array of *Indication* objects; used to store the indication values obtained from associated traffic standards.
m_Count	private: *int*	Used to count the number of vehicle passing through the approach.
m_SpeedLimit	private: *int*	Speed limit (in km/h) associated with the approach.

3.3.2.3 Approach Methods

Table 12 Approach class-methods

Method	Type	Notes
Approach ()	public:	Constructor.
~Approach ()	private abstract:	Destructor.
setAspect (Aspect)	public: *void*	param: aspect [Aspect - in]
		Mutator for attribute *m_CurrentAspect*.
getAspect ()	public: *Aspect**	Accessor used to fetch the aspect actually being displayed by the set of approach traffic standards.
getCount ()	public: *int**	Accessor for the *m_Count* attribute.
bumpCount (void)	public: *void*	Method called to increment the attribute *m_Count* by 1.
clearCount (void)	public: *void*	Method called to set the attribute *m_Count* to 0.
onPedestrianRequest ()	public: *void*	Event triggered by a valid pedestrian crossing request from one of the pedestrian request pushbuttons associated with the approach.
onEntryStateSet (void)	public: *void*	Event triggered when the vehicle detector attribute *m_State* is set.
onEntryStateCleared (void)	public: *void*	Event triggered when the vehicle detector attribute *m_State* is cleared.

3.3.3 IntersectionStandard Class (Pedestrian Traffic & Vehicle Traffic Standard)

This is the programmatic representation of a traffic control signal.

The IntersectionStandard class is responsible for managing the following functions:

1. Displaying the commanded aspect from the Intersection Controller.
2. Determining the aspect actually displayed.
3. Checking for discrepancies between commanded and displayed aspects.
4. Raising an error event if there is an aspect discrepancy.

Figure 17 illustrates the attributes, methods, and events of the IntersectionStandard class.

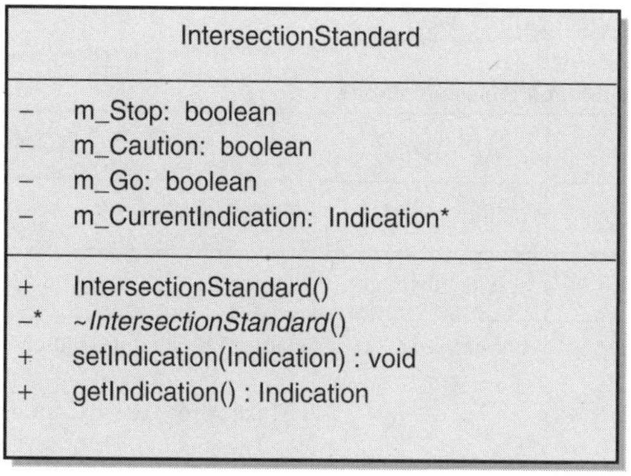

IntersectionStandard
− m_Stop: boolean − m_Caution: boolean − m_Go: boolean − m_CurrentIndication: Indication*
+ IntersectionStandard() −* ~IntersectionStandard() + setIndication(Indication) : void + getIndication() : Indication

Figure 17   IntersectionStandard class.

3.3.3.1   IntersectionStandard Relationships

- Association link from class *Approach*
- Association link to class *Indication*
- Association link from class *Approach*

3.3.3.2   IntersectionStandard Attributes

Table 13   **IntersectionStandard class-attributes**

Attribute	Type	Notes
m_Stop	private: *boolean*	A boolean value indicating that the signal is commanded to show a Stop signal (corresponding to an Indication value of R).
m_Caution	private: *boolean*	A boolean value indicating that the signal is commanded to show a Caution signal (corresponding to an Indication value of Y).
m_Go	private: *boolean*	A boolean value indicating that the signal is commanded to show a Go signal (corresponding to an Indication value of G).
m_CurrentIndication	private: *Indication*	An instance of the Indication enumerated class indicating the current traffic signal to be displayed.

3.3.3.3 IntersectionStandard Methods

Table 14 **IntersectionStandard class-methods**

Method	Type	Notes
IntersectionStandard ()	public:	Constructor.
~IntersectionStandard ()	private abstract:	Destructor.
setIndication (Indication)	public: *void*	param: indication [Indication - in] Mutator for the m_CurrentIndication attribute. The method performs the following: 1. Check whether the commanded aspect is valid. If not, raise an error. 2. If the commanded aspect is valid, display it.
getIndication ()	public: *Indication*	Accessor for determining the value of the indication actually being displayed.

3.3.3.4 Correspondence between Indications and Actual Displayed Signals Since this class is used for both the Vehicle and Pedestrian Traffic Standard objects, it is necessary to define the relationship between the attribute values and the actual displayed signal; this is shown in Table 15.

Table 15 **Attribute and signal correspondence**

m_CurrentIndication	m_Stop	m_Caution	m_Go	Vehicle Standard	Pedestrian Standard
R	True	False	False	Red	DON'T WALK
Y	False	True	False	Amber	Flashing DON'T WALK
G	False	False	True	Green	WALK

3.3.4 OnOffSensor This class represents the pedestrian crossing request pushbuttons located on opposite sides of the crosswalk associated with an approach.

Objects of the OnOffSensor class are responsible for managing the following functions:

1. Filtering of pushbutton service requests.
2. Generation of Pedestrian Service Request event.

Figure 18 illustrates the attributes, methods, and events of the OnOffSensor class.

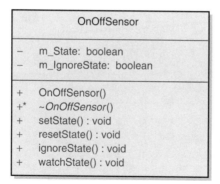

Figure 18 OnOffSensor class.

3.3.4.1 OnOffSensor Relationships

- Association link from class *Approach*
- Generalization link from class *VehicleDetector*

3.3.4.2 OnOffSensor Attributes

Table 16 OnOffSensor class-attributes

Attribute	Type	Notes
m_State	private: *boolean*	Indicates whether or not a valid pedestrian service request has been made since the last time the value was reset.
m_IgnoreState	private: *boolean*	A value that indicates whether subsequent pedestrian service requests should raise an event or simply be ignored.

3.3.4.3 OnOffSensor Methods

Table 17 OnOffSensor class-methods

Method	Type	Notes
OnOffSensor ()	public:	Constructor.
~ OnOffSensor ()	public abstract:	Destructor.
setState ()	public: *void*	Sets the object's m_State attribute to True indicating that a pedestrian service request is pending.
resetState ()	public: *void*	Sets the object's state attribute to False to indicate that any previous pedestrian service requests have been completed.

(continued)

Table 17 *(continued)*

Method	Type	Notes
ignoreState ()	public: *void*	Sets the object's m_IgnoreState attribute to True indicating that subsequent pedestrian requests are to be ignored.
watchState ()	public: *void*	Sets the object's m_IgnoreState attribute to False indicating that subsequent pedestrian requests are to be processed.

3.3.4.4 OnOffSensor Behavioral Details

Name: IntersectionController- Pedestrian Request
Author: Team 2
Version: 1.0
Created: 09-Dec-2002 12:15:35
Updated: 09-Dec-2002 14:05:38

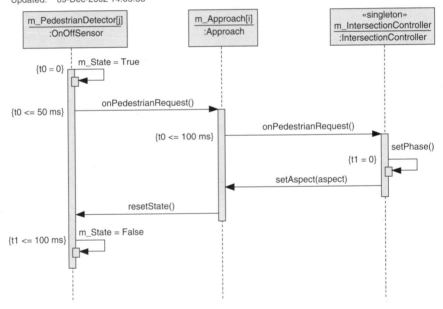

Figure 19   OnOffSensor sequence diagram.

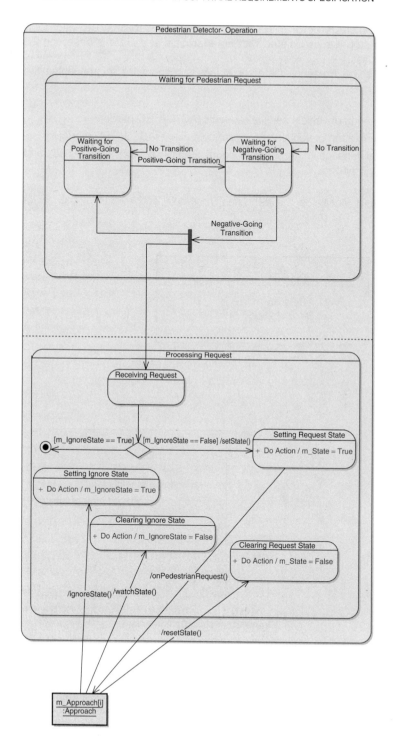

Figure 20   OnOffSensor statechart.

3.3.5 VehicleDetector   This class represents the proximity detection loop located near the stop line associated with an approach. The class is based on the OnOff-Sensor class.

The Vehicle Presence Detector object is responsible for managing the following functions:

1. Filtering of vehicle service requests (ACTUATED mode).
2. Generation of vehicle service request event (ACTUATED mode).
3. Maintenance of the vehicle count statistic (FIXED, ACTUATED, and ADAPTIVE modes).

Figure 21 illustrates the attributes, methods, and events of the VehicleDetector class.

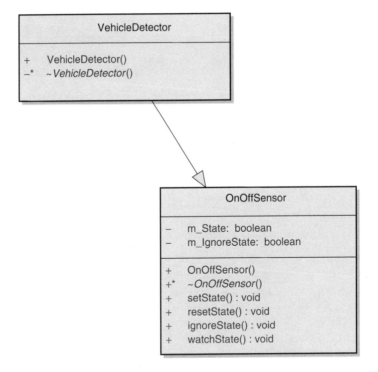

Figure 21   VehicleDetector class.

3.3.5.1   VehicleDetector Relationships

- Association link from class *Approach*
- Generalization link to class *OnOffSensor*

3.3.5.2   VehicleDetector Attributes   Inherited from superclass.

3.3.5.3 VehicleDetector Methods Inherited from superclass. Overridden methods are described in Table 18.

Table 18 VehicleDetector class-attributes

Method	Type	Notes
VehicleDetector ()	public:	Constructor.
~ VehicleDetector ()	private abstract:	Destructor.
setState ()	public: *void*	Set the *m_State* attribute and trigger the *onVehicleEntry* event.
resetState ()	public: *void*	Clear the *m_State* attribute and trigger the *onVehicleExit* event.

3.3.5.4 VehicleDetector Behavioral Details

296

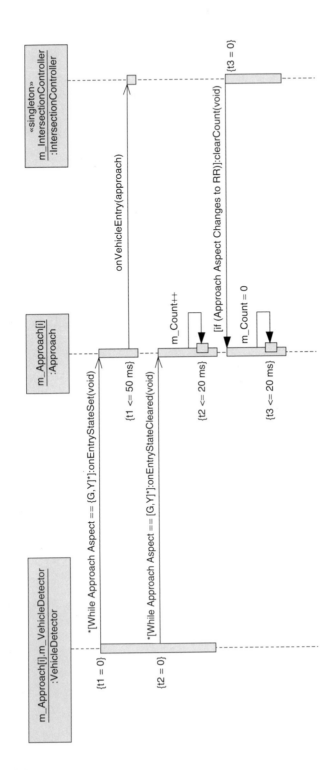

Figure 22 VehicleDetector class sequence diagram.

Name: IntersectionController- Vehicle Detector
Author: Team 2
Version: 1.0
Created: 07-Dec-2002 21:54:19
Updated: 09-Dec-2002 13:58:06

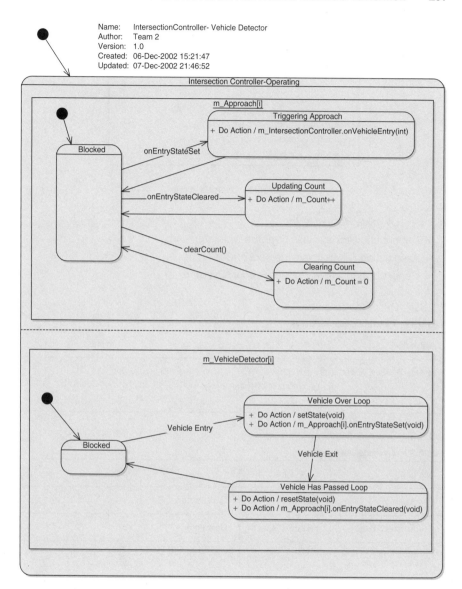

Figure 23 VehicleDetector class statechart.

3.3.6 *Override* This is class represents the set of pushbuttons on the manual override console.

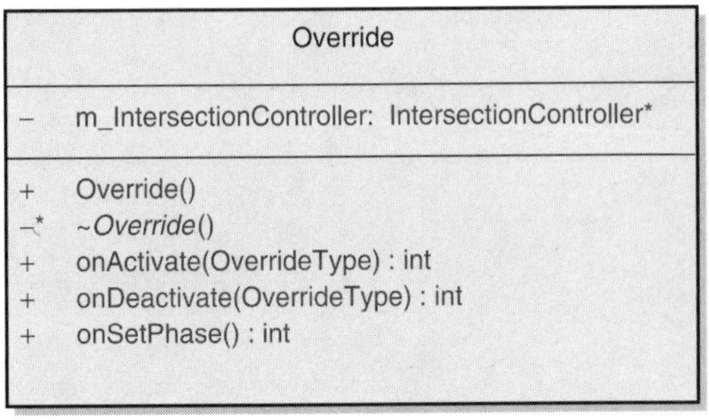

Override
−   m_IntersectionController: IntersectionController*
+   Override()
−*   ~Override()
+   onActivate(OverrideType) : int
+   onDeactivate(OverrideType) : int
+   onSetPhase() : int

Figure 24   Override class.

3.3.6.1   Override Relationships

- Dependency link to class *OverrideType*
- Association link to class *IntersectionController*
- Generalization link from class *RemoteOverride*

3.3.6.2   Override Attributes

Table 19   Override class-attributes

Attribute	Type	Notes
m_IntersectionController	private: *IntersectionController*	Pointer to the m_IntersectionController object.

3.3.6.3   Override Methods

Table 20   Override class-methods

Method	Type	Notes
Override ()	public:	Constructor.
~ Override ()	private abstract:	Destructor.
onActivate (*OverrideType*)	public: *int*	param: type [OverrideType - in] Event triggered by receipt of an activation command from the local override console.
onDeactivate (*OverrideType*)	public: *int*	param: type [OverrideType - in] Event triggered by receipt of an deactivation command from the local override console.
onSetPhase ()	public: *int*	Event triggered by receipt of an advance phase command from the local override console.

3.3.6.4 *Override Behavioral Details*

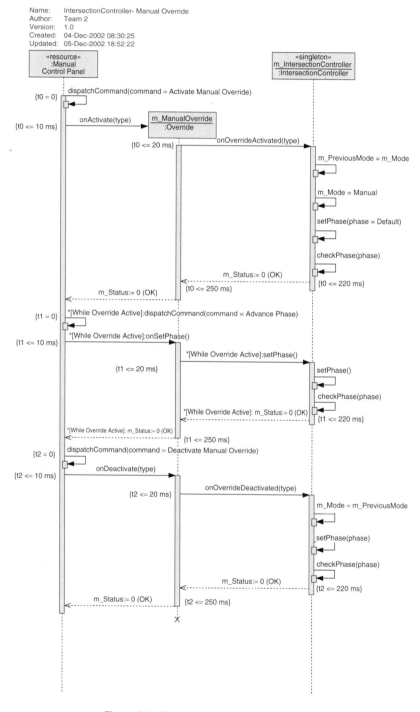

Figure 25 Override class sequence diagram.

3.3.7 RemoteOverride This class represents the commands available on the Remote Software console. Additionally, the class provides an interface for remote access to and update of intersection traffic data and cycle parameters for coordinated intersection control (option).

The RemoteOverride class is responsible for managing the following functions:

1. Triggering the appropriate mode change.
2. Generation and handling of events required to control intersection phase.
3. Acting as a substitute for the Calculate Cycle Parameters method of the Intersection Control object (in coordinated mode, not covered by this specification).

Figure 26 illustrates the attributes, methods, and events of the Remote Override class.

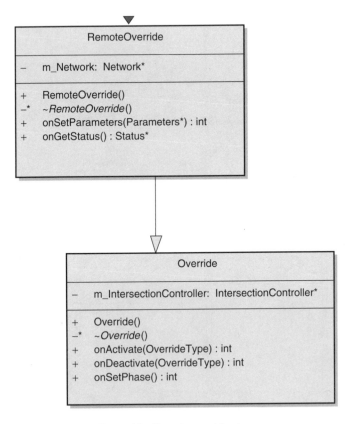

Figure 26   Remote override class.

3.3.7.1 Remote Override Relationships

- Dependency link to class *Status*
- Generalization link to class *Override*

- Association link from class *IntersectionController*
- Association link to class *Network*

3.3.7.2 RemoteOverride Attributes In addition to those inherited from the super-class Override, RemoteOverride attributes are as listed in Table 21.

Table 21 RemoteOverride class-attributes

Attribute	Type	Notes
m_Network	private: *Network*	Pointer to the m_Network object.

3.3.7.3 RemoteOverride Methods In addition to those inherited from the super-class Override, RemoteOverride methods are as listed in Table 22.

Table 22 RemoteOverride class-methods

Method	Type	Notes
RemoteOverride ()	public:	Constructor.
~ RemoteOverride ()	private abstract:	Destructor.
onSetParameters (*Parameters**)	public: *int*	param: parameters [Parameters* - in] Event triggered under coordinated control; used to set the intersection timing parameters under remote control. Completes within 100 ms.
onGetStatus ()	public: *Status**	Event triggered under coordinated control; used to get the intersection timing parameters under remote control. Completes within 100 ms.

3.3.7.4 RemoteOverride Behavioral Details Behavior of the RemoteOverride class is identical to that of the Override class for methods inherited from the superclass.

3.3.8 PreEmpt This class manages the wireless transponder interface to authorized emergency vehicles and accesses the m_IntersectionControl object in order to display the correct traffic signals, allowing the emergency vehicle priority access to the intersection.

The PreEmpt class is responsible for managing the following functions:

1. Triggering the appropriate mode change.
2. Reception of emergency vehicle preemption requests.
3. Decryption and validation of emergency vehicle preemption requests.
4. Generation and handling of events required to control intersection phase.

Figure 27 illustrates the attributes, methods, and events of the PreEmpt class.

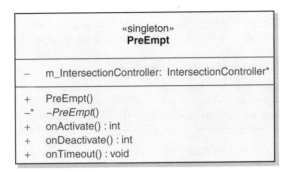

Figure 27   PreEmpt class.

3.3.8.1   PreEmpt Relationships

- Association link from class *IntersectionController*
- Association link to class *IntersectionController*

3.3.8.2   PreEmpt Attributes

Table 23   **PreEmpt class-attributes**

Attribute	Type	Notes
m_IntersectionController	private: *IntersectionController*	Pointer to the m_Intersection controller object.

3.3.8.3   PreEmpt Methods

Table 24   **PreEmpt class-methods**

Method	Type	Notes
PreEmpt ()	public:	Constructor.
~PreEmpt ()	private abstract:	Destructor.
onActivate ()	public: *int*	Event triggered by receipt of an activate signal from the emergency vehicle transponder.
onDeactivate ()	public: *int*	Event triggered by receipt of an deactivate signal from the emergency vehicle transponder.
onTimeout ()	public: *void*	Event triggered if a deactivate signal is not received after the timeout interval has elapsed.

3.3.8.4   *PreEmpt Behavioral Details*

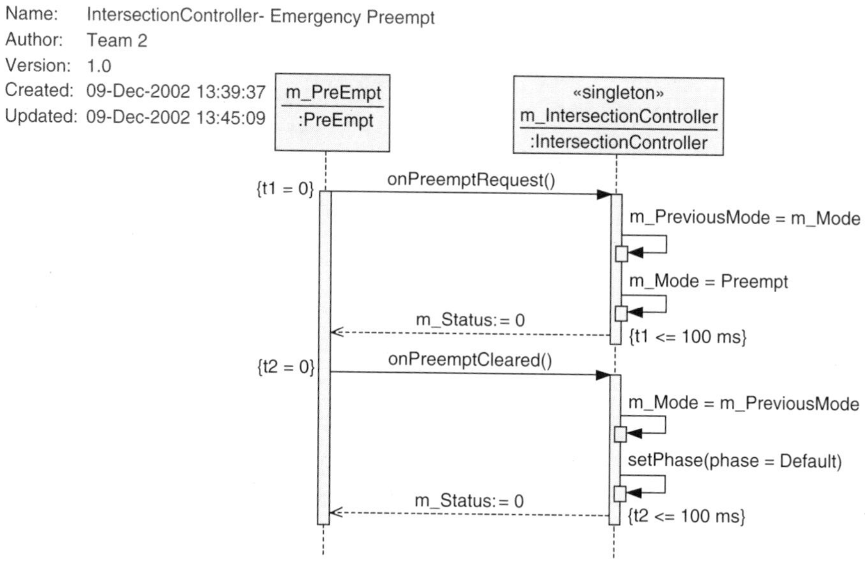

Name: IntersectionController- Emergency Preempt
Author: Team 2
Version: 1.0
Created: 09-Dec-2002 13:39:37
Updated: 09-Dec-2002 13:45:09

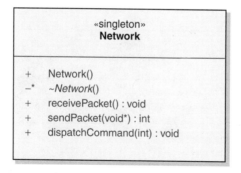

Figure 28   PreEmpt sequence diagram.

3.3.9   *Network*   This class manages communication via the Ethernet port.

Figure 29 below illustrates the attributes, methods, and events of the Network Interface class.

```
                    «singleton»
                     Network

     +      Network()
     −*     ~Network()
     +      receivePacket() : void
     +      sendPacket(void*) : int
     +      dispatchCommand(int) : void
```

Figure 29   Network class.

3.3.9.1   *Network Relationships*

- Association link from class *IntersectionController*
- Association link from class *Maintenance*
- Association link from class *RemoteOverride*

3.3.9.2 Network Methods

Table 25 Network class-methods

Method	Type	Notes
Network ()	public:	Constructor.
~ Network ()	private abstract:	Destructor.
receivePacket ()	public: *void*	Method responsible for receiving network SNMP packets.
sendPacket (*void**)	public: *int*	param: packet [void* - in] Method responsible for sending network SNMP packets.
dispatchCommand (*int*)	public: *void*	param: command [int - in] Interprets the received SNMP packet and invokes the appropriate method in response.

3.3.10 Maintenance This class provides a maintenance interface to the intersection controller, accessible either from the local maintenance Ethernet port or the DOT WAN.

The Maintenance class is responsible for managing the following functions:

1. Retrieval of database information.
2. Retrieval of current intersection controller status.

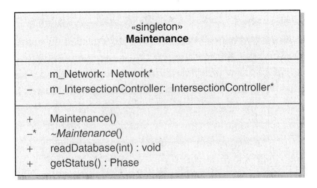

Figure 30 Maintenance class.

3.3.10.1 Maintenance Relationships

- Association link to class *IntersectionController*
- Association link to class *Network*

3.3.10.2 Maintenance Attributes

Table 26 Maintenance class-attributes

Attribute	Type	Notes
m_Network	private: *Network*	Pointer to the *m_Network* object.
m_IntersectionController	private: *IntersectionController*	Pointer to the *m_IntersectionController* object.

3.3.10.3 Maintenance Methods

Table 27 Maintenance class-methods

Method	Type	Notes
Maintenance ()	public:	Constructor.
~ Maintenance ()	private abstract:	Destructor.
readDatabase (*int*)	public: *void*	param: database [int - in] Method to read the contents of the database specified by the parameter *database*.
getStatus ()	public: *Phase*	Method to get the intersection status.

3.3.11 Database (Traffic History; Incident Log) Instances of this class are used to store the Traffic History and the Incident Log for the intersection being controlled. The Traffic History object is responsible for managing the following functions:

1. Storage and retrieval of traffic history database records.
2. Clearing of traffic history in response to a command from a remote host.

Figure 31 illustrates the attributes, methods, and events of the Traffic History class.

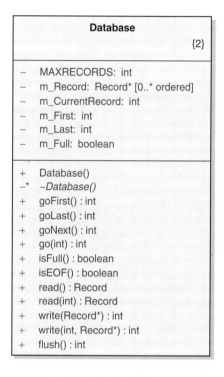

Figure 31 Database class.

3.3.11.1 Database Relationships

- Association link to class *Record*
- Association link from class *IntersectionController*
- Association link from class *IntersectionController*

3.3.11.2 Database Attributes

Table 28 Database class-attributes

Attribute	Type	Notes
MAXRECORDS	private: *int*	Constant defining the maximum number of records permitted.
m_Record	private: *Record*	Pointer to database records, which are of type Record.
m_CurrentRecord	private: *int*	Position (index) of current record.
m_First	private: *int*	Position (index) of first (least recent) record in FIFO database structure.

Table 28 (*continued*)

Attribute	Type	Notes
m_Last	private: *int*	Position (index) of last (most recent) record in FIFO database structure.
m_Full	private: *boolean*	True if data is being overwritten.

3.3.11.3   Database Methods

Table 29   Database class-methods

Method	Type	Notes
Database ()	public:	Constructor.
~ Database ()	private abstract:	Destructor.
goFirst ()	public: *int*	Move cursor to first (least recent) record. Completes in 40 ms.
goLast ()	public: *int*	Move cursor to last (most recent) record. Completes in 40 ms.
goNext ()	public: *int*	Move cursor to the next record. Completes in 40 ms.
go (*int*)	public: *int*	param: record [int - in] Move cursor to the specified record. Completes in 40 ms.
isFull ()	public: *boolean*	True if the database is full. Subsequent writes will overwrite oldest data (FIFO).
isEOF ()	public: *boolean*	True when the cursor is at the last record.
read ()	public: *Record*	Read record at current position. Completes in 10 ms.
read (*int*)	public: *Record*	param: position [int - in] Read record at specified position; update current record to specified position. Completes in 50 ms.
write (*Record**)	public: *int*	param: record [Record* - inout] Add new record to end of database. If isFull() is True, data will be overwritten. Completes in 50 ms.
write (*int, Record**)	public: *int*	param: position [int - in] param: record [Record* - inout]

(*continued*)

Table 29 (*continued*)

Method	Type	Notes
		Overwrite record at specified position; update current record to specified position. Completes in 50 ms.
flush ()	public: *int*	Clear all records by setting first and last logical record positions to zero; move cursor to first physical record position. Completes in 200 ms.

3.3.12 *Record*   This class defines the attributes and methods used by records contained in object instances of the Database class.

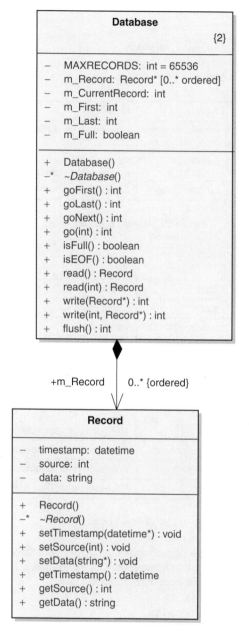

Figure 32   Record class.

3.3.12.1 Record Relationships

- Association link from class *Database*

3.3.12.2 Record Attributes

Attribute	Type	Notes
timestamp	private: *datetime*	Date and time of the incident or traffic history entry.
source	private: *int*	Integer value representing the object that is the source of the database record.
data	private: *string*	String of bytes containing the actual data.

Figure 33 Record class-attributes.

3.3.12.3 Record Methods

Method	Type	Notes
Record ()	public:	Constructor.
~Record ()	private abstract:	Destructor.
setTimestamp (*datetime*)	public: *void*	param: timestamp [datetime - inout] Mutator for m_Timestamp attribute.
setSource (*int*)	public: *void*	param: source [int - in]
setData (*string*)	public: *void*	param: data [string - inout] Mutator for m_Data attribute.
getTimestamp()	public: *datetime*	Accessor for m_Timestamp attribute.
getSource()	public: *int*	Accessor for m_Source attribute.
getData()	public: *string*	Accessor for m_Data attribute.

Figure 34 Record class-methods.

3.3.13 ErrorHandler This class handles all errors generated by the application. All errors are generated by the IntersectionController class, in response either to internal errors or error returns from method calls.

3.3.13.1 ErrorHandler Relationships

- Association link from class *IntersectionController*

3.3.13.2   *ErrorHandler Methods*

Method	Type	Notes
ErrorHandler()	public:	Constructor.
~ErrorHandler()	private abstract:	Destructor.
onNonCriticalError (*int*)	public: *void*	param: error [int - in] Logs the error incident and resumes normal operation.
onCriticalError (*int*)	public: *void*	param: error [int - in] Attempts to set the intersection to the default phase. If unsuccessful, attempts a reset. If this fails or the error occurs again immediately after reset, the watchdog timer will override software error handling. Logs the error and sends a network message to the DOT central office via the DOT WAN.

Figure 35   ErrorHandler class-methods.

3.3.13.3   *ErrorHandler Behavioral Details*

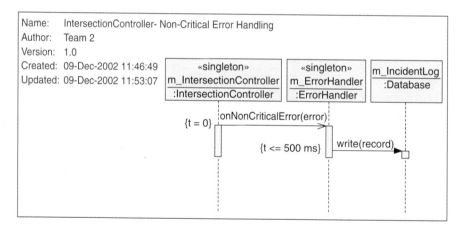

Figure 36   Noncritical error sequence diagram.

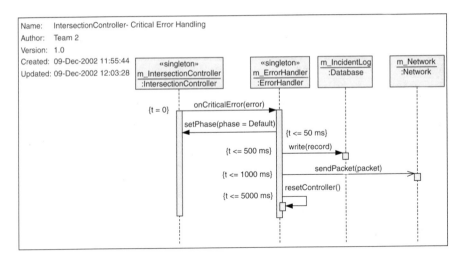

Name: IntersectionController- Critical Error Handling
Author: Team 2
Version: 1.0
Created: 09-Dec-2002 11:55:44
Updated: 09-Dec-2002 12:03:28

Figure 37   Critical error sequence diagram.

3.3.14 Support Classes   These comprise the structures and enumerated classes used to define attributes in the classes detailed above.

3.3.14.1   Split

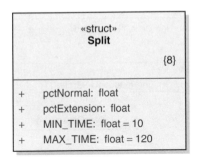

Figure 38   Split class.

Percentage of cycle time per phase. Comprises the nominal phase time plus the calculated extension due to traffic volume.

The values are determined as follows:

1. In FIXED mode, the nominal times are used (i.e., the extensions are set to zero).
2. In ACTUATED mode, the extensions contain fixed values at the start of each cycle. These values are modified in response to Vehicle Entry and Pedestrian Request events.
3. In ADAPTIVE mode, the extensions are updated prior to the start of each cycle as determined by the calculateParameters() method of the m_IntersectionController object.

3.3.14.1.1 SPLIT RELATIONSHIPS

* Association link from class *Parameters*

3.3.14.2 Parameters

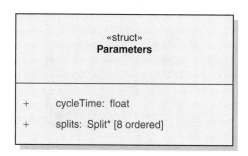

Figure 39 Parameters class.

3.3.14.2.1 PARAMETERS RELATIONSHIPS

* Association link from class *Status*
* Association link to class *Split*
* Association link from class *IntersectionController*

3.3.14.3 Status

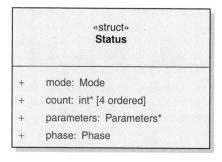

Figure 40 Status class.

3.3.14.3.1 STATUS RELATIONSHIPS

* Association link to class *Parameters*
* Association link to class *IntersectionController*
* Dependency link from class *RemoteOverride*
* Association link to class *Mode*
* Association link to class *Phase*

3.3.14.4   Phase

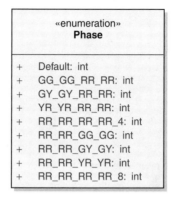

Figure 41   Phase class.

3.3.14.4.1   PHASE RELATIONSHIPS

- Association link from class *IntersectionController*
- Association link from class *Status*

3.3.14.5   Aspect

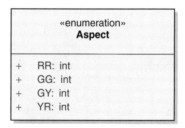

Figure 42   Aspect class.

3.3.14.5.1   ASPECT RELATIONSHIPS

- Association link from class *Approach*

3.3.14.6   Indication

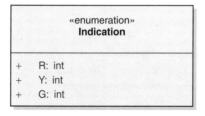

Figure 43   Indication class.

3.3.14.6.1 INDICATION RELATIONSHIPS

- Association link from class *IntersectionStandard*

3.3.14.7 Mode

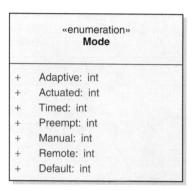

Figure 44 Mode class.

3.3.14.7.1 MODE RELATIONSHIPS

- Association link from class *Status*
- Association link from class *IntersectionController*

3.3.14.8 OverrideType

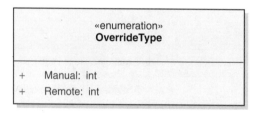

Figure 45 OverrideType class.

3.3.14.8.1 OVERRIDETYPE RELATIONSHIPS

- Dependency link from class *Override*

4 REQUIREMENTS TRACEABILITY

Table 30, Table 31, and Table 32 illustrate SDD compliance with the SRS requirements.

Table 30 **Architectural requirements**

Section Reference for SRS Requirement	SDD Section Demonstrating Compliance	Comments
2.5(10)	2.2.2	Application software to be written in high-order OO language; C++ selected
2.6(2)	2.2.1	Commercial RTOS
2.6(3)	2.2.3	Resource managers

Table 31 **Functional requirements**

Section Reference for SRS Requirement	SDD Section Demonstrating Compliance	Comments
2.6(1)	3.3.2.2	SI units; speed limit is in km/h
3.1.1 (Figure 3)	3.1, Figure 3	Use cases and collaborations
3.2, Figure 4	3.2; Figure 10	Class model
3.2.1	3.1, 3.3.1	Requirements for Intersection Controller class
3.2.2	3.1, 3.3.2	Requirements for Approach class
3.2.3	3.1, 3.3.3	Requirements for Pedestrian Traffic Standard class
3.2.4	3.1, 3.3.3	Requirements for Vehicle Traffic Standard class
3.2.5	3.1, 3.3.4	Requirements for Pedestrian Service Button class
3.2.6	3.1, 3.3.5	Requirements for Vehicle Presence Detector class
3.2.7	3.1, 3.3.6	Requirements for Manual Override class
3.2.8	3.1, 3.3.7	Requirements for Remote Override class
3.2.9	3.1, 3.3.7.4	Requirements for Emergency Vehicle Interface class
3.2.10	3.1, 3.3.8.4, 3.3.10	Requirements for Network Interface class
3.2.11	3.1, 3.3.11	Requirements for Traffic History class
3.2.12	3.1, 3.3.11	Requirements for Incident Log class

Table 32   Timing requirements

Section Reference for SRS Requirement	SDD Section Demonstrating Compliance	Comments
3.3.1.1, Table 14 (1)	3.3.1.4, Figure 13	Initialization
3.3.1.1, Table 14 (2)	3.3.1.4, Figure 13	Set Default Phase
3.3.1.1, Table 14 (3)	3.3.1.4, Figure 13	Start Normal Operation
3.3.1.1, Table 14 (4)	3.3.1.4, Figure 13	Advance Phase – Normal
3.3.1.1, Table 14 (5)	3.3.6.4, Figure 25	Advance Phase – Local
3.3.1.1, Table 14 (6)	3.3.6.4, Figure 25	Advance Phase – Remote
3.3.1.1, Table 14 (7)	3.3.1.4, Figure 13	Calculate Cycle Parameters – Actuated
3.3.1.1, Table 14 (8)	3.3.1.4, Figure 13	Calculate Cycle Parameters – Adaptive
3.3.1.1, Table 14 (9)	3.3.13.3, Figure 37	Critical Error – Display Defaults
3.3.1.1, Table 14 (10)	3.3.13.3, Figure 37	Critical Error – Alarm
3.3.1.1, Table 14 (11)	3.3.13.3, Figure 37	Critical Error – Reset
3.3.1.1, Table 14 (12)	3.3.13.3, Figure 37, Figure 36	Write Error Log
3.3.1.1, Table 14 (13)	3.3.1.4, Figure 13	Set Phase
3.3.1.1, Table 14 (14)	3.3.1.4, Figure 13	Get Phase
3.3.1.1, Table 14 (15)	3.3.1.4, Figure 13	Check Phase
3.3.1.1, Table 14 (16)	3.3.4.4, Figure 19	Pedestrian Request Latching
3.3.1.1, Table 14 (17)	3.3.4.4, Figure 19	Pedestrian Request Reset
3.3.1.1, Table 14 (18)	3.3.4.4, Figure 19	Pedestrian Request Processing
3.3.1.1, Table 14 (19)	3.3.5.4, Figure 22	Vehicle Entrance
3.3.1.1, Table 14 (20)	3.3.5.4, Figure 22	Vehicle Exit
3.3.1.1, Table 14 (21)	3.3.5.4, Figure 22	Vehicle Request Processing
3.3.1.1, Table 14 (22)	3.3.5.4, Figure 22	Vehicle Reset Request State

(*continued*)

Table 32 *(continued)*

Section Reference for SRS Requirement	SDD Section Demonstrating Compliance	Comments
3.3.1.1, Table 14 (23)	3.3.5.4, Figure 22	Vehicle Count Update
3.3.1.1, Table 14 (24)	3.3.1.4, Figure 13	Vehicle Count Fetch
3.3.1.1, Table 14 (25)	3.3.5.4, Figure 22	Vehicle Count Reset
3.3.1.1, Table 14 (26)	3.3.7.3	Get Cycle Parameters
3.3.1.1, Table 14 (27)	3.3.7.3	Update Cycle Parameters
3.3.1.1, Table 14 (28)	3.3.8.4, Figure 28	Process Message
3.3.1.1, Table 14 (29)	3.3.8.4, Figure 28	Process Command
3.3.1.1, Table 14 (30)	3.3.8.4, Figure 28	Process Message
3.3.1.1, Table 14 (13)	3.3.11.3, Table 29	Fetch Database
3.3.1.1, Table 14 (32)	3.3.11.3, Table 29	Add Record
3.3.1.1, Table 14 (33)	3.3.11.3, Table 29	Clear Database
3.3.1.1, Table 14 (34)	3.3.11.3, Table 29	Add Record
3.3.1.1, Table 14 (35)	3.3.11.3, Table 29	Clear Database

5.7 EXERCISES

5.1 Why is it that there is no one, universally accepted strategy for software design modeling?

5.2 How would you handle the situation in which the Software Requirements Specification contains numerous, if not excess, design specifications?

5.3 Who should you write the design specification?

5.4 What are the differences between object-oriented modeling and using data flow diagrams?

5.5 Using a data flow diagram, capture the data and functional requirements for monitoring the entry, exit, and traversal of planes in an airspace. Planes entering the space are sensed by the **Radar** input; the **Comm** input identifies planes that leave the space. The current contents of the space are maintained in the data area **AirspaceStatus**. A log or history of the space is kept in the **AirspaceLog** storage. An air traffic controller can request the display of the status of a particular plane through the **Controller}** input.

5.6 Why is it that the code, even though it is a model of behavior, is insufficient in serving as either a software requirements document or a software design document

5.7 Why is it important that the code be traceable to the Software Design Specification and in turn, to the Software Requirements Specification? What happens, or should happen, if it is not?

5.8 Redraw the inertial measurement system context diagram in Figure 4.14 to take into account calibration and diagnostic modes.

5.9 For each of the following systems

    (a) Inertial measurement unit

    (b) Nuclear monitoring system

    (c) Airline reservations system

    (d) Pasta sauce bottling system

    what design approach would you use?

6

PROGRAMMING LANGUAGES AND THE SOFTWARE PRODUCTION PROCESS

6.1 INTRODUCTION

Misuse of the underlying programming language can be the single greatest source of performance deterioration and missed deadlines in real-time systems. Moreover the increased use of object-oriented languages such as C++, Java, and sometimes Ada[1] in real-time systems can make the problem more insidious. But these languages are rapidly displacing the lower-level languages like C and assembly language in real-time embedded systems programming, and it is probably a good thing because of the benefits that accompany these languages.

A programming language represents the nexus of design and structure. Hence, because the actual "build" of software depends on tools to compile, generate binary code, link, and create binary objects, "coding" should take relatively little time if the design is solid. Nevertheless, coding (or programming) is more craftlike than mass production, and as with any craft, the best practitioners are known for the quality of their tools and their skill with them.

The main tool in the software production process is the language compiler. Real-time systems have been built with a wide range of programming languages,

Some of this chapter has been adapted from Phillip A. Laplante, *Software Engineering for Image Processing*, CRC Press, Boca Raton, FL, 2003.

[1] Ada has also been called "object based" to recognize is uniqueness as both object oriented and procedural.

Real-Time Systems Design and Analysis, By Phillip A. Laplante
ISBN 0-471-22855-9 © 2004 Institute of Electrical and Electronics Engineers

including various dialects of C, C++, C#, Java, Fortran, Pascal, Ada 95, assembly language, and even Visual BASIC, and Basic.

Of this list C++, Java, and C# are object-oriented, while the others are procedural. Ada 95 has elements of both object-oriented and procedural languages, and can be used either way, depending on the skill and preferences of the programmer. Applicative languages such as LISP and Prolog have also been used to build real-time systems [Allard91], but this is rather unusual and this discussion of languages is confined to those just listed.

An appropriate question is: "What is the fitness of a language for real-time applications and what criteria can be used to measure such fitness?" To answer this consider the following (informal) metrics [Cardelli96]:

- *Economy of Execution* How fast does a program run?
- *Economy of Compilation* How long does it take to go from sources to executables?
- *Economy of Small-Scale Development* How hard must an individual programmer work?
- *Economy of Large-Scale Development* How hard must a team of programmers work?
- *Economy of Language Features* How hard is it to learn or use a programming language?

Each programming language offers its own strengths and weaknesses with respect to real-time systems, and these criteria can be used to calibrate the features of a particular language for apples-to-apples comparison for a given application.

The focus of this chapter is on those language features that minimize the final code execution time and that lend themselves to performance prediction.

The compile time prediction of execution time performance is known as a schedulability analysis. In the design of modern real-time languages, the emphasis is on eliminating those constructs that render the language nonanalyzable, for example, unbounded recursion and unbounded while loops. Most so-called "real time languages" strive to eliminate all of these.

6.2 ASSEMBLY LANGUAGE

Although lacking most of the features of high-level languages, assembly language does have certain advantages for use in real-time systems, in that it provides more direct control of the computer hardware. Unfortunately, assembly language is unstructured and has limited abstraction properties. It varies widely from machine to machine. Coding in assembly language is usually difficult to learn, tedious, and error prone. Finally, the resulting code is nonportable. Therefore its use in embedded real-time systems is discouraged.

Until just a few years ago, the best programmers could generate assembly code that was more efficient than the code generated by a compiler. But with improvements in optimizing compilers, this is rarely the case. Thus, the need to write assembly code exists only in cases where the compiler does not support certain macroinstructions, or when the timing constraints are so tight that hand tuning is needed to produce optimal code. In any case, a system will likely find that 99% of the code, if not all, will be written in the high-order language.

In cases where complex prologues and epilogues are needed to prepare an assembly language program, often a shell of the program is written in the high-order language and compiled to an assembly file, which is then massaged to obtain the desired effect. Some languages, such as Ada[2] and versions of Pascal, provide a pragma pseudo-op, which allows for assembly code to be placed in-line with the high-order language code.

In terms of Cardelli's criteria, assembly languages have excellent economy of execution, and vacuously, of compilation because they are not compiled. Assembly languages, however, have poor economies of small- and large-scale development and of language features. Therefore, assembly language programming should be limited to use in very tight timing situations or in controlling hardware features that are not supported by the compiler.

6.3 PROCEDURAL LANGUAGES

Procedural languages such as C, Fortran, Basic, Ada 95, Modula-2, and Pascal, are those in which the action of the program is defined by a series of operations executed in sequence. These languages are characterized by facilities that allow for instructions to be grouped together into subprograms or procedures (modules). Appropriate structuring of the subprograms allow for achievement of desirable properties of the software (e.g., modularity, reliability, reuse).

There are several programming language features that stand out in procedural languages that are of interest in real-time systems, particularly:

- Versatile parameter passing mechanisms
- Dynamic memory allocation facilities
- Strong typing
- Abstract data typing
- Exception handling
- Modularity

These language features help promote the desirable properties of software and best real-time engineering practices.

---

[2] From here on, when referring to "Ada" it is assumed that "Ada 95" is meant, unless otherwise noted.

6.3.1   Parameter Passing Techniques

There are several methods of parameter passing, including the use of parameter lists and global variables. While each of these techniques has preferred uses, each has a different performance impact. Note that these parameter passing mechanisms are also found in object-oriented programming languages.

6.3.2   Call-by-Value and Call-by-Reference

The two most widely found parameter passing methods are call-by-value and call-by-reference.[3] In call-by-value parameter passing, the value of the actual parameter in the subroutine or function call is copied into the procedure's formal parameter. Since the procedure manipulates the formal parameter, the actual parameter is not altered. This technique is useful either when a test is being performed or the output is a function of the input parameters. For example, in passing accelerometer readings from the 10-ms cycle to the 40-ms cycle, the raw data need not be returned to the calling routine in changed form. When parameters are passed using call-by-value they are copied onto a run-time stack, at additional execution time cost.

In call-by-reference or call-by-address the address of the parameter is passed by the calling routine to the called procedure so that it can be altered there. Execution of a procedure using call-by-reference can take longer than one using call-by-value, since in call-by-reference indirect mode instructions are needed for any calculations involving the variables passed. However, in the case of passing large data structures such as buffers between procedures it is more desirable to use call-by-reference, since passing a pointer is more efficient than passing the data by byte.

Parameter lists are likely to promote modular design because the interfaces between the modules are clearly defined. Clearly defined interfaces can reduce the potential of untraceable corruption of data by procedures using global access. However, both call-by-value and call-by-reference parameter passing techniques can impact performance when the lists are long, since interrupts are frequently disabled during parameter passing to preserve the integrity of the data passed. Moreover, call-by-reference can introduce subtle function side effects, depending on the compiler.

Before deciding on a set of rules concerning parameter passing for optimum performance, it is advisable to construct a set of test cases that exercise different variations. These test cases need to be rerun every time the compiler, hardware, or application (because this can very the instruction mix) changes in order to update the rules.

---

[3] There are three historically interesting parameter passing mechanisms; call-by-constant, which was removed almost immediately from the Pascal language; call-by-value-result, which is used in Ada; and call-by-name, which was a mechanism peculiar to Algol-60.

6.3.3 Global Variables

Global variables are variables that are within the scope of all code. This usually means that references to these variables can be made in direct mode, and thus are faster than references to variables passed via parameter lists. For example, in many image processing applications, global arrays are defined to represent images, hence allowing costly parameter passing to be avoided.

Global variables are dangerous because references to them can be made by unauthorized code, thus introducing subtle faults. Even in languages like Fortran, where blocks of global variables can be defined via named COMMON declarations, access is still not well controlled. For this and other reasons, unwarranted use of global variables is to be avoided. Global parameter passing is only recommended when timing warrants, or if the use of parameter passing leads to convoluted code. In any case, the use of global variables must be clearly documented.

The decision to use one method of parameter passing or the other represents a trade-off between good software engineering practice and performance needs. For example, often timing constraints force the use of global parameter passing in instances when parameter lists would have been preferred for clarity and maintainability.

6.3.4 Recursion

Most programming languages provide recursion in that a procedure can either call itself or use itself in its construction. While recursion is elegant and is often necessary, its adverse impact on performance must be considered. Procedure calls require the allocation of storage on one or more stacks for the passing of parameters and for storage of local variables. The execution time needed for the allocation and deallocation, and for the storage of those parameters and local variables can be costly. In addition, recursion necessitates the use of a large number of expensive memory and register indirect instructions. Moreover, precautions need to be taken to ensure that the recursive routine will terminate, otherwise the run-time stack will overflow. The use of recursion often makes it impossible to determine the size of run-time memory requirements. Thus, iterative techniques such as while and for loops must be used where performance and determinism are crucial or in those languages that do not support recursion.

6.3.5 Dynamic Memory Allocation

The ability to dynamically allocate memory is important in the construction and maintenance of many data structures needed in real-time systems. While dynamic allocation can be time-consuming, it is necessary, especially in the construction of interrupt handlers, memory managers, and the like. Linked lists, trees, heaps and other dynamic data structures can benefit from the clarity and economy introduced by dynamic allocation. Furthermore, in cases where just a pointer is used to pass a data structure, the overhead for dynamic allocation can be quite reasonable. When writing real-time systems, however, care should be taken to ensure that the compiler will pass pointers to large data structures and not the data structure itself.

Languages that do not allow dynamic allocation of memory, for example, primitive high-order languages or assembly language require data structures of fixed size. While this may be faster, flexibility is sacrificed and memory requirements must be predetermined. Languages such as C, C++, Ada, and Java have dynamic allocation facilities, while BASIC and old versions of Fortran do not.

6.3.6 Typing

Typed languages require that each variable and constant be of a specific type (e.g., Boolean, integer, and real), and that each be declared as such before use. Strongly typed languages prohibit the mixing of different types in operations and assignments, and thus force the programmer to be precise about the way data are to be handled. Precise typing can prevent corruption of data through unwanted or unnecessary type conversion. Moreover, compiler type-checking is an important way to find errors at compile time, rather than at run time, when they are more costly to repair. Hence, strongly typed languages are desirable for real-time systems.

Generally, high-level languages provide integer and floating-point types, along with Boolean, character, and string types. In some cases, abstract data types are supported. These allow programmers to define their own types along with the associated operations. Use of abstract data types, however, may incur an execution time penalty, as complicated internal representations are often needed to support the abstraction.

Some languages are typed, but do not prohibit mixing of types in arithmetic operations. Since these languages generally perform mixed calculations using the type that has the highest storage complexity, they must promote all variables to that type. For example, in C, the following code fragment illustrates automatic promotion and demotion of variable types:

```
int  x,y;
float k,l,m;
  .
  .
j = x*k+m;
```

Here the variable x will be promoted to a float (real) type and then multiplication and addition will take place in floating point. Afterward, the result will be truncated and stored in j. The performance impact is that hidden promotion and more time-consuming arithmetic instructions can be generated, with no additional accuracy. In addition, accuracy can be lost due to the truncation, or worse, an integer overflow can occur if the floating-point value is larger than the allowable integer value. Programs written in languages that are weakly typed need to be scrutinized for such effects. Most C compilers can be tuned to catch type mismatches in function parameters, preventing unwanted type conversions.

6.3.7 Exception Handling

Certain languages provide facilities for dealing with errors or other anomalous conditions that arise during program execution. These conditions include the obvious, such as floating-point overflow, square root of a negative, divide-by-zero, and user-defined ones. The ability to define and handle exceptional conditions in the high-level language aids in the construction of interrupt handlers and other code used for real-time event processing. Moreover poor handling of exceptions can degrade performance. For example, floating-point overflow errors can propagate bad data through an algorithm and instigate time-consuming error-recovery routines.

For example, in ANSI-C, the raise and signal facilities are provided. A signal is a type of software interrupt handler that is used to react to an exception indicated by the `raise` operation. Both are provided as function calls, which are typically implemented as macros.

The following prototype can be used as the front end for an exception handler to react to signal S.

```
void (*signal (int S, void (*func) (int)))(int);
```

When signal S is set, function `func` is invoked. This function represents the interrupt handler. The prototype:

```
int raise (int S);
```

`raise` is used to invoke the process that reacts to signal S.

ANSI-C includes a number of predefined signals needed to handle anomalous conditions such as overflow, memory access violations and illegal instruction, but these signals can be replaced with user-defined ones. The following C code portrays a generic exception handler that reacts to a certain error condition.

```
#include <signal.h>

main ()
{
    void handler (int sig);
    :
    :
    signal (SIGINT, handler);   /*set up to handle SIGINT */
    :                           /*do some processing      */
    if (error) raise (SIGINT);  /*anomaly detected        */
    :                           /*continue processing     */
}

void handler (int sig)
{
                                /*handle error here       */
}
```

In C, the `signal` library function call is used to construct interrupt handlers to react to a signal from external hardware, and to handle certain traps, such as floating-point overflow, by replacing the standard C library handlers. This situation was illustrated in the case study in the Appendix of Chapter 2.

Of all the procedural languages discussed in this chapter, Ada has the most explicit exception handling facility. Exception handling in Ada looks somewhat different. Consider an Ada exception handler to determine whether a matrix is singular (its determinant is 0). Assume that a matrix type has been defined, and it can be determined that the matrix is singular. An associated code fragment might be:

```
begin
--
--calculate determinant
-- ...
--
    exception
    when SINGULAR : NUMERIC/ERROR => PUT ("MATRIX IS SINGULAR ");
    When others =>PUT ("FATAL Error");
    raise ERROR;
end;
```

Here the `exception` keyword is used to indicate that this is an exception handler and the `raise` keyword plays a role similar to that of `raise` in the C exception handler. The definition of `SINGULAR`, which represents a matrix whose determinant is zero, is defined elsewhere, such as in a header file. Finally, exception handling are explicit features of C++ and Java.

6.3.8 Modularity

Procedural languages that are amenable to the principle of information hiding tend to make it easy to construct high-integrity real-time systems. While C and Fortran both have mechanisms for this (procedures and subroutines), other languages such as Ada tend to foster more modular design because of the requirement to have clearly defined input and outputs in the module parameter lists.

In Ada 95 the notion of a package embodies the concept of Parnas information hiding exquisitely. The Ada package consists of a specification and declarations that include its public or visible interface and its invisible or private elements. In addition the package body, which has more externally invisible components, contains the working code of the package. Packages are separately compiliable entities, which further enhances their application as black boxes. In Fortran there is the notion of a `SUBROUTINE` and separate compilation of source files. These language features can be used to achieve modularity and design abstract data types. The C language also provides for separately compiled modules and other features that promote a rigorous top-down design approach that should lead to a good modular design.

While modular software is desirable, there is a price to pay in the overhead associated with procedure calls and parameter passing. This adverse effect should be considered when sizing modules.

6.3.9 Cardelli's Metrics and Procedural Languages

Taking the set of procedural languages as a whole, Cardelli considered them for use in real-time systems with respect to his criteria. His comments are paraphrased in the foregoing discussion. First, he notes that variable typing was introduced to improve code generation. This was particularly true in the case of Fortran. Hence, economy of execution is high for procedural languages provided the compiler is efficient. Further, because modules can be compiled independently compilation of large systems is efficient, at least when interfaces are stable. The more challenging aspects of system integration are thus eliminated.

Small-scale development is economical because type checking can catch many coding errors, reducing testing and debugging efforts. The errors that do occur are easier to debug, simply because large classes of other errors have been ruled out. Finally, experienced programmers adopt a coding style that causes some logical errors to show up as type checking errors hence they can use the type checker as a development tool. For example, changing the name of a type when its invariants change even though the type structure remains the same, yields error reports on all its old uses.

In addition, data abstraction and modularization have methodological advantages for large-scale code development. Large teams of programmers can negotiate the interfaces to be implemented, and then proceed separately to implement the corresponding pieces of code. Dependencies between pieces of code are minimized, and code can be locally rearranged without fear of global effects.

Finally, procedural languages are economical because some well-designed constructions can be naturally composed in orthogonal ways. For example, in C an array of arrays models two-dimensional arrays. Orthogonality of language features reduces the complexity of programming languages. The learning curve for programmers is thus reduced, and the relearning effort that is constantly necessary in using complex languages is minimized [Cardelli96].

6.4 OBJECT-ORIENTED LANGUAGES

The benefits of object-oriented techniques are well known, for example, increasing programmer efficiency, reliability, and the potential for reuse. Object-oriented languages include Smalltalk, C++, Java, C#, Eiffel, and Ada 95 when so used. Formally, object-oriented programming languages are those that support data abstraction, inheritance, polymorphism, and messaging.

Objects are an effective way to manage system complexity, as they provide a natural environment for information hiding, or protected variation and encapsulation. In encapsulation, a class of objects and methods associated with them are enclosed or encapsulated in class definitions. An object can utilize another object's encapsulated data only by sending a message to that object with the name of the method to apply. For example, consider the problem of sorting objects. A method may exist for sorting an object class of integers in ascending order. A class of people might be sorted by height. A class of objects that has an attribute of color might be sorted by that attribute. All of these objects have a comparison message method with different implementations. Therefore, if a client sends a message to compare one of these objects to another, the runtime code must resolve which method to apply dynamically – with some execution time penalty. This matter will be discussed shortly.

Object-oriented languages provide a natural environment for information hiding, for example, in image-processing systems it might be useful to define a class of type pixel, with attributes describing its position, color, and brightness; and operations that can be applied to a pixel such as add, activate, deactivate. It might also be desirable to define objects of type image as a collection of pixels with other attributes of width, height, and so on. In some cases, expression of system functionality is easier to do in an object-oriented manner.

6.4.1 Synchronizing Objects

Rather than extending classes through inheritance, in practice, it is often preferable to use composition. However, in doing so there is the need to support different synchronization policies for objects, due to different usage contexts. Specifically, consider the following common synchronization policies for objects:

Synchronized Objects   A synchronization object (e.g., mutex) is associated with an object that can be concurrently accessed by multiple threads. If internal locking is used, then on method entry each public method acquires a lock on the associated synchronization object and releases the lock on method exit. If external locking is used, then clients are responsible for acquiring a lock on the associated synchronization object before accessing the object and subsequently releasing the lock when finished.

Encapsulated Objects   When an object is encapsulated within another object (i.e., the encapsulated object is not accessible outside of the enclosing object), it is redundant to acquire a lock on the encapsulated object, since the lock of the enclosing object also protects the encapsulate object. Operations on encapsulated objects therefore require no synchronization.

Thread-Local Objects   Objects that are only accessed by a single thread require no synchronization.

Objects Migrating between Threads   In this policy, ownership of a migrating object is transferred between threads. When a thread transfers ownership of a migrating object it can no longer access it. When a thread receives ownership of a migrating object it is guaranteed to have exclusive access to it (i.e., the migrating object is local to the thread). Therefore, migrating objects require no synchronization. However, the transfer of ownership does require synchronization.

Immutable Objects   An immutable object's state can never be modified after it is instantiated. Therefore immutable objects require no synchronization when accessed by multiple threads since all accesses are read-only.

Unsynchronized Objects   Objects in a single-threaded program require no synchronization.

To illustrate the necessity of supporting parameterization of synchronization policies, consider a class library. A developer of a class library wants to ensure the widest possible audience for this library, so he or she makes all classes synchronized so that they can be used safely in both single-threaded and multithreaded applications. However, clients of the library whose applications are single-threaded are unduly penalized with the unnecessary execution overhead of synchronization that they do not need. Even multithreaded applications can be unduly penalized if the objects do not require synchronization (e.g., the object is thread-local). Therefore to promote reusability of a class library without sacrificing performance, classes in a library ideally would allow clients to select on a per-object basis which synchronization policy to use.

6.4.2   [4]Garbage Collection

Recall that garbage refers to allocated memory that is no longer being used but is not otherwise available. Excessive garbage accumulation can be detrimental and therefore garbage must be reclaimed. Garbage collection algorithms generally have unpredictable performance (although average performance may be known).

Garbage can be created in both procedural and object-oriented languages. For example, in Pascal or C, garbage can be created by allocating memory, but not deallocating it properly. But garbage is generally associated with object-oriented languages like C++ and Java. Java is noteworthy in that the standard environment incorporates garbage collection, whereas C++ does not. Therefore this discussion of garbage collection strategies centers on Java. It is an interesting case study in comparing several possible solutions to a real-time problem, involving dynamic memory management, recursion, and bounded performance.

Java garbage collection routines allocate objects in a memory area called the Java virtual machine (JVM) garbage-collection heap. Unlike traditional heaps for a process' virtual memory space, the garbage collector threads monitor objects in use. Once the garbage collection routines determine that an object is unreachable, the garbage collector automatically returns that object's memory to the system. Because the JVM specification does not specify how to free the memory, a JVM can implement one of several garbage collection strategies, including:

- Mark and sweep
- Generational collection
- Reference counting

---

[4] Much of this section is adapted from [Lo03] is adapted from [Lo03] with permission.

- Copying
- Mark-compact
- Stop-the-world garbage collection

Mark-and-sweep involves two phases, marking and sweeping. The marking phase marks reachable objects through either a depth- or breadth-first search on the object reference graph. A depth-first search scours one branch of the reference graph before proceeding to the next branch. Breadth-first searches go through only the first level of every branch, then return to search another level down.

Generational collectors divide objects into groups (young, old, and so on) based on their age. Because of infant mortality – the fact that many dynamically created objects die, that is, become garbage quickly – garbage collection for young-generation objects involves minimal marking effort. Garbage collection in long-lived objects in old generations, on the other hand, could still need full marking.

Reference counting tries to circumvent the nondeterministic marking time by amortizing time over modifications on the object reference graph. This technique associates each object with a reference count, which records the number of objects that can access the object. Once the reference count reaches zero, the corresponding object is garbage. Therefore, the time for reference counting is bounded (rather than nondeterministic) and is suitable for real-time applications. However, reference counting cannot handle cyclic structures like the other methods, which is a significant limitation.

Copy collectors trace garbage but do not mark it. Once a copying collector identifies live objects, it copies those objects to another memory space to achieve memory compaction, thus eliminating memory fragmentation. The original memory space becomes free automatically. After compaction, memory allocation is very fast. On the other hand, a noncopying collector simply maintains its garbage collecting data structure to reflect the new header configuration, which, however, can cause memory fragmentation. A copying collector typically requires a reference handle for each object. The handle decreases the effort involved in moving an object around, but consumes more memory space and access time.

Mark-compact garbage collection involves marking and compacting. The marking phase is similar to that of the mark-and-sweep algorithm. Once the marking phase finishes, live objects are moved side by side in the memory. The copying garbage collection, however, does not involve a marking phase. In the copying garbage collection scheme, the memory is divided into two spaces: a *from* space and a *to* space. The copying garbage collector recursively copies objects, and those referenced by the object, from the *from* space to the *to* space. Both mark-compact and copying algorithms achieve heap compaction through copying. Thus, memory allocation is extremely fast. However, the copying garbage collection has low memory utilization. The mark-compact yields good memory utilization, but takes some time in the marking phase.

Stop-the-world garbage collector suspends all program threads when garbage collector threads are running. This suspension is required to prevent the program threads from modifying the object reference graph. This is one way to synchronize program execution with garbage collection. However, the suspension could result in a long garbage collection pause with no progress in program execution. Typically, the shorter the garbage collection pause, the better the performance of garbage collection. Although stop-the-world garbage collection involves a garbage collection pause, it does not need a sophisticated synchronization mechanism. The mechanism that suspends the program threads when garbage collection threads are running tends to result in a smaller memory footprint. Stop-the-world garbage collection is not the only garbage collector to induce a garbage collection pause. For example, the copying collector must stop the program threads during copying, and the mark-compact algorithm requires suspending program threads in its marking and copying phases. Whether the stop-the-world approach is suitable for real-time systems depends on application requirements [Lo03].

Garbage collection algorithms generally have unpredictable performance (although average performance may be known). The loss of determinism results from the unknown amount of garbage, the tagging time of the nondeterministic data structures, and the fact that many incremental garbage collectors require that every memory allocation or deallocation from the heap be willing to service a page-fault trap handler.

6.4.3 Cardelli's Metrics and Object-Oriented Languages

Consider object-oriented languages in the context of Cardelli's metrics as paraphrased from his analysis. In terms of economy of execution, object-oriented style is intrinsically less efficient than procedural style. In pure object-oriented style, every routine is supposed to be a method. This introduces additional indirections through method tables and prevents optimizations such as inlining. The traditional solution to this problem (analyzing and compiling whole programs) violates modularity and is not applicable to libraries.

With respect to economy of compilation, often there is no distinction between the code and the interface of a class. Some object-oriented languages are not sufficiently modular and require recompilation of superclasses when compiling subclasses. Therefore, the time spent in compilation may grow disproportionately with the size of the system.

On the other hand, object-oriented languages are superior with respect to economy of small-scale development. For example, individual programmers can take advantage of class libraries and frameworks, drastically reducing their work load. When the project scope grows, however, programmers must be able to understand the details of those class libraries, and this turns out to be more difficult than understanding module libraries. The type systems of most object-oriented languages are not expressive enough; programmers must often resort to dynamic checking or to unsafe features, damaging the robustness of their programs.

In terms of economy of large-scale development, many developers are often involved in developing class libraries and specializing existing class libraries.

Although reuse is a benefit of object-oriented languages, it is also the case that these languages have extremely poor modularity properties with respect to class extension and modification via inheritance. For example, it is easy to override a method that should not be overridden, or to reimplement a class in a way that causes problems in subclasses. Other large-scale development problems include the confusion between classes and object types, which limits the construction of abstractions, and the fact that subtype polymorphism is not good enough for expressing container classes.

Object-oriented languages have low economy of language features. Smalltalk was originally intended as a language that would be easy to learn. C++ is based on a fairly simple model, inherited from Simula, but is overwhelming in the complexity of its many features. Unfortunately, what started as economical and uniform language ("everything is an object") ended up as a vast collection of class varieties. Java represents a step forward in the complexity trend, but is more complex than most people realize [Cardelli96].

6.4.4   Object-Oriented versus Procedural Languages

There is still no agreement on which is better for real-time systems – object-oriented or procedural languages. Even this author is somewhat conflicted. The benefit of an object-oriented approach to problem solving and the use of object-oriented languages are clear, and have already been described. Moreover, it is possible to imagine certain aspects of an operating system that would benefit from objectification, such as task, thread, file, or device, and certain application domains can clearly benefit from an object-oriented approach. The main arguments against object-oriented programming languages for real-time systems, however, are that they can lead to unpredictable and inefficient systems and that they are hard to optimize.

The unpredictability argument is hard to defend, however, at least with respect to object-oriented languages, such as C++, that do not use garbage collection. It is probably the case that a predictable system can be just as easily built in C++ as C. Similarly, it is probably just as easy to build an unpredictable system in C as in C++. The case for more unpredictable systems using object-oriented languages is probably easier to sustain when arguing about garbage collecting languages like Java.

In any case, the inefficiency argument against object-oriented languages is a powerful one. Generally, there is an execution time penalty in object-oriented languages in comparison to procedural languages. This "penalty" is due, in part, to late binding (resolution of memory locations at run-time rather than at compile time) necessitated by function polymorphism and inheritance. These represent considerable and often nondeterministic delay factors. Another problem results from the overhead of the garbage collection routines. One possible way to reduce these penalties is not to define too many classes and only define classes that contain coarse detail and high-level functionality.

Some anecdotal evidence illustrates that the use of object-oriented language for real-time systems presents other, more subtle, difficulties. Consider the following vignette. A design team for a particular real-time system used Unified Modeling Language (UML) diagrams and insisted that C++ be used to implement a fairly simple and straightforward requirements specification. After coding was complete, testing began. Although the system never failed, several users expressed the desire to add a few requirements that would cause the system to miss deadlines. A competing company that designed using a structured language met the new requirements by optimizing the macroinstructions. They could do this because of the close correspondence between the procedural code they had been written in C and the actual instructions. This option was not available to developers using C++.[5] Such cases are frequently cited to dispute the viability of object-oriented languages for real-time applications.

A more serious problem, perhaps, is the inheritance anomaly in object-oriented languages. The inheritance anomaly arises when an attempt is made to use inheritance as a code reuse mechanism, which does not preserve substitutability (i.e., the subclass is not a subtype). If the substitutability were preserved, then the anomaly would not occur. Since the use of inheritance for reuse has fallen out of favor in object-oriented approaches (in favor of composition), however, it seems that most inheritance anomaly rejections of object-oriented languages for real-time focus reflects an antiquated view of object orientation.

Consider the following example from an excellent text on real-time operating systems [Shaw01]:

```
BoundedBuffer
{
DEPOSIT
pre: not full

REMOVE
pre: not empty
}

MyBoundedBuffer extends BoundedBuffer
{
DEPOSIT
pre: not full

REMOVE
pre: not empty AND lastInvocationIsDeposit
}
```

Assuming that preconditions are checked and have "wait semantics" (i.e., wait for the precondition to become true), then clearly `MyBoundedBuffer` has strengthened the precondition of `BoundedBuffer`, and hence violated substitutability (and as such is a questionable use of inheritance).

Most opponents of object-oriented languages for real-time assert that concurrency and synchronization are poorly supported. However, when built-in language support for concurrency does not exist, it is standard practice to create "wrapper facade" classes to encapsulate system concurrency application program interface (APIs) for use in object-orientation (e.g., wrapper classes in C++ for POSIX

---

[5] Reported by one of the author's clients who prefers to remain anonymous.

threads). Further, it has already been noted that there are several concurrency patterns available for object-oriented real-time systems [Douglass03], [Schmidt00]. While concurrency may be poorly supported at the language level in practice, it is not an issue since developers use libraries instead.

In summary, critics of object-oriented languages for real-time systems (including, formerly, the author) seem fixated on Smalltalk and Java, ignoring C++. C++ is probably more suitable for real-time since among other things it does not have built-in garbage collection and class methods, and by default does not use "dynamic binding." In any case, there are no clear guidelines for where object-oriented approaches and languages should be used. Each situation needs to be considered separately.

6.5   BRIEF SURVEY OF LANGUAGES

For purposes of illustrating some of the aforementioned language properties it is helpful to review some of the more widely used languages in real-time systems. The languages are presented in alphabetical order, and not in any rank of endorsement.

Functional languages, such as LISP and ML, have been omitted from the discussion. This is not because they are useless in the context of real-time applications, but simply because their use in this setting is rare. The discussion also omits object-oriented scripting languages, which have become popular for writing tools, and test harnesses such as Python, Ruby, because they are generally not appropriate for embedded targets.

6.5.1   Ada 95

Ada was originally intended to be the mandatory language for all U.S. Department of Defense projects, which included a high percentage of embedded real-time systems. The first version, which became standardized in 1983, had significant problems. The programming language community had long been aware of the problems with the first release of the Ada standard, and practically since the first delivery of an Ada 83 compiler, had sought to resolve them, which resulted in a new version. The new language, now called "Ada 95," was the first internationally standardized object-oriented programming language. However, Ada's original intent has been consistently undermined by numerous exceptions that were granted, and it seems inevitable that Ada is not destined to fulfill its original intent.

Ada was intended to be designed specifically for embedded real-time systems, but systems builders have typically found the language to be too bulky and inefficient. Moreover, significant problems were found when trying to implement multitasking using the limited tools supplied by the language, such as the roundly criticized rendezvous mechanism.

Three pragmas were introduced in Ada 95 to resolve some of the uncertainty in scheduling, resource contention, and synchronization.

- A pragma that controls how tasks are dispatched.
- A pragma that controls the interaction between task scheduling.
- A pragma that controls the queuing policy of task/resource entry queues. First-in-first-out (FIFO) and priority queuing policies are available.

Other expansions to the language were intended to make Ada 95 an object-oriented language. These include:

- Tagged types
- Packages
- Protected units

Proper use of these constructs allows for the construction of objects that exhibit the four characteristics of object-oriented languages (abstract data typing, inheritance, polymorphism, and messaging).

However, as mentioned, Ada has never lived up to its promise of universality. Nevertheless, even though the number of available Ada developers continues to dwindle, the language is staging somewhat of a minicomeback, particularly because of the availability of open-source versions of Ada for Linux (Linux is an open-source derivative of the Unix operating system).

6.5.2 C

The C programming language, invented around 1971, is a good language for "low-level" programming. The reason for this is that it is descended from the language, BCPL (whose successor, C's parent, was "B"), which supported only one type, the machine word. Consequently, C supported machine-related objects like characters, bytes, bits, and addresses, which could be handled directly in high-level language. These entities can be manipulated to control interrupt controllers, CPU registers, and other hardware needed by a real-time system. Indeed, C is also often used as a high-level cross-platform assembly language.

C provides special variable types, such as `register`, `volatile`, `static`, and `constant`, which allows for control of code generation at the high-order language level. For example, declaring a variable as a register type indicates that it will be used frequently. This encourages the compiler to place such a declared variable in a register, which often results in smaller and faster programs. C supports call-by-value only, but call-by-reference can be implemented by passing a pointer to anything as a value.

Variables declared as type `volatile` are not optimized by the compiler. This is useful in handling memory-mapped I/O and other instances where the code should not be optimized.

Automatic coercion refers to the implicit casting of data types that sometimes occurs in C. For example, a float value can be assigned to an `int` variable, which can result in a loss of information. C provides functions, such as `printf`, that take a variable number of arguments. Although this is a convenient feature, it is impossible for the compiler to thoroughly type check the arguments, which means problems can mysteriously arise at runtime.

The C language provides for exception handling through the use of signals (see Section 3.5.6), and two other mechanisms, setjmp and longjmp, are provided to allow a procedure to return quickly from a deep level of nesting, a useful feature in procedures requiring an abort. The setjmp procedures call, which is really a macro (but often implemented as a function), saves environment information that can be used by a subsequent longjmp library function call. The longjmp call restores the program to the state at the time of the last setjmp call. For example, suppose a process is called to do some processing and error checking. If an error is detected, a longjmp can be used to transfer to the first statement after the setjmp.

Overall, the C language is good for embedded programming, because it provides for structure and flexibility without complex language restrictions.

6.5.3 C++

C++ is a hybrid object-oriented programming language that was originally implemented as a macro-extension of C. Today, C++ stands as a separately compiled language, although strictly speaking, C++ compilers should accept standard C code. C++ exhibits all characteristics of an object-oriented language and promotes better software engineering practice through encapsulation and better abstraction mechanisms than C.

C++ compilers implement a preprocessing stage that basically performs an intelligent search and replace on identifiers that have been declared using the #define or #typedef directives. Although most advocates of C++ discourage the use of the preprocessor, which was inherited from C, it is still widely used. Most of the processor definitions in C++ are stored in header files, which complement the actual source code files. The problem with the preprocessor approach is that it provides an easy way for programmers to inadvertently add unnecessary complexity to a program. An additional problem with the preprocessor approach is that it has weak type checking and validation.

Most developers agree that the misuse of pointers causes the majority of bugs in C/C++ programming. Previously C++ programmers used complex pointer arithmetic to create and maintain dynamic data structures, particular during string manipulation. Consequently, C++ programmers spent a lot of time hunting down complex bugs for simple string management. Today, however, standard libraries of dynamic data structures are available. For example the Standard Template Language (STL), is now a standard library of C++, and it has both a string and wstring data type for wide character strings. While these data types are not as flexible as, say, Java's String class, they neutralize any arguments against C++ based on string manipulation issues.

There are three types of complex data types in C++: classes, structures, and unions. C++ has no built-in support for text strings. The standard technique is to use null-terminated arrays of characters to represent strings.

C code is organized into functions, which are global subroutines accessible to a program. C++ added classes and class methods, which are functions that are connected to classes. C++ class methods are very similar to Java class methods. However, because C++ still supports C, there is nothing to discourage C++ programmers from using functions. This results in a mixture of function and method use that makes for confusing programs.

Multiple inheritance is a feature of C++ that allows a class to be derived from multiple parent classes. Although multiple inheritance is indeed powerful, it is difficult to use correctly and causes many problems otherwise. It is also very complicated to implement from the compiler perspective. The unwanted goto statement has all but disappeared in C++, although technically it is still a part of the language.

Significantly, more embedded systems are being constructed in C++ and many practitioners are asking, "Should I implement the system in C or C++?" The answer is always "it depends." Choosing C in lieu of C++ in real-time embedded applications is, roughly speaking, a trade-off between a "lean and mean" C program that will be faster and easier to predict but harder to maintain, and a C++ program that will be slower and unpredictable but potentially easier to maintain.

C++ still allows for low-level control, for example, it can use inline methods rather than a runtime call. This kind of implementation is not completely abstract, nor completely low-level, but is acceptable in embedded environments.

To its detriment, there is some tendency to take existing C code and objectify it by wrapping the procedural code into objects with little regard for the best practices of object-orientation. This kind of approach is to be avoided because it has the potential to incorporate all of the disadvantages of C++ and none of the benefits. C++ also does not provide automatic garbage collection, which means dynamic memory must be managed manually or garbage collection must be home grown.

6.5.4 C#

C#[6] is a C++-like language that, along with its operating environment, has similarities to the JVM and Java, respectively. C# is associated with Microsoft's. NET Framework for scaled-down operating systems like Windows CE 3.0. Windows CE 3.0 is highly configurable, capable of scaling from small, embedded system footprints (350 KB) and upwards (e.g., for systems requiring user interface support). The minimum kernel configuration provides basic networking support, thread management, dynamic link library support, and virtual memory management. While a detailed discussion is outside the scope of this text, it is clear that Windows CE 3.0 was intended as a real-time operating system for the. NET Platform.

C# supports "unsafe code," allowing pointers to refer to specific memory locations. Objects referenced by pointers must be "pinned," disallowing the garbage

---

[6] Much of this discussion is adapted from [Lutz03].

collector from altering their location in memory. The garbage collector collects pinned objects, it just does not move them. This capability would tend to increase schedulability, and it also allows for DMA device access to write to specific memory locations, a necessary capability in embedded real-time systems. .NET currently offers a generational approach to garbage collection intended to minimize thread blockage during mark and sweep. For instance, a means to create a thread at a particular instant, and guarantee the thread completes by a particular point in time, is not supported. C# supports many thread synchronization mechanisms, but none with this level of precision. C# supports an array of thread synchronization constructs: (1) lock, (2) monitor, (3) mutex, and (4) interlock. A Lock is semantically identical to a critical section – a code segment guaranteeing entry into itself by only one thread at a time. Lock is shorthand notation for the monitor class type. A mutex is semantically equivalent to a lock, with the additional capability of working across process spaces. The downside to mutexes is their performance cost. Finally, interlock, a set of overloaded static methods, is used to increment and decrement numerics in a thread safe manner in order to implement the priority inheritance protocol.

Timers that are similar in functionality to the existing Win32 timer exist in C#. When constructed, timers are told how long to wait in milliseconds before their first invocation, and are also supplied an interval, again in milliseconds, specifying the period between subsequent invocations. The accuracy of these timers is machine dependent, and not guaranteed, reducing their usefulness for real-time systems.

C# and the. NET platform may not be appropriate for hard real-time systems for several reasons, including the unbounded execution of its garbage collected environment and its lack of threading constructs to adequately support schedulability and determinism. It is possible that C# and. NET might be appropriate for some soft and firm real-time systems, provided that these shortcomings are not critical.

Indeed, C#'s ability to interact with operating system API's, shield developers from complex memory-management logic, and floating-point performance approaching C make it a platform that is possibly appropriate for certain soft and even firm applications. However, disciplined programming would be required [Lutz03].

6.5.5 Fortran

The Fortran[7] language is the oldest high-order language used in modern real-time systems (developed circa 1955). Because in its earlier versions it lacked recursion and dynamic allocation facilities, embedded systems written in this language typically included a large portion of assembly language code to handle interrupts and scheduling, and communication with external devices was through the use of memory-mapped I/O, direct memory access (DMA), and I/O instructions.

---

[7] Although Fortran is an acronym for Formula Translator, it is often written as "Fortran" because the word has entered the mainstream in the same way that the acronyms "laser" and "sonar" have.

Later versions of the language included such features as reentrant code, but even today, an embedded Fortran system requires some assembly language code to accompany it.

Fortran was developed in an era when efficient code was essential to optimizing performance in small, slow machines. As a result, the language constructs were selected for efficiency, and early Fortran code generators were unusually so.[8]

To its detriment, Fortran is weakly typed, but because of the subroutine construct and the if-then-else construct, it can be used to design highly structured code. Fortran has no built-in exception handling or abstract data types. Fortran is still used in many legacy and even new real-time applications.

6.5.6 Java

Java is an interpreted language, that is, the code compiles into machine-independent code that runs in a managed execution environment. This environment is a virtual machine (Figure 6.1), which executes "object" code instructions as a series of program directives. The advantage of this arrangement is that the Java code can run on any device that implements the virtual machine. This "write once, run anywhere" philosophy has important applications in embedded and portable computing, such as in cell phones, and smart cards, as well as in Web-based computing.

There are also native-code Java compilers, which allow Java to run directly "on the bare metal," that is, the compilers convert Java directly to assembly code or object code. For example, beginning with Java 2 (JDK 1.2 and later), Java Virtual Machines support "Hot Spot" compilers that compile into machine code for several standard architectures. And there are Java microprocessors which directly execute Java byte code in hardware.

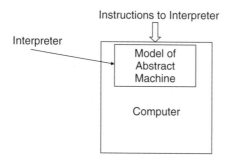

Figure 6.1 The Java interpreter as a model of a virtual machine.

---

[8] Sheldon Best is credited by Fortran inventor John Backus with introducing the notion of registerization of frequently used variables. This optimization approach is the basis of RISC computing and of real-time performance enhancements. In the 1980s the author briefly worked alongside Best, who was still writing state-of-the-art compilers for avionics computers.

Java is an object-oriented language that looks very similar to C++. Like C, Java supports call-by-value, but call-by-reference can be simulated, which will be discussed shortly.

Java does provide a preprocessor. Constant data members are used in place of the `#define` directive, and class definitions are used in lieu of the `#typedef` directive. The result is that Java source code is much more consistent and easier to read than C++ source code. The Java compiler builds class definitions directly from the source-code files, which contain both class definitions and method implementations. However, there are performance penalties for the resultant portability.

The Java language does not support pointers, however, it provides similar functionality via references. Java passes all arrays and objects by reference, which prevents common errors due to pointer mismanagement. The lack of pointers might seem to preclude implementation of data structures, such as dynamic arrays. However, any pointer task can be carried out with references, with the safety provided by the Java run-time system, such as boundary checking on all array indexing operations – all with performance penalty.

Java only implements one complex data type: classes. Java programmers use classes when the functionality of structures and unions is desired. This consistency comes at the cost of increased execution time over simple data structures.

Java does not support stand-alone functions. Instead Java requires programmers to bundle all routines into class methods again with significant cost.

Java has no direct support for multiple inheritance. Interfaces, however, allow for implementation of multiple inheritance. Java interfaces provide object method descriptions, but contain no implementations.

In Java, strings are implemented as first-class objects (`String` and `StringBuffer`), meaning that they are at the core of the Java language. Java's implementation of strings as objects provides several advantages. First, string creation and access is consistent across all systems. Next, because the Java string classes are defined as part of the Java language strings function predictably every time. Finally, the Java string classes perform extensive run-time checking, which helps eliminate errors. But all of this increases execution time.

The Java language specifies `goto` as a keyword, but its usage is not supported. Instead it provides labeled `breaks`, which are a more controlled form of `goto`.

Operator overloading is not supported in Java. However, in Java's string class, "+" represents concatenation of strings as well as numeric addition.

Java does not support automatic coercions. In Java, if a coercion will result in a loss of data, then it is necessary to explicitly cast the data element to the new type. Java does have implicit "upcasting." However, any instance can be upcast to `Object`, which is the parent class for all objects. Downcasting is explicit, and requires a cast. This explicitness is important to prevent hidden loss of precision.

The command-line arguments passed from the system into a Java program differ from the command-line arguments passed into a C++ program. In C and C++, the system passes two arguments to a program: `argc` and `argv`. `argc` specifies the number of arguments stored in `argv`, and `argv` is a pointer to an array of characters containing the actual arguments. In Java, the system passes a single value to a program: `args`. `args` is an array of strings that contains the command-line arguments.

6.5.6.1 *Real-Time Java*

In addition to the unpredictable performance of garbage collection, the Java specification provides only broad guidance for scheduling. For example, when there is competition for processing resources, threads with higher priority are generally executed in preference to threads with lower priority. This preference is not, however, a guarantee that the highest priority thread will always be running, and thread priorities cannot be used to reliably implement mutual exclusion. It was soon recognized that this and other shortcomings rendered standard Java inadequate for real-time systems.

In response to this problem, a National Institute of Standards and Technology (NIST) task force was charged with developing a version of Java that was suitable for embedded real-time applications. The final workshop report, published in September 1999, defines nine core requirements for the real-time specification for Java (RTSJ):

1. The specification must include a framework for the lookup and discovery of available profiles.

2. Any garbage collection that is provided shall have a bounded preemption latency.

3. The specification must define the relationships among real-time Java threads at the same level of detail as is currently available in existing standards documents.

4. The specification must include APIs to allow communication and synchronization between Java and non-Java tasks.

5. The specification must include handling of both internal and external asynchronous events.

6. The specification must include some form of asynchronous thread termination.

7. The core must provide mechanisms for enforcing mutual exclusion without blocking.

8. The specification must provide a mechanism to allow code to query whether it is running under a real-time Java thread or a non-real-time Java thread.

9. The specification must define the relationships that exist between real-time Java and non-real-time Java threads.

The RTSJ satisfies all but the first requirement, which was considered irrelevant because access to physical memory is not part of the NIST requirements, but industry input led the group to include it [Bollella00a].

6.5.6.2 Implementation of Real-Time Java

The RTSJ defines the real-time thread (RT) class to create threads, which the resident scheduler executes.[9] RT threads can access objects on the heap, and therefore can incur delays because of garbage collection.

For garbage collection, the RTSJ extends the memory model to support memory management in a way that does not interfere with the real-time code's ability to provide deterministic behavior. These extensions allow both short- and long-lived objects to be allocated outside the garbage-collection heap. There is also sufficient flexibility to use familiar solutions, such as preallocated object pools.

RTSJ uses "priority" somewhat more loosely than is traditionally accepted. "Highest priority thread" merely indicates the most eligible thread – the thread that the scheduler would choose from among all the threads ready to run. It does not necessarily presume a strict priority-based dispatch mechanism.

The system must queue all threads waiting to acquire a resource in priority order. These resources include the processor as well as synchronized blocks. If the active scheduling policy permits threads with the same priority, the threads are queued FIFO. Specifically, the system (1) orders waiting threads to enter synchronized blocks in a priority queue; (2) adds a blocked thread that becomes ready to run to the end of the run-ready queue for that priority; (3) adds a thread whose priority is explicitly set by itself or another thread to the end of the run ready queue for the new priority; and (4) places a thread that performs a yield to the end of its priority queue. The Priority Inheritance Protocol is implemented by default. The specification also provides a mechanism by which systemwide default policy can be implemented.

The asynchronous event facility comprises two classes: `AsyncEvent` and `AsyncEventHandler`. An `AsyncEvent` object represents something that can happen – like a Posix signal or hardware interrupt – or it represents a computed event – like an airplane entering a specified region. When one of these events occurs, indicated by the `fire()` method being called, the system schedules associated `AsyncEventHandlers`. An `AsyncEvent` manages two things: the dispatching of handlers when the event is fired, and the set of handlers associated with the event. The application can query this set and add or remove handlers. An `AsyncEventHandler` is a schedulable object roughly similar to a thread. When the event fires, the system invokes `run()` methods of the associated handlers.

---

[9] Most of this discussion has been adapted from [Bollela00a].

Unlike other runnable objects, however, an `AsyncEventHandler` has associated scheduling, release, and memory parameters that control the actual execution read or write.

Asynchronous control transfer allows for identification of particular methods by declaring them to throw an `AsynchronouslyInterrupted Exception` (AIE). When such a method is running at the top of a thread's execution stack and the system calls `java.lang.Thread.interrupt()` on the thread, the method will immediately act as if the system had thrown an AIE. If the system calls an interrupt on a thread that is not executing such a method, the system will set the AIE to a pending state for the thread and will throw it the next time control passes to such a method, either by calling it or returning to it. The system also sets the AIE's state to pending while control is in, returns to, or enters synchronized blocks.

The RTSJ defines two classes for programmers who want to access physical memory directly from Java code. The first class, `RawMemoryAccess`, defines methods that let you build an object representing a range of physical addresses and then access the physical memory with `byte`, `word`, `long`, and multiple `byte` granularity. The RTSJ implies no semantics other than the set and get methods. The second class, `PhysicalMemory`, allows the construction of a `PhysicalMemoryArea` object that represents a range of physical memory addresses where the system can locate Java objects. For example, a new Java object in a particular `PhysicalMemory` object can be built using either the `newInstance()` or `newArray()` methods. An instance of `RawMemoryAccess` models a raw storage area as a fixed-size sequence of bytes. Factory methods allow for the creation of `RawMemoryAccess` objects from memory at a particular address range or using a particular memory type. The implementation must provide and set a factory method that interprets these requests accordingly. A full complement of get and set methods lets the system access the physical memory area's contents through offsets from the base – interpreted as `byte`, `short`, `int`, or `long` data values – and copy them to or from `byte`, `short`, `int`, or `long` arrays.

6.5.7 Occam 2

Occam 2 is a language based on the communicating sequential processes (CSP) formalism. The name derives from English philosopher, William of Occam, who propounded Occam's Razor, that is, the fewer assumptions an explanation of a phenomenon depends on, the better it is.

The basic entity in Occam 2 is the process, of which there are four fundamental types, assignment, input, output, and wait. More complex processes are constructed from these by specifying sequential or parallel execution or by associating a process with an input from a channel. The process whose channel inputs first is executed. The branch constructor is IF with a list of conditions and associated processes. The process executed is the one with the first true condition in textual order. There is no operator precedence.

The Occam 2 language was designed to support concurrency on transputers, but compilers are available for other architectures. It has found some practical implementation in the UK.

6.5.8   Special Real-Time Languages

A number of other specialized languages for real-time have appeared and disappeared over the last 30 years to handle real-time applications. Briefly, these include:

- *PEARL*   The Process and Experiment Automation Real-time Language developed in the early 1970s by a group of German researchers. PEARL uses the augmentation strategy and has fairly wide application in Germany, especially in industrial controls settings.
- *Real-Time Euclid*   An experimental language that enjoys the distinction of being one of the only languages to be completely suited for schedulability analysis. This is achieved through language restriction. Unfortunately, the language never found its way into mainstream application.
- *Real-Time C*   Actually a generic name for any of a variety of C macroextension packages. These macroextensions typically provide timing and control constructs that are not found in standard C.
- *Real-Time C++*   A generic name for one of several object-class libraries specifically developed for C++. These libraries augment standard C++ to provide an increased level of timing and control.

There are, of course, many other real-time languages/operating environments, including MACH, Eiffel, MARUTI, ESTEREL. Many of these languages are used for highly specialized applications or in research only.

6.5.9   Know the Compiler and Rules of Thumb

Understanding the mapping between high-order language source and assembly language translation for a particular compiler is essential in generating code that is optimal in either execution time or memory utilization. The easiest and most reliable way to learn about any compiler is to run a series of tests on specific language constructs. For example, in many compilers the case statement is efficient only if more than three cases are to be compared, otherwise nested if statements should be used. Sometimes the code generated for a case statement can be quite convoluted, for example, a jump through a register, offset by the table value. This process can be time-consuming.

It has already been mentioned that procedure calls are costly in terms of the passing of parameters via the stack. The software engineer should determine whether the compiler passes the parameters by byte or by word.

Other language constructs that may need to be considered include:

- Use of while loops versus for loops or do-while loops.
- When to "unroll" loops, that is, to replace the looping construct with repetitive code (thus saving the loop overhead as well as providing the compiler the opportunity to use faster, direct, or single indirect mode instructions).
- Comparison of variable types and their uses (e.g., when to use short integer in C versus Boolean, when to use single precision versus double precision floating point, and so forth).
- Use of in-line expansion of code via macros versus procedure calls.

This is, by no means, an exhaustive list.

While good compilers should provide optimization of the assembly language code output so as to, in many cases, make the decisions just listed, it is important to discover what that optimization is doing to produce the resultant code. For example, compiler output can be affected by optimization for speed, memory and register usage, jumps, and so on, which can lead to inefficient code, timing problems, or critical regions. Thus, real-time systems engineers must be masters of their compilers. That is, at all times the engineer must know what assembly language code will be output for a given high-order language statement. A full understanding of each compiler can only be accomplished by developing a set of test cases to exercise it. The conclusions suggested by these tests can be included in the set of coding standards to foster improved use of the language and, ultimately, improved system performance.

When building real-time systems, no matter which language, bear in mind these rules of thumb:

- Avoid recursion (and other nondeterministic constructs where possible).
- Avoid unbounded while loops and other temporally unbounded structures.
- Avoid priority inversion situations.
- Avoid overengineering/gold-plating.
- Know your compiler!

6.6 CODING STANDARDS

Coding standards are different from language standards. A language standard, for example, ANSI C, embodies the syntactic rules of the language. A program violating those rules will be rejected by the compiler. Conversely, a coding standard is a set of stylistic conventions. Violating the conventions will not lead to compiler rejection. In another sense, compliance with language standards is mandatory, while compliance with coding standards is voluntary.

Adhering to language standards fosters portability across different compilers and, hence, hardware environments. Complying with coding standards will not foster portability, but rather in many cases, readability and maintainability. Some even contend that the use of coding standards can increase reliability. Coding standards may also be used to foster improved performance by encouraging or mandating the use of language constructs that are known to generate more efficient code. Many agile methodologies, for example, eXtreme Programming, embrace coding standards.

Coding standards involve standardizing some or all of the following elements of programming language use:

- Header format.
- Frequency, length, and style of comments.
- Naming of classes, methods, procedures, variable names, data, file names, and so forth.
- Formatting of program source code, including use of white space and indentation.
- Size limitations on code units, including maximum and minimum lines of code, and number of methods.
- Rules about the choice of language construct to be used; for example, when to use `case` statements instead of nested `if-then-else` statements.

While it is unclear if conforming to these rules fosters improvement in reliability, clearly close adherence can make programs easier to read and understand and likely more reusable and maintainable.

There are many different standards for coding that are language independent, or language specific. Coding standards can be teamwide, companywide, user-group specific (for example, the Gnu software group has standards for C and C++), or customers can require conformance to a specific standard that they own. Still other standards have come into the public domain. One example is the Hungarian notation standard, named in honor of Charles Simonyi, who is credited with first promulgating its use. Hungarian notation is a public domain standard intended to be used with object-oriented languages, particularly C++. The standard uses a complex naming scheme to embed type information about the objects, methods, attributes, and variables in the name. Because the standard essentially provides a set of rules about naming variables, it can be and has been used with other languages, such as C++, Ada, Java, and even C. Another example is in Java, which, by convention, uses all uppercase for constants such as PI and E. Further, some classes use a trailing underscore to distinguish an attribute like x_ from a method like x().

One problem with standards like the Hungarian notation is that they can create mangled variable names, in that they direct focus on how to name in Hungarian rather than a meaningful name of the variable for its use in code. In other words, the desire to conform to the standard may not result in a particularly meaningful variable name. Another problem is that the very strength of a coding standard can be its own undoing. For example, in Hungarian notation what if the type information embedded in the object name is, in fact, wrong? There is no way for a compiler to check this. There are commercial rules wizards, reminiscent of lint, that can be tuned to enforce the coding standards, but they must be programmed to work in conjunction with the compiler.

Finally, adoption of coding standards is not recommended midproject. It is much easier to start conforming than to be required to change existing code to comply. The decision to use coding standards is an organizational one that requires significant forethought and debate.

6.7 EXERCISES

6.1 Which of the languages discussed in this chapter provide for some sort of goto statement? Does the goto statement affect performance? If so, how?

6.2 It can be argued that in some cases there exists an apparent conflict between good software engineering techniques and real-time performance. Consider the relative merits of recursive program design versus interactive techniques, and the use of global variables versus parameter lists. Using these topics and an appropriate programming language for examples, compare and contrast real-time performance versus good software engineering practices as you understand them.

6.3 What other compiler options are available for your compiler and what do they do?

6.4 In the object-oriented language of your choice, design and code an "image" class that might be useful across a wide range of projects. Be sure to follow the best principles of object-oriented design.

6.5 In a procedural language of your choice develop an abstract data type called "image" with associated functions. Be sure to follow the principle of information hiding.

6.6 Write a set of coding standards for use with any of the real-time applications introduced in Chapter 1 for the programming language of your choice. Document the rationale for each provision of the coding standard.

6.7 Develop a set of tests to exercise a compiler to determine the best use of the language in a real-time processing environment. For example, your tests should determine such things as when to use case statements versus nested if-then-else statements; when to use integers versus Boolean variables for conditional branching; whether to use while or for loops, and when; and so on.

6.8 How can misuse or misunderstanding of a software technology impede a software project? For example, writing structured C code instead of classes in C++, or reinventing a tool for each project instead of using a standard one.

6.9 Compare how Ada95 and Java handle the goto statement. What does this indicate about the design principles or philosophy of each language?

6.10 Java has been compared to Ada95 in terms of hype and "unification" – defend or refute the arguments against this.

6.11 Are there language features that are exclusive to C/C++? Do these features provide any advantage or disadvantage in embedded environments?

6.12 What programming restrictions should be used in a programming language to permit the analysis of real-time applications?

7

PERFORMANCE ANALYSIS
AND OPTIMIZATION

7.1 THEORETICAL PRELIMINARIES

Of all the places where theory and practice never seem to coincide, none is more obvious than in performance analysis. For all the well-written and well-meaning research on real-time performance analysis, those that have built real systems know that practical reality has the annoying habit of getting in the way of theoretical results. Neat little formulas that ignore resource contention, use theoretically artificial hardware, or have made the assumption of zero context switch time are good as abstract art, but of little practical use. These observations, however, do not mean that theoretical analysis is useless or that there are no useful theoretical results. It only means that there are far less realistic, cookbook approaches than might be desired.

7.1.1 NP-Completeness

The complexity class P is the class of problems that can be solved by an algorithm that runs in polynomial time on a deterministic machine. The complexity class NP is the class of all problems that cannot be solved in polynomial time by a deterministic machine, although a candidate solution can be verified to be correct by a polynomial time algorithm. A decision or recognition problem is NP-complete if it is in the class NP and all other problems in NP are polynomial

Some of this chapter has been adapted from Phillip A. Laplante, *Software Engineering for Image Processing*, CRC Press, Boca Raton, FL, 2003.

Real-Time Systems Design and Analysis, By Phillip A. Laplante
ISBN 0-471-22855-9 © 2004 Institute of Electrical and Electronics Engineers

transformable to it. A problem is NP-hard if all problems in NP are polynomial transformable to that problem, but it hasn't been shown that the problem is in the class NP.

The Boolean Satisfiability Problem, for example, which arose during requirements consistency checking in Chapter 4 is NP-complete. NP-complete problems tend to be those relating to resource allocation, which is exactly the situation that occurs in real-time scheduling. This fact does not bode well for the solution of real-time scheduling problems.

7.1.2   Challenges in Analyzing Real-Time Systems

The challenges in finding workable solutions for real-time scheduling problems can be seen in more than 30 years of real-time systems research. Unfortunately most important problems in real-time scheduling require either excessive practical constraints to be solved or are NP-complete or NP-hard. Here is a sampling from the literature as summarized in [Stankovic95].

1. When there are mutual exclusion constraints, it is impossible to find a totally on-line optimal run-time scheduler.

2. The problem of deciding whether it is possible to schedule a set of periodic processes that use semaphores only to enforce mutual exclusion is NP-hard.

3. The multiprocessor scheduling problem with two processors, no resources, arbitrary partial-order relations, and every task having unit computation time is polynomial. A partial-order relation indicates that any process can call itself (reflexivity), if process A calls process B, then the reverse is not possible (antisymmetry), and if process A calls process B and process B calls process C, than process A can call process C (transitivity).

4. The multiprocessor scheduling problem with two processors, no resources, independent tasks, and arbitrary computation times is NP-complete.

5. The multiprocessor scheduling problem with two processors, no resources, independent tasks, arbitrary partial order, and task computation times of either 1 or 2 units of time is NP-complete.

6. The multiprocessor scheduling problem with two processors, one resource, a forest partial order (partial order on each processor), and each computation time of every task equal to 1 is NP-complete.

7. The multiprocessor scheduling problem with three or more processors, one resource, all independent tasks, and each computation time of every task equal to 1 is NP-complete.

8. Earliest deadline scheduling is not optimal in the multiprocessing case.

9. For two or more processors, no deadline scheduling algorithm can be optimal without complete a priori knowledge of deadlines, computation times, and task start times,

It turns out that most multiprocessor scheduling problem are in NP, but for deterministic scheduling this is not a major problem because a polynomial scheduling algorithm can be used to develop an optimal schedule if the specific problem is not NP-complete [Stankovic95]. In these cases, alternative, off-line heuristic search techniques can be used. These off-line techniques usually only need to find feasible schedules, not optimal ones. But this is what engineers do when workable theories do not exist – engineering judgment must prevail.

7.1.3 The Halting Problem

The Halting Problem, simply stated, is: does there exist a computer program that takes an arbitrary program, P_i, and an arbitrary set of inputs, I_j, and determines whether or not P_i will halt on I_j (Figure 7.1). The question of the existence of such an oracle is more than a theoretical exercise, and it has important implications in the development of process monitors, program verification, and in schedulability analysis. Unfortunately, such an oracle cannot be built.[1] Thus the Halting Problem is unsolvable. There are several ways to demonstrate this surprising fact. One way is using Cantor's diagonal argument, first used to show that the real numbers are not countably denumerable.

It should be clear that every possible program, in any computer language, can be encoded using a numbering scheme in which each program is represented as the binary expansion of the concatenated source-code bytes. The same encoding can be used with each input set. Then if the proposed oracle could be built, its behavior would be described in tabular form as in Table 7.1. That is, for each program P_i and each input set I_j it would simply have to determine if program P_i halts on I_j. Such an oracle would have to account for every conceivable program and input set.

In Table 7.1, the ↑ symbol indicates that the program does not halt and the symbol ↓ indicates that the program will halt on the corresponding input. However, the table is always incomplete in that a new program P^* can be found

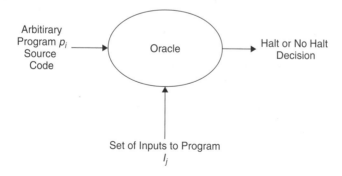

Figure 7.1   A graphical depiction of the Halting Problem.

[1] Strictly speaking, such an oracle can be built if it is restricted to a computer with fixed-size memory since, eventually, a maximum finite set of inputs would be reached, and hence the table could be completed.

Table 7.1 Diagonalization argument to show that no oracle can be constructed to solve the Halting Problem

	I_1	I_2	.	.	.	I_n	.	.	.	.
P_1	↓	↑	.	.	.	↓	.	.	.	.
P_2	↓	↑	.	.	.	↓	.	.	.	.
.				.	.		.	.	.	.
.			.		.		.	.	.	.
.			.	.			.	.	.	.
P_n	↑	↑	.	.	.	↓	.	.	.	.
.			.	.	.		.	.	.	.
.			.	.	.		.	.	.	.
.			.	.	.		.	.	.	.
P^*	↑	↓	.	.	.	↑	.	.	.	.

that differs from every other in at least the input at the diagonal. Even with the addition of a new program P^*, the table cannot be completed because a new P^* can be added that is different from every other program by using the same construction.

To see the relevance of the Halting Problem to real-time systems suppose a schedulability analyzer is to take an arbitrary program and the set of all possible inputs to that program and determine the best-, worst-, and average-case execution times for that program (Figure 7.2).

A model of the underlying machine is also needed, but this can be incorporated as part of the input set. It is easy to see that is a manifestation of the Halting Problem, since in order to determine the running time, the analyzer must know when (and hence, if) the program stops. While it is true that given a program in a specific language and a fixed set of inputs, the execution times can be found, the running times can be determined only through heuristic techniques that are not generalizable, that is, they could not work for an arbitrary and dynamic set of programs.

The Halting Problem also has implications in process monitoring. For example, is a process deadlocked or simply waiting? And also in the theory of recursive programs, for example, will a recursive program finish referencing itself?

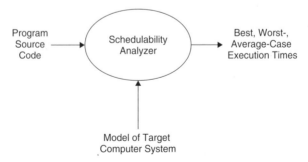

Figure 7.2 A schedulability analyzer whose behavior is related to the Halting Problem.

7.1.4 Amdahl's Law

Amdahl's Law is a statement regarding the level of parallelization that can be achieved by a parallel computer [Amdahl67].[2] Amdahl's law states that for a constant problem size, speedup approaches zero as the number of processor elements grows. It expresses a limit of parallelism in terms of speedup as a software property, not a hardware one.

Formally, let n be the number of processors available for parallel processing. Let s be the fraction of the code that is of a serial nature only, that is, it cannot be parallelized. A simple reason why a portion of code cannot be parallelized would be a sequence of operations, each depending on the result of the previous operation. Clearly $(1 - s)$ is the fraction of code that can be parallelized. The speedup is then given as the ratio of the code before allocation to the parallel processors to the ratio of that afterwards. That is,

$$\text{Speedup} = \frac{s + (1 - s)}{\left(s + \dfrac{(1 - s)}{n} \right)}$$

$$= \frac{1}{\left(s + \dfrac{(1 - s)}{n} \right)}$$

$$= \frac{1}{\left(\dfrac{ns}{n} + \dfrac{(1 - s)}{n} \right)}$$

$$= \frac{1}{\left(\dfrac{ns + 1 - s}{n} \right)}$$

$$= \frac{n}{ns + 1 - s}$$

Hence,

$$\text{Speedup} = \frac{n}{1 + (n - 1)s} \tag{7.1}$$

Clearly for $s = 0$ linear speedup can be obtained as a function of the number of processors. But for $s > 0$, perfect speedup is not possible due to the sequential component.

Amdahl's Law is frequently cited as an argument against parallel systems and massively parallel processors. For example, it is frequently suggested that "there will always be a part of the computation which is inherently sequential, [and that]

---

[2] Some of the following two sections has been adapted from Gilreath, W. and Laplante, P., *Computer Architecture: A Minimalist Perspective*, Kluwer Academic Publishers, Dordrecht, The Netherlands, 2003 [Gilreath03].

no matter how much you speed up the remaining 90 percent, the computation as a whole will never speed up by more than a factor of 10. The processors working on the 90 percent that can be done in parallel will end up waiting for the single processor to finish the sequential 10 percent of the task" [Hillis98]. But the argument is flawed. One underlying assumption of Amdahl's law is that the problem size is constant, and then at some point there is a diminishing margin of return for speeding up the computation. Problem sizes, however, tend to scale with the size of a parallel system. Parallel systems that are bigger in number of processors are used to solve very large problems in science and mathematics.

Amdahl's Law stymied the field of parallel and massively parallel computers, creating an insoluble problem that limited the efficiency and application of parallelism to different problems. The skeptics of parallelism took Amdahl's Law as the insurmountable bottleneck to any kind of practical parallelism, which ultimately impacted on real-time systems. However, later research provided new insights into Amdahl's Law and its relation to parallelism.

7.1.5 Gustafson's Law

Gustafson demonstrated with a 1024-processor system that the basic presumptions in Amdahl's Law are inappropriate for massive parallelism [Gustafson88]. Gustafson found that the underlying principle that "the problem size scales with the number of processors, or with a more powerful processor, the problem expands to make use of the increased facilities is inappropriate" [Gustafson88].

Gustafson's empirical results demonstrated that the parallel or vector part of a program scales with the problem size. Times for vector start-up, program loading, serial bottlenecks, and I/O that make up the serial component of the run do not grow with the problem size [Gustafson88].

Gustafson formulated that if the serial time, s, and parallel time, $p = (1 - s)$, on a parallel system with n processors, then a serial processor would require the time:

$$s + p \cdot n \qquad (7.2)$$

Comparing the plots of Equations 7.1 and 7.2 in Figure 7.3, it can be seen that Gustafson presents a much more optimistic picture of speedup due to parallelism than does Amdahl. Unlike the curve for Amdahl's Law, Gustafson's Law is a simple line, "one with a much more moderate slope: $1 - n$. It is thus much easier to achieve parallel performance than is implied by Amdahl's paradigm" [Gustafson88].

A different take on the flaw of Amdahl's Law can be observed as "a more efficient way to use a parallel computer is to have each processor perform similar work, but on a different section of the data... where large computations are concerned this method works surprisingly well" [Hillis98]. Doing the same task but on a different range of data circumvents an underlying presumption in Amdahl's Law, that is, "the assumption that a fixed portion of the computation... must be sequential. This estimate sounds plausible, but it turns out not to be true of most computations" [Hillis98].

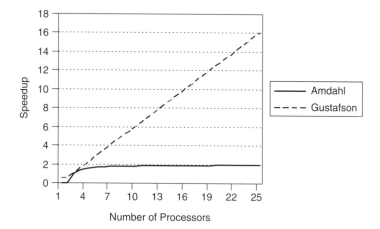

Figure 7.3 Linear speedup of Gustafson compared to "diminishing return" speedup of Amdahl with 50% of code available for parallelization. Notice as number of processors increase, speedup does not increase indefinitely for Amdahl due to serial component [Gilreath03].

7.2 PERFORMANCE ANALYSIS

It is natural to desire to analyze systems a priori to see if they will meet their deadlines. Unfortunately, in a practical sense, this is rarely possible due to the NP-completeness of most scheduling problems and constraints imposed by synchronization mechanisms. Nonetheless, it is possible to get a handle on the system's behavior through analysis. The first step in performing any kind of schedulability analysis is to determine, measure, or otherwise estimate the execution of specific code units.

The need to know the execution time of various modules and the overall system time-loading before implementation is important from both a management and an engineering perspective. Not only are CPU utilization requirements stated as specific design goals, but also knowing them a priori is important in selecting hardware and the system design approach. During the coding and testing phases, careful tracking of CPU utilization is needed to focus on those code units that are slow or whose response times are inadequate. Several methods can be used to predict or measure module execution time and CPU utilization.

7.2.1 Code Execution Time Estimation

Most measures of real-time performance require an execution-time estimate, e_i, for each task. The best method for measuring the execution time of completed code is to use the logic analyzer that is described in Chapter 8. One advantage of this approach is that hardware latencies and other delays are taken into account. The drawback in using the logic analyzer is that the system must be completely (or partially) coded and the target hardware available. Hence, the logic analyzer is usually only employed in the late stages of coding, during testing, and especially during system integration.

When a logic analyzer is not available, the code execution time can be estimated by examining the compiler output and counting macroinstructions either manually or using automated tools. This technique also requires that the code be written, an approximation of the final code exists, or similar systems are available for analysis. The approach simply involves tracing the worst-case path through the code, counting the macroinstructions along the way, and adding their execution times.

Another accurate method of code execution timing uses the system clock, which is read before and after executing code. The time difference can then be measured to determine the actual time of execution. This technique, however, is only viable when the code to be timed is large relative to the timer calls.

7.2.1.1 *Instruction Counting*

When it is too early for the logic analyzer, or if one is not available, instruction counting is the best method of determining CPU utilization due to code execution time. This technique requires that the code already be written, that an approximation of the final code exist, or that similar systems be available for inspection. The approach simply involves tracing the longest path through the code, counting the instruction types along the way, and adding their execution times.

Of course, the actual instruction times are required beforehand. They then can be obtained from the manufacturer's data sheets, by timing the instructions using a logic analyzer or simulators, or by educated guessing. If the manufacturer's data sheets are used, memory access times and the number of wait states for each instruction are needed as well. For example, consider, in the inertial measurement system. This module converts raw pulses into the actual accelerations that are later compensated for temperature and other effects. The module is to decide if the aircraft is still on the ground, in which case only a small acceleration reading by the accelerometer is allowed (represented by the symbolic constant PRE_TAKE).

Consider a time-loading analysis for the corresponding C code.

```
#define SCALE .01      /*.01 delta ft/sec/pulse is scale factor */
#define PRE_TAKE .1    /* .1 ft.sec/5ms max. allowable          */
void accelerometer (unsigned x, unsigned y, unsigned z,
float *ax, float *ay, float *az, unsigned on_ground, unsigned
*signal)

{
   *ax = (float) x*SCALE; /*covert pulses to accelerations */
   *ay = (float) y*SCALE;
   *az = (float) z*SCALE;
if(on_ground)
   if(*ax > PRE_TAKE || *ay > PRE_TAKE || *az > PRE_TAKE)
      *signal = *signal | 0x0001; /*set bit in signal */
}
```

A mixed listing combines the high-order language instruction with the equivalent assembly language instructions below it for easy tracing. A mixed listing for this code in a generic assembly language for a 2-address machine soon follows. The assembler and compiler directives have been omitted (along with some data-allocation pseudo-ops) for clarity and because they do not impact the time loading.

The instructions beginning in "F" are floating-point instructions that require 50 microseconds. The FLOAT instruction converts an integer to floating-point format. Assume all other instructions are integer and require 6 microseconds:

```
void accelerometer (unsigned x, unsigned y, unsigned z,
float *ax, float *ay, float *az, unsigned on_ground, unsigned
*signal)
{
*ax = (float) x *SCALE; /* convert pulses to accelerations */
    LOAD      R1,&x
    FLOAT     R1
    FMULT     R1,&SCALE
    FSTORE    R1,&ax,I

*ay = (float) y *SCALE;
    LOAD      R1,&y
    FLOAT     R1
    FMULT     R1,&SCALE
    FSTORE    R1,&ay,I

*az = (float) z SCALE;
    LOAD      R1,&z
    FLOAT     R1
    FMULT     R1,&SCALE
    FSTORE    R1,&az,I

if(on_ground)
    LOAD      R1,&on_ground
    CMP       R1,0
    JE @2

if(*ax > PRE_TAKE || *ay > PRE_TAKE || *az > PRE_TAKE)
    FLOAD     R1,&ax,I
    FCMP      R1,&PRE_TAKE
    JLE @1
    FLOAD     R1,&ay,I
    FCMP      R1,&PRE_TAKE
    JLE @1
    FLOAD     R1,&ay,I
    FCMP      R1,&PRE_TAKE
    JLE @1
@4:

*signal = signal | 0x0001; set bit in signal */
    LOAD      R1,&signal,I
    OR        R1,1
    STORE     R1,&signal, I
@3:
@2:
@1:
```

Tracing the worst path and counting the instructions shows that there are 12 integer and 15 floating-point instructions for a total execution time of 0.822 millisecond. Since this program runs in a 5-millisecond cycle, the time-loading is 0.822/5 = 16.5%. If the other cycles were analyzed to have a utilization as follows – 1-second cycle 1%, 10-millisecond cycle 30%, and 40-millisecond cycle 13% – then the overall time-loading for this foreground/background system would be 60.5%. Could the execution time be reduced for this module? It can, and these techniques will be discussed shortly.

In this example, the comparison could have been made in fixed point to save time. This, however, restricts the range of the variable PRE_TAKE, that is, PRE_TAKE could only be integer multiples of SCALE. If this were acceptable, then this module need only check for the pretakeoff condition and read the direct memory access (DMA) values into the variables ax, ay, and az. The compensation routines would perform all calculations in fixed point and would convert the results to floating point at the last possible moment.

As another instruction-counting example, consider the following 2-address assembly language code:

```
        LOAD    R1,&a     ; R1 <-- contents of "a"
        LOAD    R2,&a     ; R2 <-- contents of "a"
        TEST    R1,R2     ; compare R1 and R2, set condition code
        JNE     @L1       ; goto L1 if not equal
        ADD     R1,R2     ; R1 <-- R1 + R2
        TEST    R1,R2     ; compare R1 and R2, set condition code
        JGE     @L2       ; goto L2 if R1 >= R2
        JMP     @END      ; goto END
@L1     ADD     R1, R2    ; R1 <-- R1 + R2
        JMP     @END      ; goto END
@L2     ADD     R1, R2    ; R1 <-- R1 + R2
@END    SUB     R2, R3    ; R2 <-- R2 - R3
```

Calculate the following:

1. The best- and worst-case execution times.
2. The best- and worst-case execution times. Assume a three-stage instruction pipeline is used.

First, construct a branching tree enumerating all of the possible execution paths:

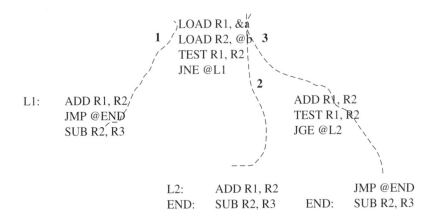

Path 1 includes 7 instructions @ 6 microseconds each = 42 microseconds. Path 2 and 3 include 9 instructions @ 6 microseconds each = 54 microsends. These are the best- and worst-case execution times.

For the second part, assume that a three-stage pipeline consisting of fetch, decode, and execute stages is implemented and that each stage takes 2 microseconds. For each of the three execution paths, it is necessary to simulate the contents of the pipeline, flushing the pipeline when required. To do this, number the instructions for ease of reference:

```
1. LOAD     R1, @a     ; R1 <-- contents of "a"
2. LOAD     R2, @b     ; R2 <-- contents of "b"
3. TEST     R1,R2      ; compare R1 and R2, set condition code
4. JNE      @L1        ; goto L1 if not equal
5. ADD      R1,R2      ; R1 <-- R1 + R2
6. TEST     R1,R2      ; compare R1 and R2, set condition code
7. JGE      @L2        ; goto L2 if R1 >= R2
8. JMP      @END       ; goto END
9. ADD      R1, R2     ; R1 <-- R1 + R2
10.JMP      @END       ; goto END
11.ADD      R1, R2     ; R1 <-- R1 + R2
12.SUB      R2, R3     ; R2 <-- R2 - R3
```

If "Fn," "Dn," and "En" indicate fetch, decode, and execution for instruction n, respectively, then for path 1, the pipeline execution trace looks like:

```
F1  D1  E1
    F2  D2  E2
        F3  D3  E3
            F4  D4  E4  (Flush)

            F5  D5
                    F9   D9   E9
                         F10  D10  E10  (flush)
                              F11  D11
                                        F12  D12  E12

2   4   6   8   10   12   14   16   18   20   22   24   26
                                        Time in microseconds
```

This yields a total execution time of 26 microseconds.

For path 2, the pipeline execution trace looks like:

```
F1   D1   E1
     F2   D2   E2
          F3   D3   E3
               F4   D4   E4   (Flush)
                    F5   D5

                         F9   D9   E9
                              F10  D10  E10  (flush)
                                   F11  D11
                                             F12  D12  E12

2    4    6    8    10   12   14   16   18   20   22   24   26
                                             Time in microseconds
```

This represents a total execution time of 26 microseconds.
For path 3, the pipeline execution trace looks like

```
F1   D1   E1
     F2   D2   E2
          F3   D3   E3
               F4   D4   E4
                    F5   D5   E5
                         F6   D6   E6
                              F7   D7   E7
                                   F8   D8   E8   (flush)
                                        F9   E9
                                                  F12  D12  E12

2    4    6    8    10   12   14   16   18   20   22   24   26
                                             Time in microseconds
```

This yields a total execution time of 26 microseconds. It is just a coincidence in this case that all three paths have the same execution time. Normally, there would be different execution times.

As a final note, the process of instruction counting can be automated if a parser is written for the target assembly language that can resolve branching.

7.2.1.2 *Instruction Execution-Time Simulators* The determination of instruction times requires more than just the information supplied in the CPU manufacturer's data books. It is also dependent on memory access times and

wait states, which can vary depending on the source region of the instruction or data in memory. Some companies that frequently design real-time systems on a variety of platforms use simulation programs to predict instruction execution time and CPU throughput. Then engineers can input the CPU types, memory speeds for each region of memory, and an instruction mix, and calculate total instruction times and throughput.

7.2.1.3 Using the System Clock

Sections of code can be timed by reading the system clock before and after the execution of the code. The time difference can then be measured to determine the actual time of execution. If this technique is used, it is necessary to calculate the actual time spent in the open loop and subtract it from the total. Of course, if the code normally takes only a few microseconds, it is better to execute the code under examination several thousand times. This will help to remove any inaccuracy introduced by the granularity of the clock. For example, the following C code can be rewritten in a suitable language to time a single high-level language instruction or series of instructions. The number of iterations needed can be varied depending on how short the code to be timed is. The shorter the code, the more iterations should be used. current_clock_time() is a system function that returns the current time. function_to_be_timed() is where the actual code to be timed is placed.

```
#include system.h
unsigned long timer(void)
{

    unsigned long time0,time1,i,j,time2,total_time,time3,
    iteration=1000000L;

    time0=current_clock_time();        /* read time now  */
    for (j=1;j<=iteration; j++);       /* run empty loop */
    time1=current_clock_time();
    loop_time=time1-time0;             /* open loop time */
    time2=current_clock_time();        /* read time now  */

    for (i=1;i<=iteration;i++)          * time function   */
        function_to_be_timed();

    time3=current_clock_time();        /* read time now  */

/* calculate instruction(s) time */
    total_time=(time 3-time2-loop_time)/iteration;
    return total_time;
}
```

Accuracy due to the clock resolution should be taken into account. For example, if 2000 iterations of the function take 1.1 seconds with a clock granularity of 18.2 microseconds, the measurement is accurate to

$$\frac{+18.2}{1.1 \times 10^6} \approx \pm 0.0017\%$$

Clearly, running more iterations can increase the accuracy of the measurement.

7.2.2 Analysis of Polled Loops

The response-time delay for a polled loop system consists of three components: the hardware delays involved in setting the software flag by some external device; the time for the polled loop to test the flag; and the time needed to process the event associated with the flag (Figure 7.4). The first delay is on the order of nanoseconds and can be ignored. The time to check the flag and jump to the handler routine can be several microseconds. The time to process the event related to the flag depends on the process involved. Hence, calculation of response time for polled loops is quite easy.

The preceding case assumes that sufficient processing time is afforded between events. However, if events begin to overlap, that is, if a new event is initiated while a previous event is still being processed, then the response time is worse. In general, if f is the time needed to check the flag and P is the time to process the event, including resetting the flag (and ignoring the time needed by the external device to set the flag), then the response time for the nth overlapping event is bounded by

$$nfP \qquad (7.3)$$

Typically, some limit is placed on n, that is, the number of events that can overlap. Two overlapping events may not be desirable in any case.

7.2.3 Analysis of Coroutines

The absence of interrupts in a coroutine system makes the determination of response time rather easy. In this case, response time is simply found by tracing the worst-case path through each of the tasks (Figure 7.5). In this case, the execution time of each phase must be determined, which has already been discussed.

7.2.4 Analysis of Round-Robin Systems

Assume that a round-robin system is such that there are n processes in the ready queue, no new ones arrive after the system starts, and none terminate prematurely.

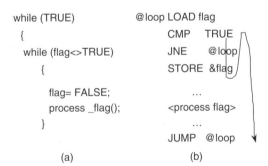

Figure 7.4 Analysis of polled-loop response time: (a) source code; (b) assembly equivalent.

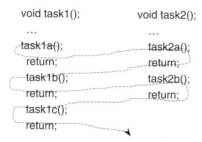

```
void task1();          void task2();
   ...                    ...
   task1a();              task2a();
   return;                return;
   task1b();              task2b();
   return;                return;
   task1c();
   return;
```

Figure 7.5 Tracing the execution path in a two-task coroutine system. The tasks are `task1()` and `task2()`. A `switch` statement in each task drives the phase-driven code (not shown). A central dispatcher calls `task1()` and `task2()` and provides intertask communication via global variables or parameter lists.

The release time is arbitrary – in other words, although all processes are ready at the same time, the order of execution is not predetermined, but is fixed.

Assume all processes have maximum end-to-end execution time, c. While this assumption might seem unrealistic, suppose that each process, i, has a different maximum execution time, c_i. Then letting $c = \max\{c_1, \ldots, c_n\}$ yields a reasonably upper bound for the system performance and allows the use of this model.

Now let the timeslice be q. If a process completes before the end of a time quantum, in practice, that slack time would be assigned to the next ready process. However, for simplicity of analysis, assume that it is not. This does not hurt the analysis because an upper bound is desired, not an analytic response-time solution.

In any case, each process, ideally, would get $1/n$ of the CPU time in chunks of q time units, and each process would wait no longer than $(n-1)q$ time units until its next time up. Now, since each process requires at most $\left\lceil \frac{c}{q} \right\rceil$ time units to complete, the waiting time will be $(n-1)q \left\lceil \frac{c}{q} \right\rceil$ (where $\lceil \; \rceil$ represents the "ceiling" function, which yields the smallest integer greater than the quantity inside the brackets). Thus, the worst-case time from readiness to completion for any task (also known as turnaround time), denoted T, is the waiting time plus undisturbed time to complete, c, or

$$T = (n-1)\left\lceil \frac{c}{q} \right\rceil q + c \qquad (7.4)$$

As an example, suppose that there is only one process with a maximum execution time of 500 ms and that the time quantum is 100 ms. Thus, $n = 1$, $c = 500$, $q = 100$, and

$$T = (1-1)\left\lceil \frac{500}{100} \right\rceil 100 + 500 = 500 \text{ ms}$$

which, is as expected.

Now suppose there are five processes with a maximum execution time of 500 ms. The time quantum is 100 ms. Hence, $n = 5, c = 500, q = 100$, which yields

$$T = (5 - 1) \left\lceil \frac{500}{100} \right\rceil 100 + 500 = 2500 \text{ ms}$$

This is intuitively pleasing, since it would be expected that five consecutive tasks of 500 ms each would take 2500 ms end-to-end to complete.

However, now assume that there is a context switching overhead, o. Now each process still waits no longer than $(n - 1)q$ until its next time quantum, but there is the additional overhead of $n \cdot o$ each time around for context switching. Again, each process requires at most $\left\lceil \dfrac{c}{q} \right\rceil$ time quanta to complete. So the worst-case turnaround time for any task is now at most

$$T = [(n - 1)q + n \cdot o] \left\lceil \frac{c}{q} \right\rceil + c \tag{7.5}$$

An assumption is that there is an initial context switch to load the first time around.

To illustrate, suppose that there is one process with a maximum execution time of 500 ms. The time quantum is 40 ms and context switch time is 1 ms. Hence, $n = 1, c = 500, q = 40, o = 1$. So,

$$T = [(1 - 1) \cdot 40 + 1 \cdot 1] \left\lceil \frac{500}{40} \right\rceil + 500$$

$$= 1 \cdot 13 + 500 = 513 \text{ ms}$$

which is expected since the context switch time to handle the round-robin clock interrupt costs 1 ms each time for the 13 times it occurs.

Next, suppose that there are six processes, each with a maximum execution time of 600 ms, the time quantum is 40 ms, and context switch time costs 2 ms. Now, $n = 6, c = 600, q = 40$, and $o = 2$. Then

$$T = [(6 - 1) \cdot 40 + 6 \cdot 2] \left\lceil \frac{600}{40} \right\rceil + 600$$

$$= [5 \cdot 40 + 10] \cdot 15 + 600 = 3750 \text{ ms}$$

which again is pleasing, because one would expect six processes of 600 ms in duration to take at least 3600 ms, without context switching costs.

In terms of the time quantum, it is desirable that $q < c$ to achieve "fair" behavior. For example, if q is very large, the round-robin algorithm is just the first-come, first-served algorithm in that each process will execute to completion, in order of arrival, within the very large time quantum.

The technique just discussed is also useful for cooperative multitasking analysis or any kind of "fair" cyclic scheduling with context switching costs.

7.2.5 Response-Time Analysis for Fixed-Period Systems

In general, utilization-based tests are not exact and provide good estimates for a very simplified task model. In this section, a necessary and sufficient condition for schedulability based on worst-case response time calculation is presented.

For the highest-priority task, its worst-case response time evidently will be equal to its own execution time. Other tasks running on the system are subjected to interference caused by execution of higher-priority tasks. For a general task τ_i, response time, R_i, is given as

$$R_i = e_i + I_i \tag{7.6}$$

where I_i is the maximum amount of delay in execution, caused by higher priority tasks, that task τ_i is going to experience in any time interval $[t, t + R_i)$. At a critical instant I_i will be maximum, that is, the time at which all higher-priority tasks are released along with task τ_i.

Consider a task τ_j of higher priority than τ_i. Within the interval $[0, R_i)$, the time of release of τ_j will be $\lceil R_i/p_j \rceil$. Each release of task τ_j is going to contribute to the amount of interference τ_i is going to face, and is expressed as:

$$\text{Maximum interference} = \lceil R_i/p_j \rceil e_j \tag{7.7}$$

Each task of higher priority is interfering with task τ_i. So,

$$I_i = \sum_{j \in hp(i)} \lceil R_i/p_j \rceil e_j \tag{7.8}$$

where $hp(i)$ is the set of higher-priority tasks with respect to τ_i. Substituting this value in $R_i = e_i + I_i$ yields

$$R_i = e_i + \sum_{j \in hp(i)} \lceil R_i/p_j \rceil e_j \tag{7.9}$$

Due to the ceiling functions, it is difficult to solve for R_i. Without getting into details, a solution is provided where the function R is evaluated by rewriting it as a recurrence relation

$$R_i^{n+1} = e_i + \sum_{j \in hp(i)} \lceil R_i^n/p_j \rceil e_j \tag{7.10}$$

where R_i^n is the response in the nth iteration.

To use the recurrence relation to find response times, it is necessary to compute R_i^{n+1} iteratively until the first value m is found such that $R_i^{m+1} = R_i^m$. R_i^m is then the response time R_i. It is important to note that if the equation does not have a solution, then the value of R_i will continue to rise, as in the case when a task set has a utilization greater than 100%.

7.2.6 Response-Time Analysis: RMA Example

To illustrate the calculation of response-time analysis for a fixed-priority scheduling scheme, consider the task set to be scheduled rate monotonically, as shown below:

τ_i	e_i	p_i
τ_1	3	9
τ_2	4	12
τ_3	2	18

The highest priority task τ_1 will have a response time equal to its execution time, so $R_1 = 3$.

The next highest priority task, τ_2 will have its response time calculated as follows. First, $R_2 = 4$. Using Equation 7.10, the next values of R_2 are derived as:

$$R_2^1 = 4 + \lceil 4/9 \rceil\, 3 = 7$$
$$R_2^2 = 4 + \lceil 7/9 \rceil\, 3 = 7$$

Since, $R_2^1 = R_2^2$, it implies that the response time of task τ_2, R_2, is 7.

Similarly, the lowest priority task τ_3 response is derived as follows. First, $R_3^0 = 5$, then use Equation 7.10 again to compute the next values of R_3:

$$R_3^1 = 2 + \lceil 2/9 \rceil\, 3 + \lceil 2/12 \rceil\, 4 = 9$$
$$R_3^2 = 2 + \lceil 9/9 \rceil\, 3 + \lceil 9/12 \rceil\, 4 = 9$$

Since, $R_3^1 = R_3^2$, the response time of the lowest priority task is 9.

7.2.7 Analysis of Sporadic and Aperiodic Interrupt Systems

Ideally, a system having one or more aperiodic or sporadic cycles should be modeled as a rate-monotonic system, but with the nonperiodic tasks modeled as having a period equal to their worst-case expected interarrival time. However, if this approximation leads to unacceptably high utilizations, it may be possible to use a heuristic analysis approach. Queuing theory can also be helpful in this regard. Certain results from queuing theory are discussed later.

The calculation of response times for interrupt systems is dependent on a variety of factors, including interrupt latency, scheduling/dispatching times, and context switch times. Determination of context save/restore times is the same as for any application code. The schedule time is negligible when the CPU uses an interrupt controller with multiple interrupts. When a single interrupt is supported in conjunction with an interrupt controller, it can be timed using instruction counting.

7.2.7.1 *Interrupt Latency* Interrupt latency is a component of response time, and is the period between when a device requests an interrupt and when the

first instruction for the associated hardware interrupt service routine executes. In the design of a real-time system, it is necessary to consider what the worst-case interrupt latency might be. Typically, it will occur when all possible interrupts in the system are requested simultaneously. The number of threads or processes also contribute to the worst-case latency. Typically, real-time operating systems need to disable interrupts while it is processing lists of blocked or waiting threads. If the design of the system requires a large number of threads or processes, it is necessary to perform some latency measurements to check that the scheduler is not disabling interrupts for an unacceptably long time.

7.2.7.2 *Instruction Completion Times*

Another contributor to interrupt latency is the time needed to complete execution of the macroinstruction that was interrupted. Thus, it is necessary to find the execution time of every macroinstruction by calculation, measurement, or manufacturer's data sheets. The instruction with the longest execution time in the code will maximize the contribution to interrupt latency if it has just begun executing when the interrupt signal is received.

For example, in a certain microprocessor, it is known that all fixed-point instructions take 10 microseconds, floating-point instructions take 50 microseconds, and other instructions, such as built-in sine and cosine functions, take 250 microseconds. The program is known to generate only one such cosine instruction when compiled. Then its contribution to interrupt latency can be as high as 250 microseconds.

The latency caused by instruction completion is often overlooked, possibly resulting in mysterious problems. Deliberate disabling of the interrupts by the software can create substantial interrupt latency, and this must be included in the overall latency calculation. Interrupts are disabled for a number of reasons, including protection of critical regions, buffering routines, and context switching.

7.2.8 Deterministic Performance

Cache, pipelines, and DMA, all designed to improve average real-time performance, destroy determinism and thus make prediction of real-time performance troublesome. In the case of cache, for example, is the instruction in the cache? From where it is being fetched has a significant effect on the execution time of that instruction. To do a worst-case performance, it must be assumed that every instruction is not fetched from cache but from in memory. However, to bring that instruction into the cache, costly replacement algorithms must be applied. This has a very deleterious effect on the predicted performance. Similarly, in the case of pipelines, one must always assume that at every possible opportunity the pipeline needs to be flushed. Finally, when DMA is present in the system, it must be assumed that cycle stealing is occurring at every opportunity, thus inflating instruction fetch times. Does this mean that these widely used architectural techniques render a system effectively unanalyzable for performance? Essentially, yes. However, by making some reasonable assumptions about the real impact of these effects, some rational approximation of performance is possible.

7.3   APPLICATION OF QUEUING THEORY

The classic queuing problem involves one or more producer processes called servers and one or more consumer processes called customers. Queuing theory has been applied to the analysis of real-time systems this way since the mid-1960s (e.g., [Martin67]), yet it seems to have been forgotten in modern real-time literature.

A standard notation for a queuing system is a three-tuple (e.g., M/M/1). The first component describes the probability distribution for the time between arrivals of customers, the second is the probability distribution of time needed to service each customer, and the third is the number of servers. The letter M is customarily used to represent exponentially distributed interarrival or service times.

In a real-time system, the first component of the tuple might be the arrival time probability distribution for a certain interrupt request. The second component would be the time needed to service that interrupt's request,. The third component would be unity for a single processing system and >1 for multiprocessing systems. Known properties of this queuing model can be used to predict service times for tasks in a real-time system.

7.3.1   The M/M/1 Queue

The simplest queuing model is the M/M/1 queue, which represents a single-server system with a Poisson arrival model (exponential interarrival times for the customers or interrupt requests with mean $1/\lambda$), and exponential service or process time with mean $1/\mu$ and $\lambda < \mu$. As suggested before, this model can be used effectively to model certain aspects of real-time systems; it is also useful because it is well known, and several important results are immediately available [Kleinrock75]. For example, let N be the number of customers in the queue. Letting $\rho = \lambda/\mu$, then the average number of customers in the queue in such a system is

$$\overline{N} = \frac{\rho}{1-\rho} \tag{7.11}$$

with variance

$$\sigma_N^2 = \frac{\rho}{(1-\rho)^2} \tag{7.12}$$

The average time a customer spends in the system is

$$T = \frac{1/\mu}{1-\rho} \tag{7.13}$$

The random variable Y for the time spent in the system has probability distribution

$$s(y) = \mu(1-\rho)e^{-\mu(1-\rho)y} \tag{7.14}$$

with $y \geq 0$.

Finally, it can be shown that the probability that at least k customers are in the queue is

$$P[\geq k \text{ in system}] = \rho^k \tag{7.15}$$

In the M/M/1 model, the probability of exceeding a certain number of customers in the system decreases geometrically. If interrupt requests are considered customers in a certain system, then two such requests in the system at the same time (a time-overloaded condition) have a far greater probability of occurrence than three or more such requests. Thus, building systems that can tolerate a single time-overload will contribute significantly to system reliability, while worrying about multiple time-overload conditions is probably futile. The following sections describe how the M/M/1 queue can be used in the analysis of real-time systems.

7.3.2 Service and Production Rates

Consider an M/M/1 system in which the customer represents an interrupt request of a certain type and the server represents the processing required for that request. In this single-processor model, waiters in the queue represent a time-overloaded condition. Because of the nature of the arrival and processing times, this condition could theoretically occur. Suppose, however, that the arrival or the processing times can vary. Varying the arrival time, which is represented by the parameter λ, could be accomplished by changing hardware or altering the process causing the interrupt. Changing the processing time, represented by the parameter μ could be achieved by optimization. In any case, fixing one of these two parameters, and selecting the second parameter in such a way as to reduce the probability that more than one interrupt will be in the system simultaneously, will ensure that time-overloading cannot occur within a specific confidence interval.

For example, suppose $1/\lambda$, the mean interarrival time between interrupt requests, is known to be 10 milliseconds. It is desired to find the mean processing time, $1/\mu$, necessary to guarantee that the probability of time overloading (more than one interrupt request in the system) is less than 1%. Use Equation 7.15 as follows:

$$P[\geq 2 \text{ in system}] = \left(\frac{\lambda}{\mu}\right)^2 \leq 0.01$$

or

$$\frac{1}{\mu} \leq \sqrt{\frac{0.01}{\lambda^2}}$$

then

$$\Rightarrow \frac{1}{\mu} \leq 0.001 \text{ seconds}$$

Thus, the mean processing time, $1/\mu$, should be no more than 1 millisecond to guarantee with 99% confidence that time overloading cannot occur.

As another example, suppose the service time, $1/\mu$, is known to be 5 milliseconds. It is desired to find the average arrival time (interrupt rate), $1/\lambda$, to guarantee that the probability of time-overloading is less than 1%. Using Equation 7.19, yields

$$\left(\frac{\lambda}{\mu}\right)^2 \le 0.01$$

or

$$\Rightarrow \frac{\mu}{\lambda} \ge 0.050 \text{ second}$$

Hence, the average interarrival time between two interrupt requests should be at least 50 milliseconds to guarantee only a 1% risk of time overloading. This result is different from the guarantee that the rate-monotonic theorem, which states that if a periodic interrupt occurs at exactly a 10-ms rate then a $1/10 = 20\%$ utilization will be realized. The result of Equation 7.15 applies if an aperiodic interrupt is arriving at an average of every 10 milliseconds.

Of course, context switching time and blocking due to semaphore waits are not incorporated in these analyses. Nevertheless, this approach can be useful in exploring the feasibility of the system with aperiodic or sporadic interrupts.

7.3.3 Some Buffer-Size Calculations

The M/M/1 queue can also be used for buffer-size calculations by portraying the "customers" as data being placed in a buffer. The "service" time is the time needed to pick up the data by some consumer task. Here the basic properties of M/M/1 queues are used to calculate the average buffer size needed to hold the data using Equation 7.11, and the average time a datum spends in the system (its age using Equation 7.13). For example, a process produces data with interarrival times given by the exponential distribution $4e^{-4t}$, and is consumed by a process at a rate given by the exponential distribution $5e^{-4t}$. To calculate the average number of data items in the buffer, use Equation 7.11:

$$\overline{N} = \frac{\rho}{1-\rho} = \frac{4/5}{1-4/5} = 4$$

A probability distribution for the random variable determining the age of the data can be found by using Equation 7.13:

$$T = \frac{1/\mu}{1-\rho} = \frac{1/5}{1-4/5} = 1 \text{ second}$$

7.3.4 Response-Time Modeling

The "average" response time for a process handling an interrupt request in the absence of other competing processes can also be computed if an M/M/1 model is assumed. In this case use Equation 7.13 to measure the average time spent in the system by an interrupt request (the response time). For example, a process is based on a sporadic interrupt that occurs with an interarrival rate given by the exponential function with mean $1/\lambda = 5$ milliseconds. A process handles the data in an amount of time determined by the exponential function with mean $1/\mu = 3$ milliseconds. The mean response time for this interrupt request is seen from Equation 7.13 to be:

$$T = \frac{3}{1 - \dfrac{1/5}{1/3}} = 7.5 \text{ milliseconds}$$

A probability distribution for the random variable determining the mean response time can be found by using Equation 7.14. Omitting context switching and inter-task interference gives

$$s(t) = \frac{1}{3}\left(1 - \frac{3}{5}\right)e^{-(1/3)(1-(3/5))t}$$

or

$$s(t) = \frac{2}{5}e^{-(2/15)t}$$

Note that the response time will be deleteriously affected if the interrupt rate is greater than the mean service rate.

7.3.5 Other Results from Queuing Theory

The M/M/1 queue can be used in a variety of other ways to model real-time systems. The only requirements are that the producer be modeled as a Poisson process and that the consumption time be exponential. Although the model assumes an infinite-length queue, confidence intervals can be fixed for modeling realistic finite-sized queues.

Systems that can be modeled to match other queuing system models can benefit from the well-known results there. For example, the M/M/1 queue with Poisson arrival (exponential interarrival) and general service time probability distributions can be used. Other results cover the general arrival and service densities. Relationships involving balking customers, those that leave the queue, can be used to represent rejected spurious interrupts or time overloads.

7.3.6 Little's Law

An important result in queuing theory, Little's Law, has some application in real-time system performance prediction. Little's Law, which appeared in 1961, states that the average number of customers in a queuing system, N_{av}, is equal to the average arrival rate of the customers to that system, r_{av}, times the average time spent in that system, t_{av} [Kleinrock75]:

$$N_{av} = r_{av}t_{av} \tag{7.16}$$

If n servers are present, then

$$N_{av} = \sum_{i=1}^{n} r_{i,av} t_{i,av} \tag{7.17}$$

where $r_{i,av}$ is the average arrival rate for customers to server i, and $t_{i,av}$ is the average service time for server i.

Viewing each process as a server and interrupt arrivals as customers, Little's Law is Equation 1.2 for CPU utilization with $e_i = t_{i,av}$ and $1/p_i = r_{i,av}$. For example, a system is known to have periodic interrupts occurring at 10, 20, and 100 milliseconds and a sporadic interrupt that is known to occur on average every 1 second. The average processing time for these interrupts is 3, 8, 25, and 30 milliseconds. Then by Little's Law the average number of customers in the queue is

$$N_{av} = (1/10)3 + (1/20)8 + (1/100)25 + (1/1000)30 = 0.98$$

This result is the same one obtained by using Equation 1.2 for CPU utilization.

7.3.7 Erlang's Formula

Another useful result of queuing theory is Erlang's Loss Formula. Suppose there are m servers (or processes) and arriving customers (interrupts). Each newly arriving interrupt is serviced by a process, unless all servers are busy (a potential time-overloaded condition). In this, case the customer (interrupt) is lost. If it is assumed that the average service (process) time is μ and the average arrival time (interrupt rate) is λ, then the fraction of time that all servers are busy (a time-overloaded condition) is given by

$$p = \frac{(\lambda/\mu)^m / m!}{\sum_{k=0}^{m} (\lambda/\mu)^k / k!} \tag{7.18}$$

This result dates back to 1917 [Kleinrock75].

Applying Erlang's Formula to the previous example gives $m = 4$, $\lambda = 380$, and $\mu = 16.5$; then

$$p = \frac{(380/16.5)^4 / 4!}{1 + (380/16.5) + (380/16.5)^2/2 + (380/16.5)^3/3! + (380/16.5)^4/4!}$$
$$= 0.834$$

This means there is a potential for time overloading 83.4% of the time. Based on the average time-loading figure of 98% and the rate-monotonic theorem, this seems reasonable.

7.4 I/O PERFORMANCE

One performance area that varies greatly owing to device dependencies is the bottleneck presented by disk and device I/O access. In many cases, disk I/O is the single greatest contributor to performance degradation. Moreover, when analyzing a system's performance through instruction counting, it is very difficult to account for disk device access times. In most cases the best approach is to assume worst-case access times for device I/O and include them in performance predictions.

In other cases, where a real-time system participates in some form of a network, for example, a local area network (LAN), loading the network can seriously degrade real-time performance and make measurement of that performance impossible. In most cases, it is necessary to assess the performance of the system assuming that the network is in the best possible state (i.e., has no other users). Then measurements of performance can be taken under varying conditions of loading, and a performance curve can be generated.

7.4.1 Basic Buffer-Size Calculation

Recall that a buffer is a set of memory locations that provide temporary storage for data that are being input or output or are being passed between two different processes. Assume that the data are being sent for some finite time called a burst period.

More precisely, if the data are produced at a rate of $P(t)$ and can be consumed at a rate of $C(t)$ (where $C(t) < P(t)$) for a burst period of T, what is the size of the buffer needed to prevent data from being lost? If both $P(t)$ and $C(t)$ are constant, denoted P and C, respectively, and if the consumptions rate C is greater than or equal to P, then no buffer is needed since the system can always consume data faster than they can be produced. If $C < P$, however, then an overflow will occur. To calculate the buffer size needed to handle the overflow for a burst of period, T, note that the total data produced is PT, while the total data consumed in that time is CT. Thus, there is an excess of $(P - C)T$ units. This is how much data must be stored in the buffer. Thus, the buffer size is

$$B = (P - C)T \qquad (7.19)$$

where C is the consumption rate, P is the production rate, and T is the burst time. For example, suppose a device is providing data to a real-time computer via DMA at 9600 bytes/second in bursts of one-second duration every 20 seconds. The computer is capable of processing the data at 800 bytes/second. Assuming there is sufficient time to empty the buffer before another burst occurs, what should the minimum buffer size be? Using Equation 7.19 yields:

$$B = (9600 - 800) \text{ bytes per second} \cdot 1 \text{ second} = 8800 \text{ bytes}$$

Handling data that occur in bursts with Equation 7.19 is possible only if the buffer can be emptied before another burst occurs. For example, emptying the buffer in the previous cases will take 11 seconds – sufficient time before the next expected burst. If bursts occur too frequently, then buffer overflow will occur. In this case the system is unstable, and either upgrading the processor or downgrading the production process is necessary to solve the problem.

7.4.2 Variable Buffer-Size Calculation

It is often not accurate to assume that burst periods are fixed; they will frequently be variable. Suppose that a process produces data at a rate given by the real-valued function $p(t)$. Also suppose that another task consumes or uses the data produced by the first task at a rate determined by the real-valued function $c(t)$. The data are produced during a burst period $T = t_1 - t_2$, where $t_1 > t_2$ represent the start and finish times of the burst. Then the buffer size needed at time t_2 can be expressed as

$$B(t_2) = \int_{t_1}^{t_2} [p(t) - c(t)] \, dt \qquad (7.20)$$

Note that when $p(t)$ and $c(t)$ are constant, Equation 7.20 becomes Equation 7.19 (see Exercise 7.9). For example, a task produces data at a rate (in bytes per second) that is determined by the function:

$$p(t) = \begin{cases} 10,000t & 0 \le t \le 1 \\ 10,000(t - 2) & 1 < t \le 2 \\ 0 & \text{elsewhere} \end{cases}$$

with t representing the burst period. The data are consumed by a task at a rate determined by the function:

$$c(t) = \begin{cases} 10,000(1/4)t & 0 \le t \le 2 \\ 10,000(1 - 1/4t) & 2 < t \le 4 \\ 0 & \text{elsewhere} \end{cases}$$

If the burst period is known to be 1.6 seconds, what is the necessary buffer size? Using Equation 7.20 yields,

$$B(1.6) = \int_0^{1.6} [p(t) - c(t)] \, dt$$

$$= 10,000 \int_0^1 [(t - 1/4t)] \, dt + 10,000 \int_1^{1.6} [(2 - t) - 1/4t] \, dt$$

$$= 10,000(3/4) \int_0^1 t \, dt + \int_1^{1.6} dt - 10,000(5/4) \int_1^{1.6} t \, dt$$

$$= 10,000(3/8)t^2 \Big|_0^1 + 10,000(2)t \Big|_1^{1.6} - 10,000(5/8)t^2 \Big|_1^{1.6}$$

$$= 10,000(3/8) + 10,000(3.2 - 2) - 10,000(5/8)(2.56 - 1)$$

$$\approx 600 \text{ bytes}$$

If the burst period is determined by the real-valued function $u(t)$, then for a burst starting at time t_1 and ending at t_2 the necessary buffer size is

$$B(t) = \int_{t_1}^{u(t_1)} [p(t) - c(t)]\, dt \tag{7.21}$$

In the previous example, if the burst starts at a time determined by the Gaussian distribution

$$u(t) = \frac{1}{\sqrt{2\pi}} e^{-(t-2)^2/2} \tag{7.22}$$

at time $t = 0$, then it is will end at time $u(0) = 0.053991$. Recalculation of the buffer size now yields

$$
\begin{aligned}
B(0.053991) &= \int_{0}^{0.053991} [p(t) - c(t)]\, dt \\
&= 10{,}000 \int_{0}^{0.053991} [t - 1/4t]\, dt \\
&= 10{,}000(3/4) \int_{0}^{0.053991} t\, dt \\
&= 10{,}000(3/8)t^2 \Big|_{0}^{0.053991} \\
&= 10{,}000(3/8)[0.002915] \\
&\approx 11 \text{ bytes}
\end{aligned}
$$

7.5 PERFORMANCE OPTIMIZATION

Identifying wasteful computation is a preliminary step in reducing code execution time, and hence, CPU utilization. Many approaches employed in compiler optimization can be used, but other methods have evolved that are specifically oriented toward real-time systems.

7.5.1 Compute at Slowest Cycle

All processing should be done at the slowest rate that can be tolerated. Checking a discrete temperature for a large room at faster than 1 second may be wasteful, for temperature typically cannot change quickly owing to thermal inertia. In the nuclear plant example, a dedicated sensor is used to monitor the temperature, which then issues a high-level priority if any overtemperature is detected.

7.5.2 Scaled Numbers

In virtually all computers, integer operations are faster than floating-point ones. This fact can be exploited by converting floating-point algorithms into scaled integer algorithms. In these so-called scaled numbers, the least significant bit (LSB) of an integer variable is assigned a real number scale factor. Scaled numbers can be added and subtracted together and multiplied and divided by a constant (but not another scaled number). The results are converted to floating point only at the last step, thus saving considerable time. For example, suppose an analog-to-digital (A/D) converter is converting accelerometer data. If the least significant bit of the two's complement 16-bit integer has value 0.0000153 ft/s^2, then any acceleration can be represented up to the maximum value of $(2^{15} - 1)*0.0000153 = 0.5013351$ ft/s^2. The 16-bit number 0000 0000 0001 011, for example, represents an acceleration of 0.0001683 ft/s^2.

A common practice is to quickly convert the integer number into its floating-point equivalent by $xf = x \cdot 0.0000153$ and then proceed to use it in calculations directly with other converted numbers; for example, $diff = xf - zf$, where zf is a similarly converted floating-point number. Instead, the calculation can be performed in integer form first and then converted to floating point: $diff = (x - z) \cdot 0.0000153$.

For applications involving the numerous addition and subtraction of like data, scaled numbers can introduce significant savings. Note, however, that multiplication and division by another scaled number cannot be performed on a scaled number as those operations change the scale factor. Finally, accuracy is generally sacrificed by excessive use of scaled numbers.

7.5.3 Binary Angular Measure

Another type of scaled number is based on the property that adding 180° to any angle is analogous to taking its two's complement. This technique, called binary angular measurement (BAM) works as follows. Consider the LSB of an n-bit word to be $2^{n-1} \cdot 180$ degrees with the most significant bit (MSB) = 180 degrees. The range of any angle θ represented this way is $0 \le \theta \le 360 - 180 \cdot 2^{-(n-1)}$ degrees. A 16-bit BAM word is shown in Figure 7.6. For more accuracy, BAM can be extended to two more words. Each n-bit word has a maximum value of

$$2^n - 2^{-(n-1)} \cdot 180° = 360° - \text{LSB} \tag{7.23}$$

with granularity

$$2^{-(n-1)} \cdot 180° = \text{LSB} \tag{7.24}$$

Consider the 16-bit BAM word:

$$0000 \ 0000 \ 10100 \ 110$$

Its binary angular measurement is $166 \cdot 180° \cdot 2^{-15} = 0.9118°$.

180	90	45	22.5	...	...	...	...	...	...	...	...	...	...	$180 \cdot 2^{-14}$	$180 \cdot 2^{-15}$

Figure 7.6 A 16-bit binary angular measurement word [Laplante03c].

BAM can be added and subtracted together and multiplied and divided by constants as if they were unsigned integers, and converted at the last stage to produce floating-point results. It is easy to show that the overflow condition for BAM numbers presents no problem as the angle simply wraps around to 0. BAM is frequently used in navigation software, robotic control, and in conjunction with digitizing imaging devices.

7.5.4 Look-Up Tables

Another variation of the scaled-number concept uses a stored table of function values at fixed intervals. Such a table, called a look-up table, allows for the computation of continuous functions using mostly fixed-point arithmetic.

Let $f(x)$ be a continuous real function and let Δx be the interval size. Suppose it is desired to store n values of f over the range $[x_0, x_0 + (n-1)\Delta x]$ in an array of scaled integers. Values for the derivative, f' may also be stored in the table. The choice of Δx represents a trade-off between the size of the table and the desired resolution of the function. A generic look-up table is given in Table 7.2.

It is well known that the table can be used for the interpolation of $x < \hat{x} < x + \Delta x$ by the formula:

$$f(\hat{x}) = f(x) + (\hat{x} - x)\frac{f(x + \Delta x) - f(x)}{\Delta x} \tag{7.25}$$

This calculation is done using integer instructions except for the final multiplication by the factor $(\hat{x} - x)/\Delta x$ and conversion to floating point. As a bonus, the look-up table has faster execution time if $\hat{x}$ happens to be one of the table values. If $f'(x)$ is also stored in the table, then the look-up formula becomes:

$$f(\hat{x}) = f(x) + (\hat{x} - x)f'(x) \tag{7.26}$$

This improves the execution time of the interpolation somewhat.

The main advantage in using look-up tables, of course, is speed. If a table value is found and no interpolation is needed, then the algorithm is much faster than the corresponding series expansion. In addition, even if interpolation is necessary,

Table 7.2   A generic function look-up table

x	$f(x)$
x_0	$f(x_0)$
$x_0 + \Delta x$	$f(x_0 + \Delta x)$
$x_0 + 2\Delta x$	$f(x_0 + 2\Delta x)$
$\vdots$	$\vdots$
$x_0 + (n-1)\Delta x$	$f(x_0 + (n-1)\Delta x)$

Table 7.3   Look-up table for trigonometric functions

Angle (rads)	Cosine	Sine	Angle (rads)	Cosine	Sine
0.000	1.000	0.000	6.981	0.766	0.643
0.698	0.766	0.643	7.679	0.174	0.985
1.396	0.174	0.985	8.378	−0.500	0.866
2.094	−0.500	0.866	9.076	−0.940	0.342
2.793	−0.940	0.342	9.774	−0.940	−0.342
3.491	−0.940	−0.342	10.472	−0.500	−0.866
4.189	−0.500	−0.866	11.170	0.174	−0.985
4.887	0.174	−0.985	11.868	0.766	−0.643
5.585	0.766	−0.643	12.566	1.000	0.000
6.283	1.000	0.000			

the algorithm is interruptible, and hence helps improve performance as compared to a series expansion.

Look-up tables are widely used in the implementation of continuous functions such as the exponential sine, cosine, and tangent functions and their inverses. For example, consider the combined look-up table for sine and cosine using radian measure shown in Table 7.3. Because these trigonometric functions and exponentials are used frequently in conjunction with the discrete Fourier transform (DFT) and discrete cosine transform (DCT), look-up tables can provide considerable savings in real-time signal processing applications.

7.5.5   Imprecise Computation

In some applications partial results can often be given in order to meet a deadline. In cases where software routines are needed to provide mathematical support (in the absence of firmware support or digital signal processing (DSP) coprocessors), complex algorithms are often employed to produce the desired calculation. For example, a Taylor series expansion (perhaps using look-up tables for function derivatives) can be terminated early, at a loss of accuracy, but with improved performance. Techniques involving early truncation of a series in order to meet deadlines are often called imprecise computation. Imprecise computation (also called approximate reasoning) is often difficult to apply, however, because it is not always easy to determine the processing that can be discarded, and its cost.

7.5.6 Optimizing Memory Usage

In modern computer architectures memory constraints are not as troublesome as they once were. Nevertheless, in embedded applications or in legacy systems (those that are being reused), often the real-time systems engineer is faced with restrictions on the amount of memory available for program storage or for scratch-pad calculations, dynamic allocation, and so on. Since there is a fundamental trade-off between memory usage and CPU utilization (with rare exceptions), when it is desired to optimize for memory usage, it is necessary to trade performance to save memory. For example in the trigonometric function just discussed, using quadrant identities can reduce the need for a large look-up table. The additional logic needed, however, represents a small run-time penalty.

Finally, it is important to match the real-time processing algorithms to the underlying architecture. In the case of the von Neumann architecture, for example, it is helpful to recognize the effects of such features as cache size and pipeline characteristics. In the case of cache size, for example, the algorithm should be chosen to optimize the cache hit ratio, that is, the percentage of time that data are found in the cache. In the case of pipelines, increasing the code locality of reference can reduce the amount of deleterious pipeline flushing.

7.5.7 Postintegration Software Optimization

After system implementation a variety of techniques can be used in conjunction with high-level languages to squeeze additional performance from the system. These techniques include the use of assembly language patches and hand-tuning compiler output. Often, however, use of these practices leads to code that is unmaintainable and unreliable because it may be poorly documented. More desirable, then, is to use coding "tricks" that involve direct interaction with the high-level language and that can be documented. These tricks improve real-time performance, but generally not at the expense of maintainability and reliability.

7.6 RESULTS FROM COMPILER OPTIMIZATION

It is important to experiment with the compiler and to know how it will react to certain high-order language constructs such as case statements versus nested if-then-else statements, integer versus character variables, and so on. As discussed in Chapter 6, a set of extensive test cases should be prepared for the high-order language in question to expose the intricacies of the compiler.

Moreover, many of the techniques used in code optimization underscore the fact that in any arithmetic expression there is no substitute for sound mathematical technique. And it is important to reformulate any algorithm or express to eliminate time-consuming function calls such as those that compute exponentials, square roots, and transcendental functions, where possible, to enhance real-time performance.

Finally, most of the code optimization techniques used by compilers can be exploited to improve real-time performance. Often these strategies will be employed invisibly by the compiler, or can be turned on or off with compiler directives or switches, but it should be known which ones are available. If a particular strategy is not being used, it can therefore be implemented at the high-order or assembly language level.

Consider some commonly used optimization techniques and their impact on real-time performance. These techniques include:

- Use of arithmetic identities
- Reduction in strength
- Common subexpression elimination
- Use of intrinsic functions
- Constant folding
- Loop invariant removal
- Loop induction elimination
- Use of revisers and caches
- Dead-code removal
- Flow-of-control optimization
- Constant propagation
- Dead-store elimination
- Dead-variable elimination
- Short-circuit Boolean code
- Loop unrolling
- Loop jamming
- Cross-jump elimination

Many of these techniques are facilitated through the use of peephole optimization. In peephole optimization a small window or peephole of assembly language code is compared against known patterns that yield optimization opportunities. These types of optimizers are easy to implement and allow for multiple optimization passes to be performed.

7.6.1 Use of Arithmetic Identifies

Good compilers should use arithmetic identities to eliminate useless code. For example, multiplication by the constant "1" or addition by the constant "0" should be eliminated from executable code, although the use of symbolic constants can obscure these situations.

7.6.2 Reduction in Strength

Reduction in strength refers to the use of the fastest macroinstruction possible to accomplish a given calculation. For example, when optimizing for speed many compilers will replace multiplication of an integer by another integer that is a power of 2 by a series of shift instructions. Shift instructions are typically faster than integer multiplication.

In some compilers, character variables are rarely loaded in registers, whereas integer variables are. This may be because it is assumed that calculations involving the integers will take place, whereas those involving characters are unlikely. Care should therefore be taken in deciding whether a variable should be a character or an integer.

Furthermore, it is well known that divide instructions typically take longer to execute than multiply instructions. Hence, it may be better to multiply by the reciprocal of a number than to divide by that number. For example, x*0.5 will be faster than x/2.0. Many compilers will not do this automatically.

7.6.3 Common Subexpression Elimination

Repeated calculations of the same subexpression in two different equations should be avoided. For example, the following C program fragment:

```
x=6+a*b;
y=a*b+z;
```

could be replaced with

```
t=a*b;
x=y+t;
y=t+z;
```

thus eliminating the additional multiplication. This can result in significant savings if a and b are floating-point numbers and the code occurs in a tight loop.

7.6.4 Intrinsic Functions

When possible, use intrinsic functions rather than ordinary functions. Intrinsic functions are simply macros where the actual function call is replaced by in-line code during compilation. This improves real-time performance because the need to pass parameters, create space for local variables, and release that space, is eliminated.

7.6.5 Constant Folding

Most compilers perform constant folding, but this should not be assumed. As an example, the statement:

```
x=2.0*x*4.0;
```

would be optimized by folding 2.0 * 4.0 into 8.0. Performing this operation manually leads to code that is easier to debug, because the programmer performs the optimization explicitly. And although the original statement may be more descriptive, a comment can be provided to explain the optimized statement.

For example, if the program uses $\pi/2$ it should be precomputed during the initialization and stored as a constant named, for example, pi_div_2. This will typically save one floating-point load and one floating-point divide instruction – potentially tens of microseconds. In a 5-millisecond real-time cycle, this can lead to time-loading savings of 0.1%. Incidentally, using this strategy again illustrates the inverse relationship between time and memory utilization: code execution time has been reduced, but extra memory is needed to store the pre-computed constant.

7.6.6   Loop Invariant Optimization

Most compilers will move computations outside loops that do not need to be performed with the loop, a process called loop invariant optimization. For example, consider the following code fragment in C:

```
x=100;
while (x>0)
    x = x-y+z;
```

it can be replaced by

```
x=100;
t=y+z;

while(x>0)
    x=x-t;
```

This moves an instruction outside the loop, but requires additional memory.

7.6.7   Loop Induction Elimination

A variable i is called an induction variable of a loop if every time i changes its value, and it is incremented or decremented by some constant. A common situation is one in which the induction variable is i and another variable, j, which is a linear function if it, is used to offset into some array. Often i is used only for a test of loop termination. Variable i can be eliminated by replacing its test for one on j. For example, consider the following C program fragment:

```
for (i=1,i<=10;i++)
    a [i+1] = 1;
```

an optimized version is

```
for (i=2,i<=11;i++)
    a[j] = 1;
```

eliminating the extra addition within the loop.

7.6.8   Use of Registers and Caches

When programming in assembly language, or when using languages that support register-type variables, such as C, it is usually advantageous to perform calculations using registers. Typically, register-to-register operations are faster than register-too-memory ones. Thus, if variables are used frequently within a module, and if enough registers are available, the compiler should be forced to generate register-direct instructions, if possible.

If the processor architecture supports memory caching, then it may be possible to force frequently used variables into the cache at the high-order language or assembly language level. Although most optimizing compilers will cache variables where possible, the nature of the source-level code affects the compiler's abilities.

7.6.9   Removal of Dead or Unreachable Code

One of the easiest methods for decreasing memory utilization is to remove dead or unreachable code – that is, code that can never be reached in the normal flow-of-control. Such code might be debug statements that are executed only if a debug flag is set, or redundant initialization instructions used. For example, consider the following C program fragment:

```
if (debug)
{
   ...
}
```

The test of the variable debug will take several microseconds, time that is consumed regardless of whether or not the code is in debug mode. Debug code should be implemented using the conditional compile facilities available with most language compilers. Thus, replace the previous fragment with

```
#ifdef DEBUG
 {
   ...
 }
#endif
```

Here, #ifdef is a compiler directive that will include the code between it and the first #endif only if the symbolic constant DEBUG is so defined. Dead code removal also increases program reliability.

7.6.10   Flow-of-Control Optimization

In flow-of-control optimization, unnecessary jump-to-jump statements are replaced by a single-jump statement. The following pseudocode illustrates the situation:

```
        goto label1;
label0  y=1;
label1  goto label2;
```

can be replaced by

```
        goto label2:
label0  y=1;
label1  goto label2;
```

Such code is not normally generated by programmers, but might result from some automatic generation or translation process and escape unnoticed.

7.6.11 Constant Propagation

Certain variable assignment statements can be changed to constant assignments, thereby permitting registerization opportunities or the use of immediate modes. In C, the following code might appear as the result of an automated translation process:

```
x=100;
y=x;
```

The corresponding 2-address assembly language code generated by a nonopti- mizing compiler might look like:

```
LOAD    R1,100
STORE   R1,&x
LOAD    R1,&x
STORE   R1,&y
```

This could be replaced by

```
x =100;
y = 100;
```

With associated 2-address assembly output:

```
LOAD    R1,100
STORE   R1,&x
STORE   R1,&y
```

Again, this type of code often appears during mechanical translation form one language to another.

7.6.12 Dead-Store Elimination

Variables that contain the same value in a short piece of code can be combined into a single temporary variable. For example,

```
t=y+z;
x=func(t);
```

Although many compilers might generate an implicit temporary location for y+z, this cannot always be relied on. By replacing the code in question with the following:

```
x=func(y+z);
```

forces the generation of a temporary location and eliminates the need for the local variable, t.

7.6.13 Dead-Variable Elimination

A variable is live at a point in a program if its value can be used subsequently; otherwise it is dead and subject to removal. The following code illustrates that z is a dead variable:

```
x=y+z;
x=y;
```

after removal of z, what is left is

```
x=y;
```

While this example appears to be trivial, again it could arise as the result of poor coding or an automated code generation or translation process.

7.6.14 Short-Circuiting Boolean Code

The test of compound Boolean expressions can be optimized by testing each subexpression separately. Consider the following:

```
if (x > 0 && y>0)
    z = 1;
```

which could be replaced by

```
if (x>0)
  if (y>0)
    z = 1;
```

In many compilers, the code generated by the second fragment will be superior to the first. ANSI-C, however, executes if(expression) constructs sequentially inside the () and drops out the first FALSE condition. That is, it will automatically short-circuit Boolean code.

7.6.15 Loop Unrolling

Loop unrolling duplicates statements executed in a loop in order to reduce the number of operations, and hence the loop overhead incurred. In the exaggerated case, the loop is completely replaced by inline code. For example,

```
for(i=1;i<=6;i++)
    a[i] = a[i]*8;
```

is replaced by

```
a[1]=a[1]*8;
a[2]=a[2]*8;
a[3]=a[3]*8;
a[4]=a[4]*8;
a[5]=a[5]*8;
a[6]=a[6]*8;
```

A less dramatic version reduces the loop overhead by a factor of 3:

```
for (i=1;i<=6;i+3)
    {
        a[i]=a[i]*8;
        a[i+1]=a[i+1]*8;
        a[i+2]=a[i+2]*8;
    };
```

7.6.16   Loop Jamming

Loop jamming or loop fusion is a technique for combining two similar loops into one, thus reducing loop overhead by a factor of 2. For example, the following C code:

```
for (i=1;i<=100;i++)
    x[i]=y[i]*8;

for (i=1;i<=100;i++)
    z[i]=x[i]*y[i];
```

can be replaced by

```
for (i=1;i<=100;i++)
{
        x[i]=y[i]*8;
        z[i]=x[i]*y[i];
};
```

7.6.16.1   Cross-Jump Elimination   If the same code appears in more than one case in a case or switch statement, then it is better to combine the cases into one. This eliminates an additional jump or cross jump. For example, the following code :

```
switch (x)
{
   case 0:  x=x+1;
            break;
   case 1:  x=x*2;
            break;
```

```
   case 2:  x=x+1;
            break;
   case 3:  x=2;
            break;
};
```

can be replaced by

```
switch (x)
{
  case 0:
  case 2:  x=x+1;
           break;
  case 1:  x=x*2;
           break;
  case 3:  x=2;
           break;
};
```

7.6.17   More Optimization Techniques

A sampling of other optimization considerations follows [Jain91]. Note that in most cases these techniques will optimize the average case, not the worst case.

- Optimize the common case. The most frequently used path should also be the most efficient. For example, arrange a series of IF statements so that the most likely value is tested first. Similarly, arrange a series of AND conditions so that the condition most likely to fail is tested first. Likewise, arrange OR conditions so the most likely to succeed is tested first. This technique is called Boolean short-circuiting.
- Arrange entries in a table so that the most frequently sought values are the first to be compared (this is just a variation of the preceding).
- Replace threshold tests on monotone (continuously nondecreasing or nonincreasing) functions by tests on their parameters, thereby avoiding evaluation of the function. For example, if exp(x) is a function computing e^x, then instead of using:

```
if  (exp(x) < exp(y)) then . . .
```

use:

```
if (x < y) then . . .
```

which will save two evaluations of the costly function exp(x).
- Link the most frequently used procedures together to maximize the locality of the reference (this only applies in paging or cached systems).
- Store redundant data elements to increase the locality of reference. This technique only applies in paging or cached systems.

- Store procedures in memory in sequence so that calling and called sub-routines (procedures) will be loaded together to increase the locality of reference. Again, this only applies in paging or cached systems.

7.6.18   Combination Effects

Although many of the optimization techniques discussed can be automated, most compilers only perform one optimization pass, overlooking opportunities that are not revealed until after at least one pass. Hence, hand optimization can provide additional execution time savings. To see the effects of multiple-pass optimization, consider the following example. The C code fragment:

```
for (j=1;i<=3;j++)
{
    a[j]=0;
    a[j]=a[j]+2*x
};

for (k=1;k<=3;k++)
  b[k]=b[k]+a[k]+2*k*k;
```

is improved by loop jamming, loop invariant removal, and removal of extraneous code (in this case the initialization of a[j]). The resultant code is:

```
t=2*x;
for (j=1;j<=3;j++)
{
    a[j]=t;
    b[j]=b[j]+a[j]+2*j*j;
  };
```

Next, loop unrolling yields:

```
t=2*x;
a[1]=t ;
b[1]=b[1]+a[1]+2*1*1;
a[2]=t ;
b[2]=b[2]+a[2]+2*2*2;
a[3]=t ;
b[3]=b[3]+a[3]+2*3*3;
```

Finally, after constant folding, the improved code is

```
t=2*x;
a[1]=t;
b[1]=b[1]+a[1]+2;
a[2]=t;
b[2]=b[2]+a[2]+8;
a[3]=t;
```

The original code involved nine additions and nine multiplications, numerous data movement instructions, and loop overheads. The improved code requires only six additions, 1 multiplication, less data movement, and no loop overhead.

It is very unlikely that any compiler would have been able to make such an optimization.

7.6.19 Speculative Execution

Speculative execution is a compiler optimization technique used in multiprocessing software systems. Speculative execution optimization is similar to that used in pipeline computer architectures (recall the prefetch of the next sequential instructions from memory). In the case of compiled code, speculative execution involves an idle processor optimistically and predictively executing code in the next process block, so long as there is no dependency in that process block on code that could be running on other processors.

The benefit of this scheme is that idle time and response times can be reduced on individual processors and for the overall system. However, there is a net penalty on the average case performance if the optimistic code execution is nullified by some activity on another processor, for example, a time penalty is incurred in order to roll back (as in pipeline flushing).

7.7 ANALYSIS OF MEMORY REQUIREMENTS

With memory becoming denser and cheaper, memory utilization analysis has become less of a concern. Still, its efficient use is important in small embedded systems and air and space applications where savings in size, power consumption, and cost are desirable.

The total memory utilization is typically the sum of the individual memory utilization for each of the memory areas. Suppose that the memory consists of the program, stack, and RAM areas (see Figure 2.10). That is,

$$M_T = M_P \cdot P_P + M_R \cdot P_R + M_S \cdot P_S \qquad (7.27)$$

where M_T is the total memory utilization, M_P, M_R, and M_S are the memory utilization for the program, RAM, and stack areas, respectively, and P_P, P_R, and P_S are percentages of total memory allocated for the program, RAM, and stack areas, respectively. Memory-mapped I/O and DMA memory are not included in the memory-utilization equation (Equation 7.27), since they are fixed in hardware.

The program area of memory is generally ROM, which contains the executable code of the real-time program, including the operating system and applications software. In addition, fixed constants can be stored in this area. Here memory utilization is calculated simply by dividing the number of used locations in the program area by the allowable locations.

$$M_P = \frac{U_P}{T_P} \qquad (7.28)$$

where M_p is the memory utilization for the program area, U_P is the number of locations used in the program area, and T_P is the total available locations in the program area. These numbers are available as output from the linker.

Although the program instructions may be stored in RAM instead of ROM for increased fetching speed and modifiability, all global variables should be stored in RAM. While the size of this area is determined at system design time, the loading factor for this area is not determined until the application programs have been completed. In any case, the memory utilization factor can be computed as

$$M_R = \frac{U_R}{T_R} \tag{7.29}$$

where M_R, is the memory utilization for the RAM area, U_R is the number of locations used in the RAM area, and T_R is the total available locations in the RAM area. Again, these numbers are available as output from the linker.

For the stack area the memory utilization factor can be computed as

$$M_S = \frac{U_S}{T_S} \tag{7.30}$$

where M_S is the memory utilization for the stack area, U_S is the number of locations used in the stack area, and T_S is the total available locations in the stack area. For example, a computer system has 64 megabytes of program memory that is loaded at 75%, 24 megabytes of RAM area that is loaded at 25%, and 12 megabytes of stack area that is loaded at 50%. The total memory utilization is

$$M_T = 0.75 \cdot \frac{64}{100} + 0.25 \cdot \frac{24}{100} + 0.50 \cdot \frac{12}{100} = 60\%$$

7.8 REDUCING MEMORY UTILIZATION

As mentioned previously, memory utilization[3] is less of a problem than it has been in the past, but occasionally a system needs to be designed in which the available main memory is small in relation to the program size. Moreover, it is expected that this situation will arise more frequently in the future, as ubiquitous and mobile computing applications call for very small processors with tiny memories. Most of the approaches developed to reduce memory utilization date from a time when memory was at a premium and might violate the principles of software engineering. Thus, they should be used with caution.

7.8.1 Variable Selection

Memory utilization in one area can be reduced at the expense of another. For example, all automatic variables (variables that are local to procedures) increase

---

[3] Earlier editions of this book included discussions of the once necessary and accepted techniques of self-modifying code and reusing global memory. While it was possible to retain these discussions for historical and entertainment purposes, there is no place for these techniques in modern software engineering.

the loading in the stack area of memory, whereas global variables appear in the RAM area. By forcing variables to be either local or global, relief can be purchased in one area of memory at the expense of the other, thus balancing the memory load.

In addition, intermediate result calculations that are computed explicitly require a variable either in the stack or the RAM area, depending on whether it is local or global. The intermediate value can be forced into a register instead by omitting the intermediate calculation. To illustrate, consider this C code fragment that calculates one root of a quadratic:

```
discriminant =b*b-4*a*c;
root=(-b+sqrt(discriminant))*0.5/a;
```

this code could be replaced by

```
root=(-b+sqrt(b*B-4*a*c)*0.5/a;
```

which saves one floating-point variable and thus at least 4 bytes of memory. In addition, this eliminates at least one STORE macroinstruction, reducing time-loading as well.

7.8.2 Memory Fragmentation

Memory fragmentation does not impact memory utilization, but it can produce effects resembling memory overloading. In this case, although sufficient memory is available, it is not contiguous. Although compaction schemes were discussed in Chapter 3 and it was noted that they were not desirable in real-time systems, they may be necessary in serious cases of memory overutilization.

7.9 EXERCISES

7.1 Can the number of instructions executed serve as a timer mechanism in a real-time operating system? How?

7.2 Derive the look-up table for the tangent function in increments of 1 degree. Be sure to take advantage of symmetry.

7.3 Suppose x is a 16-bit BAM word representing the angle 225° and y is a 16-big BAM word representing 157.5°. Using two's complement addition show that $x + y = 22.5°$.

7.4 What is the range of an unsigned scaled 16-bit number with least significant bit -0.00043?

7.5 What are the advantages and disadvantages of writing a BAM object class in an object-oriented language?

7.6 Write a program in the language of your choice, which takes an arbitrary function and table increment and writes a look-up table functional. The arbitrary function will have to be hard-coded into the program, but the table size or table increment can be input interactively. Creation of such a program will increase the accuracy of your table look-up functions and reduce the time needed to write them.

7.7   A polled loop system polls a discrete signal every 50 microseconds. Testing the signal and vectoring to the interrupt-processing routine take 40 microseconds. If it takes 6.2 milliseconds to process the interrupt, what is the minimum response time for this interrupt? What is the maximum response time?

7.8   A consumer process can read the data in 32-bit words but only at a rate of one word every 2 microseconds. Calculate the minimum-size buffer required to avoid spillover, assuming there is enough time between bursts to empty the buffer.

7.9   Show that when the producer and consumer tasks have constant rates, then Equation 7.20 becomes Equation 7.19.

7.10   A producer process is known to be able to process data at a rate that is exponentially distributed with average service time of 3 milliseconds per datum. What is the maximum allowable average data rate if the probability of collision is to be 0.1%? Assume that the data arrive at intervals that are exponentially distributed.

7.11   Consider a foreground/background system that has three cycles: 10 millisecond, 40 millisecond, and 1 second. If the cycle completion times have been estimated at 4 milliseconds, 12 milliseconds, and 98 milliseconds, respectively, what is the CPU utilization of the system?

7.12   What is the worst-case response time for the background process in a foreground/background system in which the background task requires 100 milliseconds to complete, the foreground task executes every 50 milliseconds and requires 25 milliseconds to complete, and context switching requires no more than 100 microseconds?

7.13   Consider a preemptive priority system. The tasks in the system, time needed to complete, and priority (1 being the highest) are given below:

Task	Time Needed (ms)	Priority
Task 1	40	3
Task 2	20	1
Task 3	30	2

If the tasks arrive in the order 1, 2, 3, what is the time needed to complete task 2?

7.14   A preemptive foreground/background system has three interrupt-driven cycles, described by Table 7.4 (with context switch time ignored):

Table 7.4   Table for time-loading Exercise 7.14

Task Cycle	Actual Execution Time (ms)	Priority (1 is highest, 10 is lowest)
10 ms	4	1
20 ms	5	3
40 ms	10	2
Background	5	–

(a) Draw an execution time line for this system.

(b) What is the CPU utilization?

(c) Considering the context switch time to be 1 millisecond, redraw the execution time line for this system.

(d) What is the system time-loading (CPU utilization) factor with the context switch time included?

7.15 A periodic task τ_i with phase ϕ_i, period p_i, execution time e_i, and relative deadline D_i is represented by the 4-tuple (ϕ_i, p_i, e_i, D_i). For example, (1, 10, 3, 6) is a periodic task whose phase is 1, period is 10, execution time is 3, and relative deadline is 6. Using this notation, consider the following problem. The total utilization of the period task (7, 10, 1, 10), (0, 12, 6, 12), (0, 25, 9, 25) is 0.96. Is it schedulable by the rate-monotonic algorithm? Draw the rate-monotonic schedule.

τ_i	r_i	e_i	p_i
u_1	1	1	3
τ_2	0	1	5
τ_3	1	3	10

7.16 What characteristics of reduced instruction set computer (RISC) architectures tend to reduce the total interrupt latency time as compared to complex instruction set computer (CISC)?

7.17 A computer has instructions that require two bus cycles, one to fetch the instruction and one to fetch the data. Each bus cycle takes 250 nanoseconds and each instruction takes 500 nanoseconds (i.e., the internal processing time is negligible). The computer also has a disk with 16,512 byte sectors per track. Disk rotation time is 8.092 milliseconds. To what percentage of its normal speed is the computer reduced during DMA transfer if each DMA takes one bus cycle? Consider two cases: 8-bit bus transfer and 16-bit bus transfer.

7.18 Use optimization methods to optimize the following C code:

```
#define UNIT 1
#define FULL 1

void main(void)
{
    int a,b;

    a = FULL;
    b=a;

    if ((a==FULL)&&(b==FULL))
    {
        if (debug)
            printf("a=%d b=%d", a,b);
        a=(b * UNIT) /2;
```

```
        a= 2.0 * a * 4;
        b=d sqrt(a);
    }
}
```

7.19 What unique challenges are presented in performance analysis for the example systems described in Chapter 1? Namely:

(a)   Inertial measurement system

(b)   Nuclear monitoring system

(c)   Airline reservation system

(d)   Pasta sauce bottling system

(e)   Traffic light control

8

ENGINEERING CONSIDERATIONS

8.1 METRICS

Metrics can be used in real-time systems engineering in several ways. First, certain metrics can be used during software requirements development to assist in cost estimation. Another useful application for metrics is for benchmarking. For example, if an engineering group has a set of successful systems, then computing metrics for those systems yields a set of desirable and measurable characteristics with which to seek or compare in future systems. Most metrics can be used for testing in the sense of measuring the desirable properties of the software and setting limits on the bounds of those criteria.

Of course, metrics can be used to track project progress. In fact, some companies reward employees based on the amount of software developed per day as measured by some of the metrics to be discussed (e.g., delivered source instructions, function points, or lines of code). Finally, metrics can be used during the testing phase and for debugging purposes to help focus on likely sources of errors.

8.1.1 Lines of Code

The easiest characteristic of software that can be measured is the number of lines of finished source code. Measured as thousands of lines of code (KLOC), the "clock" metric is also referred to as delivered source instructions (DSI) or

Some of this chapter has been adapted from Phillip A. Laplante, *Software Engineering for Image Processing*, CRC Press, Boca Raton, FL, 2003.

Real-Time Systems Design and Analysis, By Phillip A. Laplante
ISBN 0-471-22855-9 © 2004 Institute of Electrical and Electronics Engineers

noncommented source-code statements (NCSS). That is, the number of executable program instructions, excluding comment statements, header files, formatting statements, macros, and anything that does not show up as executable code after compilation or cause allocation of memory, are counted. Another related metric is source lines of code (SLOC), the major difference being that a single source line of code may span several lines. For example, an if-then-else statement would be a single SLOC, but multiple delivered source instructions.

While the clock metric essentially measures the weight of a printout of the source code, thinking in these terms makes it likely that the usefulness of KLOC will be unjustifiably dismissed as supercilious. But is it not likely that 1000 lines of code is going to have more errors than 100 lines of code? Would it not take longer to develop the latter than the former? Of course, the answer is dependent on how complex the code is.

One of the main disadvantages of using lines of source code as a metric is that it can only be measured after the code has been written. While it can be estimated beforehand and during software production based on similar projects, this is far less accurate than measuring the code after the fact. Nevertheless, KLOC is a useful metric, and in many cases is better than measuring nothing. Moreover many other metrics are fundamentally based on lines of code. For example, a closely related metric is delta KLOC. Delta KLOC measures how many lines of code change over some period of time. Such a measure is useful, perhaps, in the sense that as a project nears the end of code development, Delta KLOC would be expected to be small. Other, more substantial metrics are also derived from KLOC.

8.1.2 McCabe's Metric

A valid criticism of the KLOC metric is that it does not take into account the complexity of the software involved. For example, one thousand lines of print statements probably has fewer defects than one hundred lines of a real-time kernel.

To attempt to measure software complexity, cyclomatic complexity was introduced to measure program flow-of-control [McCabe76]. This concept fits well with procedural programming, but not necessarily with object-oriented programming, though there are adaptations for use with the latter. In any case, this metric has two primary uses:

1. To indicate escalating complexity in a module as it is coded and therefore assisting the coders in determining the "size" of their modules;
2. To determine the upper bound on the number of tests that must be designed and executed.

8.1.2.1 *Measuring Software Complexity*   The cyclomatic complexity is based on determining the number of linearly independent paths in a program module, suggesting that the complexity increases with this number, and reliability decreases.

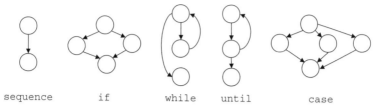

Figure 8.1 Correspondence of language statements and flow graph. (Adapted from [Pressman00].)

To compute the metric, the following procedure is followed. Consider the flow graph of a program where the nodes represent program segments and edges represent independent paths. Let e be the number of edges and n be the number of nodes. Form the cyclomatic complexity, C, as follows:

$$C = e - n + 2 \qquad (8.1)$$

This is the most generally accepted form.[1]

To get a sense of the relationship between program flow for some simple code fragments and cyclomatic complexity, refer to Figure 8.1. Here, for example, a sequence of instructions has two nodes, one edge, and hence a complexity of $C = 1 - 2 + 2 = 1$. This is intuitively pleasing, as nothing could be less complex than a simple sequence. On the other hand, the case statement shown in Figure 8.1 has six edges and five nodes with $C = 6 - 5 + 2 = 3$. The higher value for C is consistent with the notion that a case statement is somehow more complex than a simple sequence of instructions.

As a more substantial example, consider a segment of code extracted from the gyro compensation code for the inertial measurement unit. The procedure calls between modules **a**, **b**, **c**, **d**, **e**, and **f** are depicted in Figure 8.2. Here $e = 9, n = 6$, and the cyclomatic complexity of $C = 9 - 6 + 2 = 5$.

Computation of McCabe's metric can be done easily during compilation by analyzing the internal tree structure generated during the parsing phase. However, commercial tools are available to perform this analysis.

8.1.3 Halstead's Metrics

One of the drawbacks of McCabe's metric is that it measures complexity as a function of control flow. But complexity can exist internally in the way that the programming language is used. Halstead's metrics measure information content,

---

[1] There is some confusion in the literature, however, about the correct formula for C. For example, the following alternative formulations can be found: $C = e - n + 2p$, or $C = e - n + p$ (where p is the sum of degrees of freedom of the predicate nodes, that is, those with degree of 2 or greater). The confusion apparently arises from the transformation of an arbitrary directed graph to a strongly connected, directed graph obtained by adding one edge from the sink to the source node [Jorgensen02].

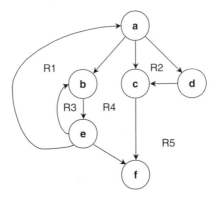

Figure 8.2 Flow graph for gyro compensation code for the inertial measurement unit [Laplante03c].

or how intensively the programming language is used. Halstead's metrics are computed using the following slightly modified algorithm:

- First, find n_1. This is essentially the number of distinct, syntactic begin–end pairs (or their equivalent), called "operators."
- Next find n_2, the number of distinct statements. A statement is determined by the syntax of the language; for example, a line terminated by a semicolon is a statement in C.
- Next count N_1, the total number of occurrences of n_1 in the program.
- Then count N_2, the total number of occurrences of operands or n_2 in the program.

From these statistics the following metrics can be computed.
    The program vocabulary, n, is defined as

$$n = n_1 + n_2 \tag{8.2}$$

The program length, N, is defined as

$$N = N_1 + N_2 \tag{8.3}$$

The program volume, V, is defined as

$$V = N \log_2 n \tag{8.4}$$

The potential volume, V^*, is defined as

$$V^* = (2 + n_2) \cdot \log_2 (2 + n_2) \tag{8.5}$$

The program level, L, is defined as

$$L = V^*/V \tag{8.6}$$

where L is a measure of the level of abstraction of the program. It is believed that increasing this number will increase system reliability.

Another Halstead metric measures the amount of mental effort required in the development of the code. The effort, E, is defined as

$$E = V/L \tag{8.7}$$

Again, decreasing the effort level is believed to increase reliability as well as ease of implementation. In principle, the program length, N, can be estimated, and therefore is useful in cost and schedule estimation. The length is also a measure of the "complexity" of the program in terms of language usage, and therefore can be used to estimate defect rates.

Halstead's metrics, though dating back almost 30 years, are still widely used and tools are available to completely automate their determination. Halstead's metrics can also be applied to requirements specifications as well as to code, by adapting the definitions of "operator" and "statements." In this way, comparative statistics can be generated and estimator effort level determined. From the software requirements specification, Halstead's metrics have also been used for related applications such as identifying whether two programs are identical except for naming changes (something that is useful in plagiarism detection or software patent infringement).

8.1.4 Function Points

Function points were introduced in the late 1970s as an alternative to metrics based on simple source line count. The basis of function points is that as more powerful programming languages are developed the number of source lines necessary to perform a given function decreases. Paradoxically, however, the cost/LOC measure indicated a reduction in productivity, as the fixed costs of software production were largely unchanged.

The solution is to measure the functionality of software via the number of interfaces between modules and subsystems in programs or systems. A big advantage of the function point metric is that it can be calculated before any coding occurs based solely on the design description.

The following five software characteristics for each module, subsystem, or system represent its function points:

- Number of inputs to the application (I)
- Number of outputs (O)
- Number of user inquiries (Q)
- Number of files used (F)
- Number of external interfaces (X)

Now consider empirical weighting factors for each aspect that reflect their relative difficulty in implementation. For example, one set of weighting factors for a particular kind of system might yield the function point (FP) value:

$$FP = 4I + 4O + 5Q + 10F + 7X \qquad (8.8)$$

The weights given in Equation 8.8 can be adjusted to compensate for factors such as application domain and software developer experience. For example, if W_i are the weighting factors, F_j are the "complexity adjustment factors," and A_i are the item counts, then FP is defined as:

$$FP = \sum(A_i \times W_i) \times \left[0.65 + 0.01x \sum F_j\right] \qquad (8.9)$$

Intuitively, the higher FP, the more difficult the system is to implement. A great advantage of the function-point metric is that it can be computed before any coding occurs.

The complexity factor adjustments can be adapted for other application domains such as embedded and real-time systems. To determine the complexity factor adjustments a set of 14 questions are answered by the software engineer(s) with responses from a scale from 0 to 5 where:

0	no influence
1	incidental
2	moderate
3	average
4	significant
5	essential

For example, in the inertial measurement unit system suppose the engineering team was queried and the following interrogatory and resulting answers to the questions were obtained.

Question 1   Does the system require reliable backup and recovery? "Yes, this is a critical system; assign a 4."

Question 2   Are data communications required? "Yes, there is communication between various components of the system over the MIL STD 1553 standard bus; therefore, assign a 5."

Question 3   Are there distributed processing functions? "Yes, assign a 5."

Question 4   Is performance critical? "Absolutely, this is a hard real-time system; assign a 5."

Question 5   Will the system run in an existing, heavily utilized operational environment? "In this case yes; assign a 5."

Question 6   Does the system require on-line data entry? "Yes via sensors; assign a 4."

Question 7 Does the on-line data entry require the input transactions to be built over multiple screens or operations? "Yes it does; assign a 4."

Question 8 Are the master files updated on-line? "Yes they are; assign a 5."

Question 9 Are the inputs, outputs, files, or inquiries complex? "Yes, they involve comparatively complex sensor inputs; assign a 4.

Question 10 Is the internal processing complex? "Clearly it is, the compensation and other algorithms are nontrivial; assign a 4."

Question 11 Is the code designed to be reusable? "Yes, there are high up-front development costs and multiple applications have to be supported for this investment to pay off; assign a 4."

Question 12 Are the conversion and installation included in the design? "In this case, yes; assign a 5."

Question 13 Is the system designed for multiple installations in different organizations? "Not organizations, but in different applications, and therefore this must be a highly flexible system; assign a 5."

Question 14 Is the application designed to facilitate change and ease of use by the user? "Yes, absolutely; assign a 5."

Then applying Equation 8.9 yields:

$$0.01 \sum F_j = 0.01 \cdot (6 \cdot 4 + 8 \cdot 5) = 0.64$$

Now suppose that it was determined from the Software Requirements Specification that the item counts were as follows:

$$A_1 = I = 5$$
$$A_2 = U = 7$$
$$A_3 = Q = 8$$
$$A_4 = F = 5$$
$$A_5 = X = 5$$

Using the weighting factors from Equation 8.8:

$$W_1 = 4$$
$$W_2 = 4$$
$$W_3 = 5$$
$$W_4 = 10$$
$$W_5 = 7$$

Table 8.1 **Programming language and lines of code per function point adapted from [Jones98]**

Language	Lines of Code per Function Point
Assembly	320
C	128
Fortran	106
Pascal	90
C++	64

and putting these into Equation 8.9, yields

$$FP = [5 \cdot 4 + 7 \cdot 4 + 8 \cdot 5 + 5 \cdot 10 + 5 \cdot 7] \, [0.65 + 0.64]$$

$$\approx 223$$

For the purposes of comparison, and as a management tool, function points have been mapped to the relative lines of source code in particular programming languages. These are shown in Table 8.1. For example, it seems intuitively pleasing that it would take many more lines of assembly language code to express functionality than it would in a high-level language like C. In the case of the inertial measurement system, with FP = 223, it might be expected that about 28.5 thousand lines of code would be needed to implement the functionality. In turn, it should take many less to express that same functionality in a more abstract language such as C++. The same observations that apply to software production might also apply to maintenance as well as to the potential reliability of software.

Real-time applications like the inertial measurement system are highly complex and they have many complexity factors rated at five, whereas in other kinds of systems, such as database applications, these factors would be much lower. This is an explicit statement about the difficulty in building and maintaining code for embedded systems versus nonembedded ones.

The function point metric has mostly been used in business processing, and not nearly as much in embedded systems. However, there is increasing interest in the use of function points in real-time embedded systems, especially in large-scale real-time databases, multimedia, and Internet support. These systems are data driven and often behave like the large-scale transaction-based systems for which function points were developed.

The International Function Point Users Group maintains a Web database of weighting factors and function point values for a variety of application domains. These can be used for comparison.

8.1.5 Feature Points

Feature points are an extension of function points developed by Software Productivity Research, Inc., in 1986. Feature points address the fact that the classic function point metric was developed for management information systems and therefore are not particularly applicable to many other systems, such as real-time, embedded, communications, and process control software. The motivation is that these systems exhibit high levels of algorithmic complexity, but relatively sparse inputs and outputs.

The feature point metric is computed in a similar manner to the function point, except that a new factor for the number of algorithms, A, is added. The empirical weightings are:

$$W_1 = 3$$
$$W_2 = 4$$
$$W_3 = 5$$
$$W_4 = 4$$
$$W_5 = 7$$
$$W_6 = 7$$

And the feature point metric, $\overline{FP}$, is then

$$\overline{FP} = 3I + 4O + 5Q + 4F + 7X + 7A \qquad (8.10)$$

For example, in the inertial measurement, using the same item counts as computed before, and supposing that the item count for algorithms, $A = 10$, and using the same complexity adjustment factor, $\overline{FP}$ would be computed as follows:

$$\overline{FP} = [5 \cdot 3 + 7 \cdot 4 + 8 \cdot 5 + 10 \cdot 4 + 5 \cdot 7 + 10 \cdot 7] \, [0.65 + 0.64]$$
$$\approx 294$$

If the system were to be written in C, it could be estimated that approximately 37.6 thousand lines of code would be needed, a slightly more pessimistic estimate than that computed using the function point metric.

8.1.6 Metrics for Object-Oriented Software

While any of the previously discussed metrics can be used in object-oriented code, other metrics are better suited for this setting. For example, some of the metrics that have been used include:

- A weighted count of methods per class.
- The depth of inheritance tree.
- The number of children in the inheritance tree.
- The coupling between object classes.
- The lack of cohesion in methods.

As with other metrics, the key to use is consistency.

8.1.7 Objections to Metrics

There are many who object to the use of metrics in one or all of the ways that have been described. Several counterarguments to the use of metrics have been stated, for example, that they can be misused or that they are a costly and an unnecessary distraction. For example, metrics related to the number lines of code imply that the more powerful the language, the less productive the programmer. Hence, obsessing with code production based on lines of code is a meaningless endeavor.

Metrics can also be misused through sloppiness, which can lead to bad decision making. Finally, metrics can be misused in the sense that they can be abused to "prove a point." For example, if a manager wishes to assert that a particular member of the team is "incompetent," he or she can simplistically base his or her assertion on the lines of code produced per day without accounting for other factors.

Another objection is that measuring the correlation effects of a metric without clearly understanding the causality is unscientific and dangerous. For example, while there are numerous studies suggesting that lowering the cyclomatic complexity leads to more reliable software, there just is no real way to know why. Obviously the arguments about the complexity of well-written code versus "spaghetti code" apply, but there is just no way to show the causal relationship. So, the opponents of metrics might argue that in if a study of several companies it was shown that software written by software engineers who always wore yellow shirts had statistically significant fewer defects in their code, companies would start requiring a dress code of yellow shirts! This illustration is, of course, hyperbole, but the point of correlation versus causality is made. While it is possible that in many cases these objections may be valid, like most things, metrics can be either useful or harmful, depending on how they are used (or abused).

8.1.8 Best Practices

The objections raised about metrics however, suggest that best practices need to be used in conjunction with metrics. These include establishing the purpose, scope, and scale if the metrics. In addition, metrics programs need to be incorporated into the management plan by setting solid measurement objectives and plans and embedded measurement throughout the process. Also, it is important to create a culture where honest measurement and collection of data is encouraged and rewarded.

8.2 FAULTS, FAILURES, AND BUGS

There is more than a subtle difference between the terms fault, failure, bug, and defect. Use of "bug" is, in fact, discouraged, since it somehow implies that an error crept into the program through no one's action. The preferred term for an error in requirement, design, or code is "error" or "defect." The manifestation of a defect during the operation of the software system is called a fault. A fault that causes the software system to fail to meet one of its requirements is a failure.[2]

8.2.1 The Role of Testing

From 1985 to 1987, faulty software in a Therac-25 radiation treatment system made by Atomic Energy of Canada Limited (AECL) resulted in several cancer patients receiving lethal doses of radiation. A subsequent investigation found that the basic mistakes involved poor testing and debugging. Clearly, such a real-time system in which human life is at risk, verification and validation of the software is crucial [Cnet00].

Verification determines whether the products of a given phase of the software development cycle fulfill the requirements established during the previous phase. Verification answers the question, "Am I building the product right?"

Validation determines the correctness of the final program or software with respect to the user's needs and requirements. Validation answers the question, "Am I building the right product?"

Testing is the execution of a program or partial program with known inputs and outputs that are both predicted and observed for the purpose of finding faults or deviations from the requirements.

Although testing will flush out errors, this is just one of its purposes. The other is to increase trust in the system. Perhaps once, software testing was thought of as intended to remove all errors. But testing can only detect the presence of errors, not the absence of them, therefore, it can never be known when all errors have been detected. Instead, testing must increase faith in the system, even though it still may contain undetected faults, by ensuring that the software meets its requirements. This objective places emphasis on solid design techniques and a well-developed requirements document. Moreover, a formal test plan must be developed that provides criteria used in deciding whether the system has satisfied the requirements.

8.2.2 Testing Techniques

There is a wide range of testing techniques for unit- and system-level testing, desk checking, and integration testing. Some techniques are often interchangeable, while others are not. Any one of these test techniques can be either insufficient or not computationally feasible for real-time systems. Therefore, some combination of testing techniques is almost always employed. Recently, commercially and open-source user-guided test-case generators have emerged. These tools (e.g., X Unit) can greatly facilitate many of the testing strategies to be discussed.

---

[2] Some define a fault as an error found prior to system delivery and a defect as an error found post delivery.

8.2.2.1 Unit Level Testing   Several methods can be used to test individual modules or units. These techniques can be used by the unit author and by the independent test team to exercise each unit in the system. These techniques can also be applied to subsystems (collections of modules related to the same function).

Black-Box Testing   In black-box testing, only inputs and outputs of the unit are considered; how the outputs are generated based on a particular set of inputs is ignored. Such a technique, being independent of the implementation of the module, can be applied to any number of modules with the same functionality. But this technique does not provide insight into the programmer's skill in implementing the module. In addition, dead or unreachable code cannot be detected.

For each module a number of test cases need to be generated. This number depends on the functionality of the module, the number of inputs, and so on. If a module fails to pass a single-module-level test, then the error must be repaired, and all previous module-level test cases are rerun and passed to prevent the repair from causing other errors.

Some widely used black-box testing techniques include:

- Exhaustive testing
- Boundary-value testing
- Random test generation
- Worst-case testing

An important aspect of using black-box testing techniques is that clearly defined interfaces to the modules are required. This places additional emphasis on the application of Parnas Partitioning principles to module design.

Exhaustive Testing   Brute-force or exhaustive testing involves presenting each code unit with every possible input combination. Brute-force testing can work well in the case of a small number of inputs, each with a limited input range, for example, a code unit that evaluates a small number of Boolean inputs. A major problem with brute-force testing, however, is the combinatorial explosion in the number of test cases. For example, for the code that will deal with raw accelerometer data $3 \cdot 2^{16}$, test cases would be required, which could be prohibitive.

Boundary-Value Testing   Boundary-value or corner-case testing solves the problem of combinatorial explosion by testing some very tiny subset of the input combinations identified as meaningful "boundaries" of input. For example, consider a code unit with five different inputs, each of which is a 16-bit signed integer. Approaching the testing of this code unit using exhaustive testing would require $2^{16} \cdot 2^{16} \cdot 2^{16} \cdot 2^{16} \cdot 2^{16} = 2^{80}$ test cases. However, if the test inputs are restricted to every combination of the min, max, and average values for each input, then the test set would consist of $3^5 = 243$ test cases. A test set of this size can be handled easily with automatic test-case generation.

Random Test-Case Generation   Random test-case generation, or statistically based testing, can be used for both unit- and system-level testing. This kind of testing involves subjecting the code unit to many randomly generated test cases over some period of time. The purpose of this approach is to simulate execution of the software under realistic conditions.

The randomly generated test cases are based on determining the underlying statistics of the expected inputs. The statistics are usually collected by expert users of similar systems or, if none exist, by educated guessing. The theory is that system reliability will be enhanced if prolonged usage of the system can be simulated in a controlled environment. The major drawback of such a technique is that the underlying probability distribution functions for the input variables may be unavailable or incorrect. In addition, randomly generated test cases are likely to miss conditions with low probability of occurrence. Precisely this kind of condition is usually overlooked in the design of the module. Failing to test these scenarios is an invitation to disaster.

Worst-Case Testing   Worst-case or pathological-case testing deals with those test scenarios that might be considered highly unusual and unlikely. It is often the case that these exceptional cases are exactly those for which the code is likely to be poorly designed, and therefore, to fail. For example, in the inertial measurement system, while it might be highly unlikely that the system will achieve the maximum accelerations that can be represented in a 16-bit scaled number, this worst case still needs to be tested.

8.2.2.2   *White-Box Testing*   One disadvantage of black-box testing is that it can often bypass unreachable or dead code. In addition, it may not test all of the control paths in the module. Another away to look at this is that black-box testing only tests what is expected to happen, not what was not intended. White-box or clear-box testing techniques can be used to deal with this problem.

Whereas black-box tests are data driven, white-box tests are logic driven, that is, they are designed to exercise all paths in the code unit. For example, in the nuclear plant monitoring system, all error paths would need to be tested, including those pathological situations that deal with simultaneous and multiple failures.

White-box testing also has the advantage that it can discover those code paths that cannot be executed. This unreachable code is undesirable because it is likely a sign that the logic is incorrect, because it wastes code space memory, and because it might inadvertently be executed in the case of the corruption of the computer's program counter.

Code Inspections   Group walkthroughs or code inspections are a kind of white-box testing in which code is inspected line-by-line. Walkthroughs have been shown to be much more effective than testing.

In code inspections, the author of some collection of software presents each line of code to a review group, which can detect errors as well as discover ways for improving the implementation. This audit also provides excellent control of the coding standards. Finally, unreachable code can be discovered.

Formal Methods in Testing   Formal program proving is a kind of white-box testing using formal methods in which the code is treated as a theorem and some form of calculus is used to prove that the program is correct.

A program is said to be partially correct if it produces the correct output for each input if it terminates. It is said to be correct if it is partially correct and it terminates. Hence to verify a program is correct, partial correctness must be demonstrated, and then it must be demonstrated that the program terminates. Recall that the halting problem was shown to be unsolvable, that is, there is no way to write a program that can answer the problem of program termination automatically. That is, it must be shown manually.

To casually illustrate formal program verification, consider the following example. It is casual because some of the more rigorous mathematics are omitted for ease of understanding. Consider a function to compute the power a^b, where a is a floating-point number and b is a nonnegative integer (type and range checking are omitted from the verification because it is assumed that this is done by the run-time library).

```
float power(float real, unsigned b)
{
  if (b==0)
    return 1;
  else
    return a*power(a,b-1);
}
```

In a real-time sense it is more important to show that this program terminates, that is, unbounded recursion does not occur. To show this, note that int b is a loop invariant and that b is monotonically decreasing. Hence, b will eventually become 0, which is the return (termination) condition.

To demonstrate partial correctness, note that $a^b = (\Pi_{i=1}^{b} a) \cdot 1$. Recognizing that the program calls itself b times through the else condition and once through the if condition, yields the equality shown. In its most rigorous form, formal verification requires a high level of mathematical sophistication and is appropriate, generally, only for limited, mission-critical situations because of the intensity of activity.

Testing Object-Oriented Software   A test process that complements object-oriented design and programming can significantly increase reuse, quality, and productivity. There are three issues in testing object-oriented software:

- Testing the base class.
- Testing external code that uses a base class.
- Dealing with inheritance and dynamic binding.

Without inheritance, testing object-oriented code is not very different from simply testing abstract data types. Each object has some data structure, such as an array, and a set of member functions to operate. There are also member functions to operate on the object. These member functions are tested like any other using black-box or white-box techniques.

In a good object-oriented design there should be a well-defined inheritance structure. Therefore, most of the tests from the base class can be used for testing the derived class, and only a small amount of retesting of the derived class is required. On the other hand, if the inheritance structure is bad, for example, if there is inheritance of implementation (where code is used from the base class), then additional testing will be necessary. Hence, the price of using inheritance poorly is having to retest all of the inherited code. Finally, dynamic binding requires that all cases have to be tested for each binding possibility.

Effective testing is guided by information about likely sources of error. The combination of polymorphism, inheritance, and encapsulation is unique to object-oriented languages, presenting opportunities for error that do not exist in conventional languages. The main rule here is that if a class is used in a new context, then it should be tested as if it were new.

Test First Coding Test first coding (or test-driven design) is a code production approach normally associated with eXtreme Programming. In test first coding the test cases are designed by the software engineer who will eventually write the code. The advantage of this approach is that it forces the software engineer to think about the code in a very different way that involves focusing on "breaking down" the software. Those who use this technique report that, while it is sometimes difficult to change their way of thinking, once the test cases have been designed, it is actually easier to write the code, and debugging becomes much easier because the unit-level test cases have already been written. Test first coding is not really a testing technique, it is a design and analysis technique, and it does not obviate the need for testing.

8.2.2.3 *Determining the Limit on Number of Test Cases* As it turns out, cyclomatic complexity measures the number of linearly independent paths through the code, and hence, provides an indication of the minimum number of test cases needed to exercise every code path and provide total code coverage. To determine the linear independent paths, McCabe developed an algorithmic procedure (called the baseline method) to determine a set of basis paths.

First, a clever construction is followed to force the complexity graph to look like a vector space by defining the notions of scalar multiplication and addition along paths. The basis vectors for this space are then determined. The method proceeds with the selection of a baseline path, which should correspond to some "ordinary" case of program execution along one of the basis vector paths. McCabe advises choosing a path with as many decision nodes as possible. Next the baseline path is retraced, and in turn, each decision is reversed, that is, when a node of outdegree of greater than two is reached, a different path must be taken. Continuing in this way until all possibilities are exhausted, it generates a set of paths representing the test set [Jorgensen02]. For example, consider Figure 8.2. Here the cyclomatic complexity was computed to be 5, indicating that there are five linearly independent test cases. Tracing through the graph, the first path is adcf. Following McCabe's procedure yields the paths **acf, abef, abeb, . . .** , and **abea . . .** . The ellipses indicate that the path includes one or more iterations through paths or subpaths that were already traced.

Function points can also be used to determine the minimum number of test cases needed for coverage. The International Function Point User's Group indicates that there is a strong relationship between the number of test cases, defects, and function points, that is, they are equal. Accordingly, the number of acceptance test cases can be estimated by multiplying the number of function points by 1.2, which is the factor suggested by McCabe. For example, if a project consists of 200 function points, then 240 test cases would be needed.

8.2.2.4 Debugging   In real-time systems, testing methods often affect the systems that they test. When this is the case, nonintrusive testing should be considered. For example, when removing code during debugging, do not use conditional branching; use conditional compilation instead. Conditional branching affects timing and can introduce subtle timing problems, for example, the one discussed in Section 2.5.4.3.

Some Debugging Tips: Unit-Level Testing   Programs can be affected by syntactic or logic errors. Syntactic or syntax errors arise from the failure to satisfy the rules of the language. A good compiler will always detect syntax errors, although the way that it reports the error often can be misleading. For example, in a C program a missing } may not be detected until many lines after it should have appeared. Some compilers only report "syntax error" rather than, for example, "missing }".

In logic errors, the code adheres to the rules of the language, but the algorithm that is specified is somehow wrong. Logic errors are more difficult to diagnose because the compiler cannot detect them, but a few basic rules may help you find and eliminate logic errors.

- Document the program carefully. Ideally, each nontrivial line of code should include a comment. In the course of commenting, this may detect or prevent logical errors.
- Where a symbolic debugging is available, use steps, traces, breakpoints, skips, and so on to isolate the logic error (discussed later).
- Use automated testing where possible. Open source test generators are available, for example, the XUnit family, which includes JUnit for Java and CUnit for C++. These tools help generate test cases and are used for ongoing unit and regression testing of components or classes.
- In the case of a command line environment (such as Unix/Linux) use print statements to output intermediate results at checkpoints in the code. This may help detect logic errors.
- In case of an error, comment out portions of the code until the program compiles and runs. Add in the commented-out code, one feature at a time, checking to see that the program still compiles and runs. When the program either does not compile or runs incorrectly, the last code added is involved in the logic error.

Finding and eliminating errors in real-time systems is as much art than science, and the software engineer develops these skills over time with practice. In many cases, code audits or walkthroughs can be quite helpful in finding logic errors.

Symbolic Debugging   Source-level debuggers are software programs that provide the ability to step through code at either a macroassembly or high-order language level. They are extremely useful in module-level testing. They are less useful in system-level debugging, because the real-time aspect of the system is necessarily disabled or affected.

Debuggers can be obtained as part of compiler support packages or in conjunction with sophisticated logic analyzers. For example, sdb is a generic name for a symbolic debugger associated with Unix and Linux. sdb allows the engineer to single step through the source language code and view the results of each step.

In order to use the symbolic debugger, the source code must be compiled with a particular option set. This has the effect of including special run-time code that interacts with the debugger. Once the code has been compiled for debugging, then it can be executed "normally." For example, in the Unix/Linux environment, the program can be started normally from the sdb debugger at any point by typing certain commands at the command prompt. However, it is more useful to single step through the source code. Lines of code are displayed and executed one at a time by using the step command. If the statement is an output statement, it will output to the screen accordingly. If the statement is an input statement, it will await user input. All other statements execute normally. At any point in the single-stepping process, individual variables can be set or examined. There are many other features of sdb, such as breakpoint setting. In more sophisticated operating environments, a graphical user interface (GUI) is also provided, but essentially, these tools provide the same functionality.

Very often when debugging a new program, the Unix operating system will abort execution and indicate that a core dump has occurred. This is a signal that some fault has occurred. A core dump creates a rather large file named core, which many programs simply remove before proceeding with the debugging. But core contains some valuable debugging information, especially when used in conjunction with sdb. For example, core contains the last line of the program that was executed and the contents of the function call stack at the time of the catastrophe. sdb can be used to single step up to the point of the core dump to identify its cause. Later on, breakpoints can be used to quickly come up to this line of code.

When removing code during debugging, it is inadvisable to use conditional branching. Conditional branching affects timing and can introduce subtle timing problems. Conditional compilation, is more useful in these instances. In conditional compilation, selected code is included only if a compiler directive is set and does not affect timing in the production system.

8.2.3 System-Level Testing

Once individual modules have been tested, then subsystems or the entire system need to be tested. In larger systems, the process can be broken down into a series of subsystem tests, and then a test of the overall system.

System testing treats the system as a black box so that one or more of the black-box testing techniques can be applied. System-level testing always occurs after all modules pass their unit test. At this point the coding team hands the software over to the test team for validation. If an error occurs during system-level testing, the error must be repaired then every test case involving the changed module must be rerun and all previous system-level tests must be passed in succession. The collection of system test cases is often called a system test suite.

Burn-in testing is a type of system-level testing that seeks to flush out those failures appearing early in the life of the system, and thus to improve the reliability of the delivered product. System-level testing may be followed by alpha testing, which is a type of validation consisting of internal distribution and exercise of the software. This testing is followed by beta testing, where preliminary versions of validated software are distributed to friendly customers who test the software under actual use. Later in the life cycle of the software, if corrections or enhancements are added, then regression testing is performed.

Regression testing, which can also be performed at the module level, is used to validate the updated software against the old set of test case that has already been passed. Any new test case needed for the enhancements are then added to the test suite, and the software is validated as if it were a new product. Regression testing is also an integral part of integration testing as new modules are added to the tested subsystem.

8.2.3.1 *Cleanroom Testing*

The principal tenant of cleanroom software development is that given sufficient time and with care, error-free software can be written. Cleanroom software development relies heavily on group walkthroughs, code inspections, and formal program validation. It is taken for granted that software specifications exist that are sufficient to completely describe the system. In this approach, the development team is not allowed to test code as it is being developed. Rather, syntax checkers, code walkthroughs, group inspections, and formal verifications are used to ensure code integrity. Statistically based testing is then applied at various stages of product development by a separate test team. This technique reportedly produces documentation and code that are more reliable and maintainable and easier to test than other development methods.

The program is developed by slowly "growing" features into the code, starting with some baseline of functionality. At each milestone an independent test team checks the code against a set of randomly generated test cases based on a set of statistics describing the frequency of use for each feature specified in the requirements. This group tests the code incrementally at predetermined milestones, and either accepts or returns it to the development team for correction. Once a functional milestone has been reached, the development team adds to the "clean" code, using the same techniques as before. Thus, like an onion skin, new layers of functionality are added to the software system until it has completely satisfied the requirements.

Numerous projects have been developed in this way, in both academic and industrial environments. In any case, many of the tenets of cleanroom testing can be incorporated without completely embracing the methodology.

8.2.3.2 Stress Testing In another type of testing, stress testing, the system is subjected to a large disturbance in the inputs (for example, a large burst of interrupts), followed by smaller disturbances spread out over a longer period of time. One objective of this kind testing is to see how the system fails (gracefully or catastrophically).

Stress testing can also be useful in dealing with cases and conditions where the system is under heavy load. For example, in testing for memory or processor utilization in conjunction with other application and operating system resources, stress testing can be used to determine if performance is acceptable. An effective way to stress test, for example, is to generate a configurable number of threads in a test program and subject the software to them. Running such tests for long periods of time also has the benefit of checking for memory leaks.

8.2.3.3 Test of Partially Implemented Systems One of the challenges in testing real-time systems is dealing with partially implemented systems. Many of the problems that arise are similar to those found in dealing with prototype hardware. There are numerous straightforward strategies involving creating stubs and drivers to deal with missing components at the interface. Commercial and open-source test generators can be helpful in these cases. But the strategies involved for testing real-time systems are nontrivial.

8.2.4 Design of Testing Plans

The test plan should follow the requirement to document item by item, providing criteria that are used to judge whether the required item has been met. A set of test cases is then written which is used to measure the criteria set out in the test plan. Writing such test cases can be extremely difficult when a user interface is part of the requirements.

The test plan includes criteria for testing the software on a module-by-module or unit level, and on a system or subsystem level; both should be incorporated in a good testing scheme. The system-level testing provides criteria for the hardware/software integration process.

Other documentation may be required, particularly in Department of Defense (DoD)-style software development, where preliminary and final documents are required and where additional documentation such as a hardware integration plan and software integration plan may be required. Many software systems that interact directly or indirectly with humans also require some form of users manual to be developed and tested.

8.3 FAULT-TOLERANCE

Fault-tolerance is the tendency to function in the presence of hardware or software failures. In real-time systems, fault-tolerance includes design choices that transform hard real-time deadlines into soft ones. These are often encountered in interrupt-driven systems, which can provide for detecting and reacting to a missed deadline.

Fault-tolerance designed to increase reliability in embedded systems can be classified as either spatial or temporal. Spatial fault-tolerance includes methods involving redundant hardware or software, whereas temporal fault-tolerance involves techniques that allow for tolerating missed deadlines. Of the two, temporal fault-tolerance is the more difficult to achieve because it requires careful algorithm design.

8.3.1 Spatial Fault-Tolerance

The reliability of most hardware can be increased using some form of spatial fault-tolerance using redundant hardware. In one common scheme, two or more pairs of redundant hardware devices provide inputs to the system. Each device compares its output to its companion. If the results are unequal, the pair declares itself in error and the outputs are ignored. An alternative is to use a third device to determine which of the other two is correct. In either case, the penalty is increased cost, space, and power requirements.

Voting schemes can also be used in software to increase algorithm robustness. Often like inputs are processed from more than one source and reduced to some sort of best estimate of the actual value. For example, an aircraft's position can be determined via information from satellite positioning systems, inertial navigation data, and ground information. A composite of these readings is made using either simple averaging or a Kalman filter.

8.3.1.1 Checkpoints One way to increase fault-tolerance is to use checkpoints. In this scheme, intermediate results are written to memory at fixed locations in code for diagnostic purposes (Figure 8.3). These locations, called checkpoints, can be used during system operation and during system verification. If the checkpoints are used only during testing, then this code is known as a test probe. Test probes can introduce subtle timing errors, which are discussed later.

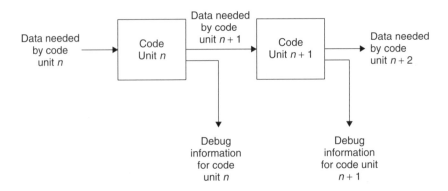

Figure 8.3 Checkpoint implementation [Laplante03c].

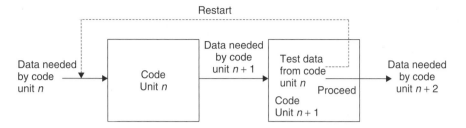

Figure 8.4 Recovery-block implementation [Laplante03c].

8.3.1.2 Recovery-Block Approach

Fault-tolerance can be further increased by using checkpoints in conjunction with predetermined reset points in software. These reset points mark recovery blocks in the software. At the end of each recovery block, the checkpoints are tested for "reasonableness." If the results are not reasonable, then processing resumes with a prior recovery block (Figure 8.4). The point, of course, is that some hardware device (or another process that is independent of the one in question) has provided faulty inputs to the block. By repeating the processing in the block, with presumably valid data, the error will not be repeated.

In the process-block model, each recovery block represents a redundant parallel process to the block being tested. Although this strategy increases system reliability, it can have a severe impact on performance because of the overhead added by the checkpoint and repetition of the processing in a block.

8.3.2 Software Black Boxes

The software black box is related to checkpoints and is used in certain mission-critical systems to recover data to prevent future disasters. The objective of a software black box is to recreate the sequence of events that led to the software failure for the purpose of identifying the faulty code. The software black-box recorder is essentially a checkpoint that records and stores behavioral data during program execution, while attempting to minimize any impact on that execution.

The execution of program functionalities results in a sequence of module transitions such that the system can be described as modules and their interaction. When software is running, it passes control from one module to the next. Exchanging control from one module to the next is considered a transition. Call graphs can be developed from these transitions graphically using an $N \times N$ matrix, where N represents the number of modules in a system.

When each module is called, each transition is recorded in a matrix, incrementing that element in a transition frequency matrix. From this, a posteriori probability of transition matrix can be derived that records the likeliness that a transition will occur. The transition frequency and transition matrices indicate the number of observed transitions and the probability that some sequence is missing in these data.

Recovery begins after the system has failed and the software black box has been recovered. The software black-box decoder generates possible functional scenarios based on the execution frequencies found in the transition matrix. The generation process attempts to map the modules in the execution sequence to functionalities, which allows for the isolation of the likely cause of failure.

8.3.3 *N*-Version Programming

In any system, a state can be entered where the system is rendered ineffective or locks up. This is usually due to some untested flow-of-control in the software for which there is no escape. That is to say that event determinism has been violated.

In order to reduce the likelihood of this sort of catastrophic error, redundant processors are added to the system. These processors are coded to the same specifications, but by different programming teams. It is therefore highly unlikely that more than one of the systems can lock up under the same circumstances. Since each of the systems usually resets a watchdog timer, it quickly becomes obvious when one of them is locked up, because it fails to reset its timer. The other processors in the system can then ignore this processor, and the overall system continues to function. This technique is called *N*-version programming, and it has been used successfully in a number of projects, including the space shuttle general-purpose computer (GPC).

The redundant processors can use a voting scheme to decide on outputs, or, more likely, there are two processors, master and slave. The master processor is on-line and produces the actual outputs to the system under control, whereas the slave processor shadows the master off-line. If the slave detects that the master has become hung up, then the slave goes on-line.

8.3.4 Built-In-Test Software

Built-in-test software (BITS) can enhance fault-tolerance by providing ongoing diagnostics of the underlying hardware for processing by the software. BITS is especially important in embedded systems. For example, if an I/O channel is functioning incorrectly as determined by its onboard circuitry, the software may be able to shut off the channel and redirect the I/O. Although BITS is an important part of embedded systems, it adds significantly to the worst-case time-loading analysis. This must be considered when selecting BITS and when interpreting the CPU utilization contributions that result from the additional software.

8.3.5 CPU Testing

In an embedded system the health of the CPU should be checked regularly. A set of carefully constructed tests can be performed to test the efficacy of its instruction set in all addressing modes. Such a test suite will be time-consuming and thus should be relegated to background processing. Interrupts should be disabled during each subtest to protect the data being used.

There is a catch-22 involved in using the CPU to test itself. If, for example, the CPU detects an error in its instruction set, can it be believed? If the CPU does not detect an error that is actually present, then this, too, is a paradox. This contradiction should not be cause for omitting the CPU instruction set test, because in any case, it is due to some failure either in the test or in the underlying hardware.

8.3.6 Memory Testing

All types of memory, including nonvolatile memories, can be corrupted via electrostatic discharge, power surging, vibration, or other means. This damage can manifest either as a permutation of data stored in memory cells or as permanent damage to the cell. Corruption of both RAM and ROM by randomly encountered charged particles is a particular problem in space. These single-event upsets do not usually happen on earth because either the magnetosphere deflects the offending particle or the mean free path of the particle is not sufficient to reach the surface.

Damage to the contents of memory is a soft error, whereas damage to the cell itself is a hard error. Chapter 2 discusses some of the characteristics of memory devices, and refers to their tolerance to upset. The embedded-systems engineer is particularly interested in techniques that can detect an upset to a memory cell and then correct it.

8.3.7 ROM

The contents of ROM are often checked by comparing a known checksum. The known checksum, which is usually a simple binary addition of all program-code memory locations, is computed at link time and stored in a specific location in ROM. The new checksum can be recomputed in a slow cycle or background processing, and compared against the original checksum. Any deviation can be reported as a memory error.

Checksums are not a very desirable form of error checking because errors to an even number of locations can result in error cancellation. For example, an error to bit 12 of two different memory locations may cancel out in the overall checksum, resulting in no error being detected. In addition, although an error may be reported, the location of the error in memory is unknown.

A reliable method for checking ROM memory uses a cyclic redundancy code (CRC). The CRC treats the contents of memory as a stream of bits and each of these bits as the binary coefficient of a message polynomial. A second binary polynomial of much lower order (for example, 16 for the Comité Consultatif International Télégraphique et Téléphonique (CCITT) or CRC-16 standards), called the generator polynomial, is divided (modulo-2) into the message, producing a quotient and a remainder. Before dividing, the message polynomial is appended with a 0 bit for every term in the generator. The remainder from the modulo-2 division of the padded message is the CRC check value. The quotient is discarded.

The CCITT generator polynomial is

$$X^{16} + X^{12} + X^5 + 1 \qquad (8.11)$$

whereas the CRC-16 generator polynomial is

$$X^{16} + X^{15} + X^2 + 1 \qquad (8.12)$$

A CRC can detect all 1-bit errors and virtually all multiple-bit errors. The source of the error, however, cannot be pinpointed. For example, ROM consists of 64 kilobytes of 16-bit memory. CRC-16 is to be employed to check the validity of the memory contents. The memory contents represent a polynomial of at most order $65,536 \cdot 16 = 1,048,576$. Whether the polynomial starts from high or low memory does not matter as long as consistency is maintained. After appending the polynomial with 16 zeroes, the polynomial is at most of order 1,048,592. This so-called message polynomial is then divided by the generator polynomial $X^{16} + X^{15} + X^2 + 1$, producing a quotient, which is discarded, and the remainder, which is the desired CRC check value.

In addition to checking memory, the CRC can be employed to perform nonvisual validation of screens by comparing a CRC of the actual output with the CRC of the desired output. The CRC of the screen memory is called a screen signature. The CRC calculation is CPU-intensive, and should only be performed in background or at extremely slow rates.

8.3.8 RAM

Because of the dynamic nature of RAM, checksums and CRCs are not viable. One way of protecting against errors to memory is to equip it with extra bits used to implement a Hamming code. Depending on the number of extra bits, known as the syndrome, errors to one or more bits can be detected and corrected. Such coding schemes can be used to protect ROM memory as well.

Chips that implement Hamming code error detection and correction (EDC chip) are available commercially. Their operation is of some interest. During a normal fetch or store, the data must pass through the chip before going into or out of memory. The chip compares the data against the check bits and makes corrections if necessary. The chip also sets a readable flag, which indicates that either a single- or multiple-bit error was found. Realize, however, that the error is not corrected in memory during a read cycle, so if the same erroneous data are fetched again, they must be corrected again. When data are stored in memory, however, the correct check bits for the data are computed and stored along with the word, thereby fixing any errors. This process is called RAM scrubbing.

In RAM scrubbing, the contents of a RAM location are simply read and written back. The error detection and correction occurs on the bus, and the corrected data are reloaded into a register. Upon writing the data back to the memory location, the correct data and syndrome are stored. Thus, the error is corrected in memory as well as on the bus. RAM scrubbing is used in the space shuttle inertial measurement unit computer [Laplante93]. The EDC chip significantly reduces the number of soft errors, which will be removed upon rewriting to the cell, and hard errors, which are caused by stuck bits or permanent physical damage to the memory.

The disadvantages of EDC are that additional memory is needed for the scheme (6 bits for every 16 bits), and an access time penalty of about 50 nanoseconds per access is incurred if an error correction is made. Finally, multiple-bit errors cannot be corrected.

In the absence of error detecting and correcting hardware, basic techniques can be used to test the integrity of RAM memory. These tests are usually run upon initialization, but they can also be implemented in slow cycles if interrupts are appropriately disabled. For example, suppose a computer system has 8-bit data and address buses to write to 8-bit memory locations. It is desired to exercise the address and data buses as well as the memory cells. This is accomplished by writing and then reading back certain bit patterns to every memory location. Traditionally, the following hexadecimal bit patterns are used:

AA	00
55	FF

The bit patterns are selected so that any cross talk between wires can be detected. Bus wires are not always laid out consecutively, however, so that other cross-talk situations can arise. For instance, the preceding bit patterns do not check for coupling between odd-numbered wires. The following set of hexadecimal patterns also checks for odd bit coupling:

AA	00
55	FF
0F	33

This test set, however, does not isolate the problem to the offending wire (bit). For complete coverage of 8 bits $7 + 6 + 5 + 4 + 3 + 2 + 1 = 28$ unique 2-bit combinations are needed. Since 8-bit words are available, $28/4 = 7$ of these combinations are available per test. Thus, seven 8-bit patterns are needed. These are given in hexadecimal in the following table:

AA	00
55	FF
0F	33
CC	

In general, for n-bit data and address buses writing to n-bit memory, where n is a power of 2, a total of $m(n-1)/2$ patterns of 2 are needed, which can be implemented in $n-1$ patterns of n bits each.

If walking ones and zeros[3] are used, there are 32 different test cases for each of the 2^{16} memory cells. Another common scheme is to test each cell with the hex patterns; 0000, FFFF, AAAA, and 5555. This test is faster than the walking ones or zeros, but still checks for cross talk between data wires and stuck-at faults.

8.3.9 Other Devices

In real-time embedded systems, A/D converters, D/A converters, MUXs, I/O cards, and the like need to be tested continually. Many of these devices have built-in watchdog timer circuitry to indicate that the device is still on-line. The software can check for watchdog timer overflows and either reset the device or indicate failure.

In addition, the built-in test software can rely on the individual built-in tests of the devices in the system. Typically, these devices will send a status word via DMA to indicate their health. The software should check this status word and indicate failures as required.

8.3.10 Spurious and Missed Interrupts

Extraneous and unwanted interrupts not due to time-loading are called spurious interrupts. Spurious interrupts can destroy algorithmic integrity and cause run-time stack overflows or system crashes. Spurious interrupts can be caused by noisy hardware, power surges, electrostatic discharges, or single-event upset. Missed interrupts can be caused in a similar way. In either case, hard real-time deadlines can be compromised, leading to system failure. It is the goal, then, to transform these hard errors into some kind of tolerable soft error.

8.3.11 Handling Spurious and Missed Interrupts

Spurious interrupts can be tolerated by using redundant interrupt hardware in conjunction with a voting scheme. Similarly, the device issuing the interrupt can issue a redundant check, such as using direct memory access (DMA) to send a confirming flag. Upon receiving the interrupt, the handler routine checks the redundant flag. If the flag is set, the interrupt is legitimate. The handler should then clear the flag. If the flag is not set, the interrupt is bogus and the handler routine should exit quickly and in an orderly fashion. The additional overhead of checking redundant flag is minimal relative to the benefit derived. Of course, extra stack space should be allocated to allow for at least one spurious interrupt per cycle to avoid stack overflow. Stack overflow caused by repeated spurious interrupts is called a death spiral.

Missed interrupts are more difficult to deal with. Software watchdog timers can be constructed that must be set or reset by the routine in question. Routines

---

[3] The sequences of bit patterns: 00000001, 00000010, 00000100, ... and 11111110, 11111101, 11111100, ...

running a higher priority or at a faster rate can check these memory locations to ensure that they are being accessed properly. If not, the dead task can be restarted or an error indicated. The surest method for sustaining integrity in the face of missed interrupts is through the design of robust algorithms.

8.3.12 The Kalman Filter

The Kalman filter is used to estimate the state variables of a multivariable feedback control system subject to stochastic disturbances caused by noisy measurements of input variables. It can also be used to provide fault-tolerance for embedded real-time systems in the face of noisy input data.

The Kalman filtering algorithm works by combining the information regarding the system dynamics with probabilistic information regarding the noise. The filter is very powerful in that it supports estimations of past, present, and even future states and, in particular, can do so even when the precise nature of the noise is unknown.

The Kalman filter estimates a process using a form of feedback control – the filter estimates the process state at some time and then obtains feedback in the form of noisy measurements. There are two kinds of equations for the Kaman filter: time update equations and measurement update equations. The time update equations project forward in time the current state and error covariance estimates to obtain the a priori estimates for the next time step. The measurement update equations are responsible for the feedback in that they incorporate a new measurement into the a priori estimate to obtain an improved estimate (Figure 8.5). As an example, in the inertial measurement system it is desired to protect against spurious noise in the accelerometer readings that could lead to unwanted interpretation of a sudden acceleration. The Kalman filter can also be used to deal

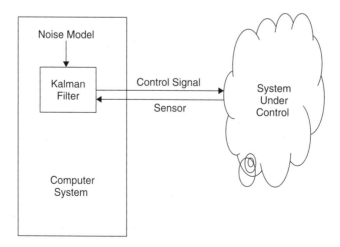

Figure 8.5 Using a Kalman filter for real-time control in the presence of noisy sensor data.

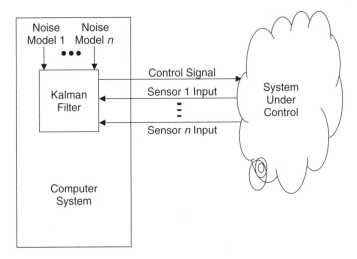

Figure 8.6 A Kalman filter used to control a real-time system involving multiple sensor sources, each with its own noise model.

with sensor fusion in a way that is less sensitive to the subtle spikes than a simple voting scheme.

In a typical mission-critical system, two or more of the sensors may measure the same process. This is done to provide redundancy and fault-tolerance. This goes beyond the simple failure of the sensor (for which the others provide backup). It helps to compensate for differing types of error in the sensor itself (Figure 8.6). For example, in the inertial measurement unit one accelerometer type may have errors that are known to have a large correlation over time, while a redundant accelerometer has a smaller error, but that exhibits no correlation. Fusing the sensor readings can provide overall improved measurements. The design of Kalman filters is beyond the scope of the text, but it is usually a topic covered in control systems texts.

8.4 SYSTEMS INTEGRATION

Integration is the process of combining partial functionality to form the overall system functionality. Because real-time systems are usually embedded, the integration process involves both multiple software units and hardware. Each of these parts potentially has been developed by different teams or individuals within the project organization. Although it is presumed that they have been rigorously tested and verified separately, the overall behavior of the system, and conformance to most of the software requirements, cannot be tested until the system is wholly integrated. Software integration can be further complicated when both hardware and software are new.

8.4.1 Goals of System Integration

The software integration activity has the most uncertain schedule and is typically the cause of project cost overruns. Moreover, the stage has been set for failure or success at this phase, by the specification, design, implementation, and testing practices used throughout the software project life cycle. Hence, by the time of software integration, it may be very difficult to fix problems. Indeed, many modern programming practices were devised to ensure arrival at this stage with the fewest errors in the source code. For example, light-weight methodologies, such as eXtreme programming, tend to reduce these kinds of problems.

8.4.2 System Unification

Fitting the pieces of the system together from its individual components is a tricky business, especially for real-time systems. Parameter mismatching, variable name mistyping, and calling sequence errors are some of the problems possibly encountered during system integration. Even the most rigorous unit-level testing cannot eliminate these problems completely.

The system unification process consists of linking together the tested software modules drawn in an orderly fashion from the source-code library. During the linking process, errors are likely to occur that relate to unresolved external symbols, memory assignment violations, page link errors, and the like. These problems must, of course, be resolved. Once resolved, the loadable code or load module, can be downloaded from the development environment to the target machine. This is achieved in a variety of ways, depending on the system architecture. In any case, once the load module has been created and loaded into the target machine, testing of timing and hardware/software interaction can begin.

8.4.3 System Verification

Final system testing of embedded systems can be a tedious process, often requiring days or weeks. During system validation a careful test log must be kept indicating the test case number, results, and disposition. Table 8.2 is a sample of such a test log for the inertial measurement system. If a system test fails, it is imperative, once the problem has been identified and presumably corrected, that all affected tests be rerun. These include

1. All module-level test cases for any module that has been changed.
2. All system-level test cases.

Even though the module-level test cases and previous system-level test cases have been passed, it is imperative that these be rerun to ensure that no side effects have been introduced during error repair.

Table 8.2 Sample test log for inertial measurement unit

Test Number	Reference Requirements Number	Test Name	Pass/Fail	Date	Tester
S121	3.2.2.2	Compensate accelerometer data 1a	Pass	5/16/03	P.L.
S122	3.2.2.2	Compensate accelerometer data 1b	Pass	5/16/03	P.L.
S123	3.2.2.2	Compensate accelerometer data 1c	Fail	5/16/03	P.L.

8.4.4 System Integration Tools

As mentioned before, it is not always easy to identify sources of error during a system test. A number of hardware and software tools are available to assist in the validation of embedded systems. Test tools make the difference between success and failure – especially in deeply embedded systems.

8.4.4.1 Multimeter The use of a multimeter in the debugging of real-time systems may seem odd nowadays, but it is an important tool in embedded systems where the software controls or reads analog values through hardware. The multimeter measures voltage, current, or power, and can be used to validate the analog input to or output from the system.

8.4.4.2 Oscilloscope An oscilloscope, like a multimeter, is not always regarded as a software-debugging tool, but it is useful in embedded software environments. Oscilloscopes range from the basic single-trace variety to storage oscilloscopes with multiple traces. Oscilloscopes can be used for validating interrupt integrity, discrete signal issuance, and receipt, and for monitoring clocks. The more sophisticated storage oscilloscopes with multiple inputs can often be used in lieu of logic analyzers, by using the inputs to track the data and address buses and synchronization with an appropriate clock.

8.4.4.3 Logic Analyzer The logic analyzer is an important tool for debugging software, especially in embedded real-time systems. The logic analyzer can be used to capture data or events, to measure individual instruction times, or to time sections of code. Moreover, the introduction of programmable logic analyzers with integrated debugging environments has further enhanced the capabilities of the system integrator.

More sophisticated logic analyzers include built-in dissemblers and compilers for source-level debugging and performance analysis. These integrated

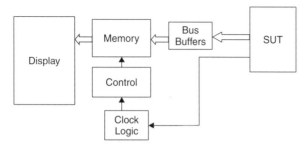

Figure 8.7 Basic logic analyzer structure connected to the system under test (SUT).

environments typically are found on more expensive models, but they make the identification of performance bottlenecks particularly easy.

No matter how elaborate, all logic analyzers have the same basic functionality. This is shown in Figure 8.7. The logic analyzer is connected to the system under test by connecting probes that sit directly on the memory and data buses. A clock probe connects to the memory-access synchronization clock. Upon each memory access, the data and address are captured by the logic analyzer and stored in buffers for transfer to the logic analyzer's main memory, from which it can be processed for display. Using the logic analyzer, the software engineer can capture specific memory locations and data for the purposes of timing or for verifying execution of a specific segment of code.

Timing Instructions The logic analyzer can be used to time an individual macroinstruction, segments of code, or an entire process. To time an individual instruction, the engineer finds a memory location in the code segment of memory containing the desired instruction. Then the logic analyzer is set to trigger on this opcode at the desired location, and on the opcode and location of the next instruction. The trace is set for absolute time. The logic analyzer will then display the difference in time between the fetch of the first instruction (the target) and the next instruction. This is the most accurate means for determining the instruction execution time. For example, suppose the pasta sauce bottling plant system contains a 30-millisecond frame buffering task. It is known from the linker output that instructions and data are found at memory location 4356 through 464B (hexadecimal). Then the memory location, corresponding numerical opcode (in hex), and symbolic equivalent will appear in the display of the logic analyzer as follows:

Location (hex)	Opcode (hex)	Instruction
4356	2321	STORE R2, R1
4357	4701 1000	LOAD R1, 1000
4359	2401 FC32	
⋮		
464B	6300 2000	JUMP 2000

The code represents part of an interrupt handler in which the first instruction is to disable all interrupts and the last instruction is to enable all interrupts. If the logic analyzer is set to trigger on address = 4357, data = 4701, and capture-only address = 4357 and data = 4701, the time to complete the LOAD (4701) will be displayed. In this case, the "data" is the instruction opcode.

Timing Code   The logic analyzer also provides an accurate method for measuring time-to-complete for any periodic task. To measure the total elapsed time for any task in the system, set the logic analyzer to trigger on the starting and ending address and opcode for the first instruction of that task. It should be the first instruction of the interrupt handler, usually a disable interrupt instruction. Disable the interrupts for all higher-priority cycles and set the trace for absolute time. The time displayed is the total time of that task cycle.

Consider the code shown in the previous example. If the logic analyzer is set to trigger on address = 4356, data = 2321, and capture-only address = 464B and data = 6300, the absolute time to execute all instructions in the module will be measured. Suppose the elapsed time is measured as 3 milliseconds for a 10-millisecond rate. Then the utilization contribution from this code is 33.33%. This approach can be used to time one or several modules within a cycle, or even sections of code within a module.

8.4.4.4 In-Circuit Emulator

During module-level debugging and system integration of embedded systems, the ability to single-step the computer, set the program counter, and insert into and read from memory is extremely important. This capability in conjunction with the symbolic debugger is the key to the proper integration of real-time systems. In an embedded environment, however, this capability is provided by an in-circuit emulator. In-circuit emulation (ICE) uses special hardware in conjunction with software to emulate the target CPU while providing the aforementioned features. Typically, the in-circuit emulator plugs into the chip carrier or card slot normally occupied by the CPU. External wires connect to an emulation system. Access to the emulator is provided directly or via a secondary computer.

In-circuit emulators are useful in software patching and for single-stepping through critical portions of code. In-circuit emulators are not typically useful in timing tests, however, because subtle timing changes can be introduced by the emulator. In certain ICE systems, the symbol table may be too large to load. Privatization of certain global variables can be used to reduce the size of the symbol table. For example, in C, judicious use of the static data type during testing can reduce the number of variables in the global symbol table. This aids the debugging process.

8.4.4.5 Software Simulators

When integrating and debugging embedded systems, software simulators are often needed to stand in for hardware or inputs that do not exist or that are not readily available, for example, to generate simulated accelerometer or gyro readings where real ones are unavailable at the time. The author of the simulator code has a task that is by no means easy. The software must be written to mimic exactly the hardware specification, especially in timing characteristics. The simulator must be rigorously tested (unfortunately, this is sometimes not the case). Many systems have been successfully validated and integrated with software simulators, only to fail when connected to the actual hardware.

8.4.4.6 Hardware Prototypes   In the absence of the actual hardware system under control, simulation hardware may be preferable to software simulators. These devices might be required when the software is ready before the prototype hardware, or when it would be impossible to test the software on the actual hardware, such as in the control of a large nuclear plant. Hardware simulators simulate real-life system inputs and can be useful for integration and testing, but are not always reliable testing the underlying algorithms, for which real data from live devices are needed.

8.4.4.7 Software Integration   A deliberate approach must be used when performing system integration to ensure system integrity. Failure to do so can lead to cost escalation and frustration. Software integration approaches are largely based on experience. The following represents one simple strategy for software integration based on significant experience.

8.4.5 A Simple Integration Strategy

In any embedded operating system it is important to ensure that all tasks in the system are being scheduled and dispatched properly. Thus, the first goal in integrating the embedded system is to ensure that each task is running at its prescribed rate, and that context is saved and restored. This is done without performing any functions within those tasks; functions are added later.

As discussed before, a logic analyzer is quite useful in verifying cycle rates by setting the triggers on the starting location of each of the tasks involved. During debugging it is most helpful to establish the fact that cyclic processes are being called at the appropriate rates. Until the system cycles properly, the application code associated with each of the tasks should not be added. The success of this method depends on the fact that one change at a time is made to the system so that when the system becomes corrupted, the problem can be isolated.

The overall approach is shown in Figure 8.8. The approach involves establishing a baseline of running kernel components (no applications programs). This ensures that interrupts are being handled properly and that all cycles are running at their prescribed rates, without worry about interference from application code. Once the baseline is established, small sections of applications code are added and the cycle rates verified. If an error is detected, it is patched if possible. If the patch succeeds in restoring the cycle rates properly, then more code is added. This ensures that the system is grown incrementally, with an appropriate baseline at each stage of the integration. This approach represents a phased integration with regression testing after each step.

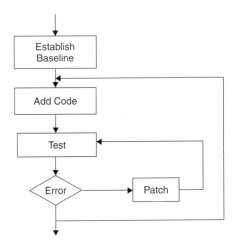

Figure 8.8 A simple integration strategy [Laplante03c].

8.4.6 Patching

The process of correcting errors in the code directly on the target machine is called patching. Patching allows minor errors detected during the integration process to be corrected directly on the target machine, without undergoing the tedious process of correcting the source code and creating a new load module. It is also useful in repairing software remotely, for example, in space-borne applications. Patching requires an expert command of the opcodes for the target machine unless a macroassembly-level patching facility is available. It also requires an accurate memory map, which includes the contents of each address in memory, and a method for inserting directly into memory. This capability is provided by many commercial development environments and by in-circuit emulators.

Patching, which is analogous to placing jumper wires on prototype hardware, typically requires only a minor change of memory contents. If the patch needed fits into the memory space accorded to the code to be changed, then it is considered an in-line patch. In Figure 8.9, for example, a 1 was supposed to be added to register 1 (R1) instead of a 0. This error can be changed easily by altering the memory location containing the LOAD R1,0 instruction to LOAD R1,1.

If the patch requires more memory than is currently occupied by the code to be replaced, it is considered an oversized patch. In this case a JUMP to some unused portion of memory is required, followed by the patched code, followed by a return JUMP to the next significant location. This technique is shown in Figure 8.10. The loading of patches during system integration can often be automated. However, a large number of patches and patches on top of others can become confusing. It is imperative that a careful record be kept of all patches made, that the patches eventually find their way back to the source code, and that a new system be generated before validation testing begins. This is essential from a maintenance standpoint. Final testing should never be performed on a patched system.

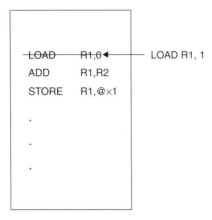

Figure 8.9 In-line patch.

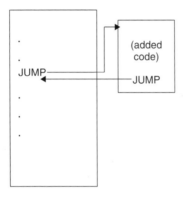

Figure 8.10 An oversized patch [Laplante03c].

Patching of software written in object-oriented languages is very difficult because of the lack of a straightforward mapping from the source code to the object code. Symbolic debuggers are quite helpful in this case, but even so, in this situation patching is risky at best.

8.4.7 The Probe Effect

The uncertainty principle, originally postulated by Werner Heisenberg in 1927, states essentially that the precise position and momentum of a particle cannot be known simultaneously. An analogy to the Heisenberg uncertainty principle applies in software integration. While software systems do not explicitly deal with electrons (except as ensemble behavior), uncertainty arises because the more closely a system is examined, the more likely the examination process will affect the system. This fact is especially true for embedded systems where test probes can affect timing. For example, an engineer is debugging the pasta sauce bottling system and discovers that a certain deadline is not being met. Some debugging code is added to print out a preliminary result to a file. But after adding the debugging code, the problem goes away. Declaring success, the engineer removes the debugging code and the problem reappears. In this case, it is clear that the debugging code somehow changed the timing of the system.

The software version of the Heisenberg uncertainty principle should be taken as a warning that testing methods often affect the systems that they test. When this is the case, nonintrusive testing should be considered, for example, using a logic analyzer. Furthermore, wherever there is an inverse correlation between two variables affecting system, Heisenberg uncertainty is suggested.

8.4.8   Fault-Tolerant Design: A Case Study

It has already been noted that single-bit errors present significant problems to control systems operating in space or in the presence of a nuclear event, such as in the control system for the nuclear energy plant. As a detailed example of a fault-tolerant design for such an environment, consider the inertial measurement system for the space shuttle [Laplante93].

8.4.8.1 Problem Description   In understanding the potential effects of single-event upsets (SEU), recall that different devices have varying susceptibilities to upset partly based on geometry, logic levels, and manufacturing process. These susceptibilities are summarized in Table 8.3. Depending on its mission, environment, technology used, and other factors, the expectation of error can be quite low. But the point here really is to illustrate a holistic approach to fault-tolerant design.

8.4.8.2 Hardware Support   Certain parts manufacturers design and test critical components for radiation hardness. In many cases, parts are preirradiated with low-level radiation to liberate any radioactive impurities in the packaging material that could have become offending particles or created latch-ups. This process may lower the part's SEU potential. However, this treatment diminishes the part's low-level radiation longevity. In addition to minimizing SEU rates through optimum part selection, error detection and correction should be used. For the space shuttle internal measurement unit (IMU) application the AMD2960 was selected to protect the 8K RAM and 4K of EEPROM used for crucial data.

8.4.8.3 Software Fault-Tolerant   Various measures can be taken during system start up, foreground processing, and background processing to enhance fault-tolerance. The software for the space shuttle IMU was particularly aggressive in this regard.

Initialization   During initialization it is necessary to perform as many system-level and device-level checks as possible. There is usually plenty of time to

Table 8.3   Device susceptibility to SEU and the possible adverse effects

Device	Susceptibility	Possible Adverse Effects
Core memory	Not susceptible	None
CMOS RAM	Very susceptible	Corruption of data
Fusible link ROM	Not susceptible	None
EEPROM	Very susceptible	Corruption of data
UVROM	Very susceptible	Corruption of data
Program counter	Very susceptible	Jump program Misexecute instruction Bad operand Latch-up
I/O circuitry	Very susceptible	Bad data
Interrupt controller	Very susceptible	Spurious interrupts Missed interrupts Misprioritized interrupts

execute the initialization routine, except in the case of a system that needs to be ready milliseconds after power-on.

For the space shuttle IMU the following procedures are performed at initialization. First, a checksum is performed on the stored program. The checksum, which is a simple sum of all words of program memory, is included in the PROM image at link time. The initialization software will sum all program memory, and compare it against the checksum. Discrepancies are flagged and recorded. If an EDC device is implemented for the ROM area of memory, then the EDC chip should be disabled during the checksum, or the checksum will always be correct regardless of errors. And, although it was noted that fusible-link PROM is virtually nonupsetable, EDC should also be implemented here to protect against hard errors not due to SEUs. The detection and correction capability of the AM2960 EDC chip, for example, can be disabled by setting a "pass through" bit with a computer-controlled discrete signal.

After checking the ROM area for errors, the RAM area of memory is fully tested by writing and reading back, several patterns of bits into each cell. The EDC chip is, of course, disabled so as not to inadvertently correct an error without flagging it. The test patterns written into memory include: all ones, alternating ones and zeros, alternating zeros and ones, and all zeros. These patterns exercise each bit independently in both states.

Once each cell of RAM has been exercised, the EDC chip is fully tested. This is accomplished by writing a word to memory (thus updating the syndrome), turning off the EDC chip and writing a different word to that cell (this changes the data but not the syndrome), turning on the EDC capability, and reading back the changed word. The EDC chip should flag and correct the error. If the EDC chip does not work as expected, an error will be indicated. This procedure is followed for various error types (single and double bit), various data patterns, and several different areas of memory. Finally, normal checking of other system hardware is performed and standard initialization of the interrupt cycle structure and variables is performed.

Foreground Processing In each of the interrupt-driven cycles, special care should be taken to ensure the validity of each interrupt. As was mentioned before, an upset to the interrupt controller can cause spurious interrupts. For the space shuttle IMU application, in order to avoid processing these false interrupts, a redundant flag (CPU-readable discrete) is generated along with each interrupt request. The software will check the redundant discrete flag when responding to an interrupt. If the interrupt is false, control can quickly be returned to the interrupted cycle. When using this technique care should be taken to allocate sufficient stack storage to allow for such an occurrence. Recall that the program stack is an area of RAM used by the computer to save the current status of the machine when interrupted. If several consecutive interrupts of increasing priority occur, the stack will grow. The size of the stack should be the amount of space needed to save the state of one cycle times the total number of cycles plus one.

For the space shuttle IMU system, which has eight asynchronous cycles, analysis was used to prove that missing any interrupt once will not compromise system performance. Another SEU-induced problem occurs when the interrupt controller wrongly prioritizes interrupts. If the status register or interrupt vector is upset, a lower priority interrupt may supersede one of higher priority. In the space shuttle IMU, this problem is solved by continually saving a copy of the previous interrupt status register in RAM. When an interrupt is serviced, the current contents of the status register are compared to the previous contents, which should always be lower. Any anomalies can be detected, and erroneous processing avoided. Finally, during real-time processing the software monitors the "error detected" flag from the EDC chip, and a count is kept and saved for postflight analysis.

Background Processing While executing in background, several operations are performed to help minimize the potential effects of an SEU. First, as in initialization, a ROM checksum is continuously formed to detect any damage to the program area of memory. Any aberrations are flagged and reported. Second, RAM "scrubbing" is performed. This protects against the small possibility that two SEUs occur in the same cell over the life of a mission, an error that cannot always be corrected by the EDC circuitry. Finally, any induced state where the interrupt controller cannot honor interrupts must be avoided. This is accomplished by continually refreshing the mask register (a bit vector containing the identity of all valid interrupt lines in the system) and by continually reenabling interrupts.

8.4.8.4 Summary In the case study the serious problem of SEU in real-time fault-tolerant control system was discussed. Moreover, the damaging effects of this phenomenon and some of the ways that these effects can be prevented or minimized were discussed. Finally, a real system, which has attempted to use some of these techniques to improve system reliability to an acceptable level, was examined. These techniques and the relative costs of the remedies are summarized in Tables 8.4 and 8.5. At the time of this writing, actual data on the relative efficiency of these techniques were unavailable, but it will be interesting to note which technique fares best.

It seems clear that the SEU problem needs careful attention. Quite a bit is understood about how SEUs can destroy data, but no system is 100% fault-tolerant against them. Nevertheless an acceptable risk level can be achieved through careful design and the use of one or more of the mechanisms discussed.

Table 8.4 SEU protection mechanisms

Adverse Effect	Remedy
Corruption of RAM data	EDC chip, RAM scrubbing
Corruption of ROM data	EDC chip
Corruption of program counter	None
CPU latch-up	Watchdog timer
I/O circuitry	None
Spurious interrupts	Confirmation flags
Missed interrupts	Watchdog timer, counters
Misprioritized interrupts	Double-check status register

Table 8.5 Costs of SEU protection mechanisms

Remedy	Cost
RAM scrubbing	None
EDC chip	Increased memory access times
Watchdog timer	Increased power, space, weight
Confirmation flags	Increased interrupt response times
Double-check status register	Increased interrupt response times

8.5 REFACTORING REAL-TIME CODE

A code smell is a popular term that refers to an indicator of poor design or coding [Fowler00]. More specifically, the term relates to visible signs that suggest the need for refactoring. Refactoring refers to a behavior preserving code transformation enacted to improve some feature of the software, which is evidenced by the code smell.

The notion of refactoring, however, is ordinarily associated with object-oriented code [Fowler00]. On the other hand, code smells are found in both object-oriented and procedurally oriented code in real-time systems, especially legacy systems. In this section a brief discussion of some of the more common code smells and their refactorings are given. Applying these refactorings appropriately can help to enhance desirable properties of the software while removing the undesirable ones.

8.5.1 Conditional Logic

There are several reasons why excessive `switch`, `if-then`, and `case` statements are an indicator of a bad design in real-time systems. First, they breed code duplication. Moreover, the code generated for a case statement can be quite convoluted, for example, a jump through a register, offset by a table value. This mechanism can be time-consuming. Furthermore, nested conditional logic can be difficult to test, especially if it is nested due to the large number of logic paths through the code. Finally, the differences between best- and worst-case execution times can be significant, leading to highly pessimistic utilization figures.

Conditional logic needs to be refactored, but there is no silver bullet here. In dealing with performance issues, often trade-offs can be made by exchanging one type of conditional logic for another. For example, in some compilers, the case statement is efficient only if more than three cases are to be compared, otherwise, nested `if` statements should be used. Reducing the switch logic (and hence, the number of conditionals) can also be accomplished through application of logical identities, and in the most complex cases, through traditional logic minimization techniques such as Karnaugh mapping and the Quine-McCluskey procedure.

8.5.2 Data Clumps

Several data items found together in lots of places are known as data clumps. In the procedural sense data clumps can arise, for example, in C from too much configuration information in `#include` files. Stewart calls this problem "`#include globals.h`" and notes that this situation is unhealthy because it leads to increased development and maintenance time and introduces circular dependencies that make reuse difficult. He suggests that to refactor, each module be defined by two files ".h" and ".c," with the former containing only that information that is to be exported by the module and ".c" file containing everything that is not exported [Stewart99].

A second manifestation of the data clump smell has to do with excessive use of `#define` statements that propagate through the code. Suppose these `#defines` are expanded in 20 places in the code. If, during debugging, it is desired to place a patch over one of the #defines, it must be done in 20 places. To refactor, place the quantity in a global variable that can be changed in one place only during debugging [Stewart99].

8.5.3 Delays as Loops

Another code smell involves timing delays implemented as while loops with zero or more instructions [Stewart99]. These delays rely on the overhead cost of the loop construct plus the execution time of the body to achieve a delay. The problem is that if the underlying architecture changes or characteristics of instruction execution (e.g., memory access time) changes, then the delay time is inadvertently altered. To refactor use a mechanism based on a timer facility provided by the operating system that is not itself based on individual instruction execution times.

8.5.4 Dubious Constraints

This code smell is particularly insidious in real-time systems where response-time constraints have questionable or no attributable source. In some cases, systems have deadlines that are imposed on them that are based on nothing less than guessing or on some forgotten and since eliminated requirement. The problem in these cases is that undue constraints may be placed on the systems. For example, suppose the response time for an interrupt is 30 ms, but no one knows why. Similarly, more than one reason given for the constraints in comments or documentation indicates a traceability conflict, which hints at other problems. This is a primary lesson in using real-time systems design to understand the basis and nature of the timing constraints, so that they can be relaxed if necessary.

In any case, to remove the code smell, some detective work is needed to discover the true reason for the constraint. If it cannot be determined, then the constraint could be relaxed and the system redesigned accordingly.

8.5.5 Duplicated Code

Obviously, duplicated code refers to the same or similar code found in more than one place. This code has an unhealthy impact on maintainability (the same change has to be propagated to each copy), and it also adversely affects memory utilization.

It is obvious that the purpose of refactoring is to assign the code to a single common code unit via better application of information hiding. While it is too easy to mock the designers of systems that contain duplicated code, it is possible that the situation arose out of a real need at the time. For example, duplicated code may have been due to legacy concerns for performance where the cost of the procedure call added too much overhead in a critical instance. Alternatively, in languages that were not reentrant, such as early versions of Fortran, duplicated code was a common means for providing utilities to each cycle in the real-time system.

8.5.6 Generalizations Based on a Single Architecture

Stewart suggests that writing software for a specific architecture, but with the intent to support easy porting to other architectures later, can lead to overgeneralizing items that are similar across architectures, while not generalizing some items that are different. He suggests developing the code simultaneously on multiple architectures and then generalizing only those parts that are different. He suggests choosing three or four architectures that are very different in order to obtain the best generalization [Stewart99]. Presumably, such an approach suggests the appropriate refactoring.

8.5.7 Large Procedures

Fowler describes two code smells, long method and large class, which are self-evident. In the procedural sense, the analogy is a large procedure. Large procedures are anathema to the divide-and-conquer principle of software engineering and need to be refactored by repartitioning the code appropriately.

8.5.8 Lazy Procedure

A lazy method is one that does not do enough to justify its existence. The procedural analogy to the lazy class/method code smells is the lazy procedure. In a real-time sense, a procedure that does too little to pay for the overhead of calling the procedure needs to be eliminated by removing its code to the calling procedure(s), or redefining the procedure to do more.

8.5.9 Long Parameter List

Long parameter lists are an unwanted remnant of the practice of using parameter lists to avoid the use of global variables. While well-defined interfaces are clearly desirable, long parameter lists can cause problems in real-time systems if interrupts are disabled during parameter passing. In this case, overly long interrupt latencies and possible missed deadlines are possible. The "long parameter list" code smell can be refactored by passing a pointer to one or more data structures that contain aggregated parameters, or by using global variables.

8.5.10 Message-Passing Overload

Stewart describes the excessive use of message passing for synchronization as another unwanted practice. He notes that this practice can lead to unpredictability (because of the synchronization necessary), the potential for deadlock, and the overhead involved. He suggests that the refactoring is to use state-based communication via shared memory with structured communication [Stewart99].

8.5.11   Self-Modifying Code

One method that can be used to save space in the code area of memory is self-modifying code. This method takes advantage of the fact that the opcodes of certain instructions may differ by only one bit. For example, by flipping a bit in a JUMP instruction, an ADD instruction is created. Although this type of programming usually arises from coincidence, stories exist about programmers who could write such code effortlessly. In fact, in some settings, it was considered a sign of cleverness to include some self-modifying code. The most important disadvantage of such coding is that it destroys determinism.

Unfortunately, many processors include on-chip caches that can obviate self-modifying code. In these cases the cache does not update the code and executes the unmodified code. Additionally, the effect of modifying code within the cache causes performance degradation. In any case, this type of programming should never be done, but was not uncommon in older systems.

8.5.12   Speculative Generality

Speculative generality relates to hooks and special cases that are built into the code to handle things that are not required (but might be needed someday). Real-time systems are no place to build in hooks for "what-if" code. Hooks lead to testing anomalies and possible unreachable code. Therefore, the refactoring is to remove hooks and special cases that are not immediately needed.

8.5.13   Telltale Comments

The telltale comment problem appears in real-time and non-real-time systems. Comments that are excessive, or that tend to explicate the code beyond a reasonable level are often indicators of some serious problem. Comments such as:

- "Do not remove this code."
- "If you remove this statement the code doesn't work, I don't know why."
- "Please see me for an explanation of this code . . . ."

are not uncommon. Humor in comment statements can sometimes be a glib way to mask the fact that writer does not know what he or she is doing. For example, "if you understand this code, could you please explain it to me?" Clearly these kinds of statements indicate that there are underlying timing errors. In the case of tell-tale comments, the refactoring involves rewriting the code so that an overly long explicating comment is not necessary.

8.5.14   Unnecessary Use of Interrupts

Stewart also suggests that indiscriminate use of interrupts is a bad code smell [Stewart99]. Interrupts can lead to deadlock, priority inversion, and the inability to make performance guarantees. Interrupt-based systems should be avoided, though as has been noted, this is possible only in the simplest of systems where real-time multitasking can be achieved with coroutines and cyclic executives implemented without interrupts. When interrupts are required to meet performance constraints, rate-monotonic or earliest deadline first scheduling should be used.

8.6 COST ESTIMATION USING COCOMO

Cost estimation is an important component of engineering real-time software systems. One of the most widely used software modeling tools is Boehm's COCOMO model, first introduced in 1981 [Boehm81]. COCOMO is an acronym for constructive cost model. That is, it is a predictive model. There are three versions of the original COCOMO: basic, intermediate and detailed, and the recently released COCOMO II model.

8.6.1 Basic COCOMO

The basic COCOMO model is based on thousands of lines of deliverable source instructions. In short, for a given piece of software, the time, T, to complete is a function of L, the number of lines of delivered source instructions (KDSI), and two additional parameters, a and b, which will be explained shortly. This is the effort equation for the basic COCOMO model:

$$T = aL^b \tag{8.13}$$

Dividing T by a known productivity factor, in KLOC per person month, yields the number of person months estimated to complete the project.

The parameters a and b are a function of the type of software system to be constructed. For example, if the system is organic, that is, one that is not heavily embedded in the hardware, then the following parameters are used: $a = 3.2, b = 1.05$. If the system is semidetached, that is, partially embedded, then these parameters are used: $a = 3.0, b = 1.12$. Finally, if the system is embedded, that is closely tied to the underlying hardware like the inertial measurement system, then the following parameters are used $a = 2.8, b = 1.20$. Note that the exponent for the embedded system is the highest, leading to the longest time to complete for an equivalent number of delivered source instructions.

Recall that for the inertial measurement system, using feature points, 37.6 thousand lines of C code were estimated. Hence, an effort-level estimate is obtained using COCOMO of:

$$T = 2.8 \cdot (37.6\text{K})^{1.2} = 169\text{K}$$

Suppose, then, it is known that an efficient software engineer using computer-aided software engineering (CASE) and other tools can generate 2000 lines of code per month. Then superficially, at least, it might be estimated that the project would take about 84.5 person months to complete. Not counting dependencies in the task graph, this implies that a five-person team would take about 17 months to complete the project. It would be expected, however, that more time would be needed because of task dependencies.

8.6.2 Intermediate and Detailed COCOMO

The intermediate or detailed COCOMO models dictate the kinds of adjustments used. Consider the intermediate model, for example. Once the effort level for the basic model is computed based on the appropriate parameters and number of source instructions, other adjustments can be made based on additional factors. In this case, for example, if the lines of code to be produced consist of design-modified code, code modified code, and integration modified code rather than straight code, a linear combination of these relative percentages is used to create an adaptation adjustment factor as follows.

Adjustments are then made to T based on two sets of factors, the adaptation adjustment factor, A, and the effort adjustment factor, E. The adaptation adjustment factor is a measure of the kind and proportion of code that is to be used in the system, namely, design-modified, code-modified, or integration-modified. The adaptation factor, A, is given by Equation 8.14:

$$A = 0.4 \text{ (\% design modified)} + 0.03 \text{ (\% code modified)}$$
$$+ 0.3 \text{ (\% integration modified)} \tag{8.14}$$

For new components $A = 100$. On the other hand, if all of the code is design modified, then $A = 40$, and so on. Then the new estimation for delivered source instructions, E, is given as

$$E = L \cdot A/100 \tag{8.15}$$

An additional adjustment, the effort adjustment factor, can be made to the number of delivered source instructions based on a variety of other factors, including:

- Product attributes
- Computer attributes
- Personnel attributes
- Project attributes

Each of these attributes is assigned a number based on an assessment that rates them on a relative scale. Then, a simple linear combination of the attribute numbers is formed based on project type. This gives a new adjustment factor, call it, E'. The second adjustment, effort adjustment factor, E'', is then made based on the formula

$$E'' = E' \cdot E \tag{8.16}$$

Then the delivered source instructions are adjusted, yielding the new effort equation:

$$T = aE''^b \tag{8.17}$$

The detailed model differs from the intermediate model in that different effort multipliers are used for each phase of the software life cycle.

COCOMO is widely recognized and respected as a software project management tool. It is useful even if the underlying model is not really understood. COCOMO software is commercially available and even can be found on the Web for free use. One drawback, however, is that the model does not take into account the leveraging effect of productivity tools. Finally, the model bases its estimation almost entirely on lines of code, not on program attributes, which is something that feature points do. Feature points, however, can be converted to lines of code using standard conversion formulas, as was shown.

8.6.3   COCOMO II

COCOMO II is a major revision of COCOMO that is evolving to deal with some of the original version's shortcomings. For example, the original COCOMO 81 model was defined in terms of delivered source instructions. COCOMO II uses the metric, source lines of code instead of delivered source instructions. The new model helps accommodate more expressive modern languages as well as software generation tools that tend to produce more code with essentially the same effort.

In addition, in COCOMO II some of the more important factors that contribute to a project's expected duration and cost are included as new scale drivers. These five scale drivers are used to modify the exponent used in the effort equation. These scale drivers are:

- Project novelty
- Development flexibility
- Architectural/risk resolution
- Team cohesion
- Process maturity

The first two drivers, project novelty and development flexibility, for example, describe many of the same influences found in the adjustment factors of COCOMO 81.

It is beyond the scope of this text to study COCOMO in detail. As with any metric and model, it must be used carefully and be based on practice and experience. Nevertheless, using proven models is better than using none at all, or guessing.

8.7 EXERCISES

8.1 Recalculate McCabe's metrics for the if, while, and until structures seen in Figure 8.1.

8.2 Research the use of McCabe's metric in real-time systems by searching the literature.

8.3 Recalculate the $\overline{FP}$ metric for the inertial measurement system using a set of weightings that assumes that significant off-the-shelf software (say 70%) is to be used. Make assumptions about which factors will be most influenced by the off-the-shelf software. How many lines of C code do you estimate you will need?

8.4 Do the same as Exercise 8.3, except recalculate the $\overline{FP}$ metric. How many lines of C code do you estimate will be needed?

8.5 For the inertial measurement system, which testing approaches would you use? When and why?

8.6 If the inertial measurement system were written in C++ according to the design fragment described in Chapter 5, describe the testing strategy you would use. If possible, try to design some test cases.

8.7 How much can testing and test case/suite generation be automated? What are the roadblocks to automating a test suite? In languages like Java?

8.8 A software module is to take as inputs four signed 16-bit integers and produce two outputs, the sum and average. How many test cases would be needed for a brute-force testing scheme? How many would be needed if the minimum, maximum, and average values for each input were to be used?

8.9 Describe the effect of the following BITS and reliability schemes without appropriately disabling interrupts. How should interrupts be disabled?
   (a) RAM scrubbing
   (b) CRC calculation
   (c) RAM pattern tests
   (d) CPU instruction set test

8.10 Suppose a computer system has 16-bit data and address buses. What test patterns are necessary and sufficient to test the address and data lines and the memory cells?

8.11 Write a module in the language of your choice that generates a CRC checkword for a range of 16-bit memory. The modules should take as input the starting and ending addresses of the range, and output the 16-bit checkword. Use either CCITT or CRC-16 as generator polynomials.

8.12 In N-version programming, the different programming teams code from the same set of specifications. Discuss the disadvantages of this (if any).

8.13 Can the tools available impact the software project and the code quality? How?

8.14 For the example systems already described:
   • Nuclear monitoring system
   • Inertial measurement system

- Airline reservation system
- Pasta sauce bottling system
- Traffic light control system

which testing methods would you use and why?

8.15 Consider the statement "good real-time engineering is not good for software engineering and vice versa." Do you consider this statement true or false? Justify your answer with sufficient examples.

GLOSSARY

abstract class A superclass that has no direct instances.

abstract data type A language construct where a user defines his or her own type (e.g., "pixel") along with the requisite operation that can be applied to it.

accept operation Operation on a mailbox that is similar to the pend operation, except that if no data are available, the task returns immediately from the call with a condition code rather than suspending.

access time The interval between when data are requested from the memory cell and when they are actually available.

accumulator An anonymous register used in certain computer instructions.

activity packet A special token passed between the processors in a data flow architecture. Each token contains an opcode, operand count, operands, and a list of destination addresses for the result of the computation.

actual parameter The named variable passed to a procedure or subroutine.

adaptive programming A lightweight programming methodology that offers a series of frameworks to apply adaptive principles and encourage collaboration.

address bus The collection of wires needed to access individual memory addresses.

agile programming A lightweight programming methodology that is divided into four activities – planning, designing, coding, and testing – all performed iteratively.

algorithm A systematic and precise, step-by-step procedure for solving a certain kind of problem or accomplishing a task, for instance, converting a particular kind of input data to a particular kind of output data, or controlling a machine tool. An algorithm can be executed by a machine.

alpha testing A type of validation consisting of internal distribution and exercise of the software.

Many of these terms have been adapted from *The Dictionary of Computer Science, Engineering, and Technology*, Phillip A. Laplante (editor-in-chief), CRC Press, Boca Raton, FL, 2001.

Real-Time Systems Design and Analysis, By Phillip A. Laplante
ISBN 0-471-22855-9 © 2004 Institute of Electrical and Electronics Engineers

ALU *See* arithmetic logic unit.

analog-to-digital conversion The process of converting continuous (analog) signals into discrete (digital) ones.

anonymous variable A hidden variable created by the compiler to facilitate call-by-value parameter passing.

application program Program to solve specific problems.

approximate reasoning *See* imprecise computation.

argument (1) An address or value that is passed to a procedure or function call, as a way of communicating cleanly across procedure/function boundaries. (2) A piece of data given to a hardware operator block.

arithmetic logic unit The CPU internal device that performs arithmetic and logical operations.

arithmetic operation Any of the following operations and combination thereof: addition, subtraction, multiplication, division.

artifact Any by-product of the software production process including code and documentation.

assembler A computer program that translates an assembly-code text file to an object file suitable for linking.

assemblers Software that translates assembly language to machine code.

assembly code Programs written in assembly language.

assembly language The set of symbolic equivalents to the macroinstruction set.

associative memory Memory organized so that it can be searched according to its contents.

asynchronous event An event that is not synchronous.

atomic instruction An instruction that cannot be interrupted.

attribute A named property of a class that describes a value held by each object of the class.

attribute multiplicity The possible number of values for an object–attribute combination.

Background Noninterrupt-driven processes in foreground/background systems.

BAM *See* binary angular measurement.

banker's algorithm A technique sometimes used to prevent deadlock situations.

bathtub curve A graph describing the phenomenon that in hardware components most errors occur either very early or very late in the life of the component. Some believe that it is applicable to software.

Belady's anomaly The observation that in the FIFO page replacement rule, increasing the number of pages in memory may not reduce the number of page faults.

benchmark Standard tests that are used to compare the performance of computers, processors, circuits, or algorithms.

beta testing A type of system test where preliminary versions of validated software are distributed to friendly customers who test the software under actual use.

binary angular measurement (BAM) An n-bit scaled number where the least significant bit is $2^{n-1} \cdot 180$.

binary semaphore A semaphore that can take on one of two values.

black-box testing A testing methodology where only the inputs and outputs of the unit are considered. How the outputs are generated inside the unit is ignored.

blocked The condition experienced by tasks that are waiting for the occurrence of an event.

branch instruction An instruction used to modify the instruction execution sequence of the CPU. The transfer of control to another sequence of instructions may be unconditional or conditional based on the result of a previous instruction. In the latter case, if the condition is not satisfied, the transfer of control will be to the next instruction in sequence. It is equivalent to a JUMP instruction, although the range of the transfer may be limited in a branch instruction compared to the JUMP.

branch prediction A mechanism used to predict the outcome of branch instructions prior to their execution.

breakpoint (1) An instruction address at which a debugger is instructed to suspend the execution of a program. (2) A critical point in a program, at which execution can be conditionally stopped to allow examination if the program variables contain the correct values and/or other manipulation of data. Breakpoint techniques are often used in modern debuggers, which provide nice user interfaces to deal with them.

breakpoint instruction A debugging instruction provided through hardware support in most microprocessors. When a program hits a break point, specified actions occur that save the state of the program, and then switch to another program that allows the user to examine the stored state. The user can suspend the execution of a program, examine the registers, stack, and memory, and then resume the program's execution, which is very helpful in a program's debugging.

broadcast communication In statecharts, a technique that allows for transitions to occur in more than one orthogonal system simultaneously.

buffer A temporary data storage area used to interface between, for example, a fast device and a slower process servicing that device.

built-in test software (BITS) Special software used to perform self-testing. On-line BITS assures testing concurrently with normal operation (e.g., accomplished with coding or duplication techniques). Off-line BITS suspends normal operation and it is carried out using built-in test pattern generator and test response analyzer (e.g., signature analyzer).

burn-in testing Testing that seeks to flush out those failures that appear early in the life of the part and thus improve the reliability of the delivered product.

burst period The time over which data are being passed into a buffer.

bus The wires that connect the CPU and main memory. The bus is used to exchange memory location information ("addresses") and data between the

CPU and main memory in binary-encoded form. The width of the bus is determined by the number of bits or wires provided for the binary code. Usually the address and data wires are referred to as the address bus and data bus, respectively.

bus arbitration The process of ensuring that only one device at a time can place data on the bus.

bus contention Condition in which two or more devices attempt to gain control of the main memory bus simultaneously.

bus cycle Memory fetch.

bus grant A signal provided by the DMA controller to a device, indicating that is has exclusive rights to the bus.

bus time-out A condition whereby a device making a DMA request does not receive a bus grant before some specified time.

busy wait In polled-loop systems, the process of testing the flag without success.

cache *See* memory caching.

cache hit ratio The percentage of time in which a requested instruction or data are actually in the cache.

call-by-address *See* call-by-reference.

call-by-reference Parameter passing mechanism in which the address of the parameter is passed by the calling routine to the called procedure so that it can be altered there. *Also known as* call-by-address.

call-by-value parameter passing Parameter passing mechanism in which the value of the actual parameter in the subroutine or function call is copied into the procedure's formal parameter.

calling tree *See* structure chart.

capability An object that contains both a pointer to another object and a set of access permissions that specify the modes of access permitted to the associated object from a process that holds the capability.

CASE Computer-aided software engineering.

catastrophic error An error that renders the system useless.

CCR *See* condition code register.

cellular automata A computational paradigm for an efficient description of SIMD massively parallel systems.

central processing unit In a computer it provides for arithmetic and logical operations. *Abbreviated* CPU.

chain reaction In statecharts, a group of sequential events where the nth event is triggered by the $(n-1)$th event.

checkpoint Time in the history of execution at which a consistent version of the system's state is saved so that if a later event causes potential difficulties, the system can be restarted from the state that had been saved at the checkpoint. Checkpoints are important for the reliability of a distributed system, since timing problems or message loss can create a need to "backup" to a previous state that has to be consistent in order for the overall system to operate functionally.

checkpointing Method used in rollback techniques in which some subset of the system states (data, program, etc.) is saved at specific points (checkpoints), during the process execution, to be used for recovery if a fault is detected.

checksum A value used to determine if a block of data has changed. The checksum is formed by adding all of the data values in the block together, and then finding the two's complement of the sum. The checksum value is added to the end of the data block. When the data block is examined (possibly after being received over a serial line), the sum of the data values and checksum should be zero.

circular queue *See* ring buffer.

CISC *See* complex instruction set computer.

class A group of objects with similar attributes, behavior, and relationships to other objects.

class definitions Object declarations along with the methods associated with them.

clear-box testing *See* white-box testing.

code inspection *See* group walkthrough.

coding The process of programming, generating code in a specific language. The process of translating data from a representation form into a different one by using a set of rules or tables.

collision Condition in which a device already has control of the bus and another obtains access. Also, simultaneous use of the critical resource.

compaction The process of compressing fragmented memory so that it is no longer fragmented. *Also called* coalescing.

compiler A program that translates a high-level language program into an executable machine instruction program or other lower-level form such as assembly language.

complex instruction set computers Architectures characterized by a large, microcoded instruction set with numerous addressing modes.

composition An operation applied to a reliability matrix that determines the maximum reliability between processors.

compute-bound Computations in which the number of operations is large in comparison to the number of I/O instructions.

computer simulation A set of computer programs that allows one to model the important aspects of the behavior of the specific system under study. Simulation can aid the design process by, for example, allowing one to determine appropriate system design parameters or aid the analysis process by, for example, allowing one to estimate the end-to-end performance of the system under study.

concrete class A class that can have direct instances.

condition code register Internal CPU register used to implement a conditional transfer.

conditional instruction An instruction that performs its function only if a certain condition is met.

conditional transfer A change of the program counter based on the result of a test transfer.

configuration Operation in which a set of parameters is imposed for defining the operating conditions.

constant folding An optimization technique that involves precomputing constants at compile time.

content-addressable memory *See* associative memory.

context The minimum information that is needed in order to save a currently executing task so that it can be resumed.

context switching The process of saving and restoring sufficient information for a real-time task so that it can be resumed after being interrupted.

contiguous file allocation The process of forcing all allocated file sectors to follow one another on the disk.

continuous random variable A random variable with a continuous sample space.

control flow diagram A real-time extension to data flow diagrams that shows the flow of control signals through the system.

control specifications In data flow diagrams, a finite state automation in diagrammatic and tabulator representation.

control unit CPU internal device that synchronizes the fetch–execute cycle.

cooperative multitasking system A scheme in which two or more processes are divided into states or phases, determined by a finite state automaton. Calls to a central dispatcher are made after each phase is complete.

coprocessor A second specialized CPU used to extend the macroinstruction set.

coroutine system *See* cooperative multitasking system.

correctness A property in which the software does not deviate from the requirements specification. Often used synonymously with reliability, correctness requires a stricter adherence to the requirements.

correlated data *See* time-relative data.

counting semaphore A semaphore than can take on two or more values. *Also* called a general semaphore.

CPU Central processing unit.

CPU utilization A measure of the percentage of nonidle processing.

CRC *See* cyclic redundancy code.

critical region Code that interacts with a serially reusable resource.

Crystal A lightweight programming methodology that empowers the development team to define the development process and refine it in subsequent iterations until it is stable.

CU *See* control unit.

cycle stealing   A situation in which DMA access precludes the CPU from accessing the bus.

cyclic redundancy code   A method for checking ROM memory that is superior to checksum. *Abbreviated* CRC.

cycling   The process whereby all tasks are being appropriately scheduled (although no actual processing is occurring).

cyclomatic complexity   A measure of a system complexity devised by McCabe.

daemon   A device server that does not run explicitly, but rather lies dormant waiting for some condition(s) to occur.

dangerous allocation   Any memory allocation that can preclude system determinism.

data bus   Bus used to carry data between the various components in the system.

data dependency   The normal situation in which the data that an instruction uses or produces depends upon the data used or produced by other instructions such that the instructions must be executed in a specific order to obtain the desired results.

data flow architecture   A multiprocessing system that uses a large number of special processors. Computation is performed by passing activity packs between them.

data flow diagram   A structured analysis tool for modeling software systems.

data structure   A particular way of organizing a group of data, usually optimized for efficient storage, fast search, fast retrieval, and/or fast modification.

data-oriented methodology   An application development methodology that considers data the focus of activities because they are more stable than processes.

dead code   *See* unreachable code.

deadlock   A catastrophic situation that can arise when tasks are competing for the same set of two or more serially reusable resources.

deadly embrace   *See* deadlock.

death spiral   Stack overflow caused by repeated spurious interrupts.

debug   To remove errors from hardware or software.

debug port   The facility to switch the processor from run mode into probe mode to access its debug and general registers.

debugger   (1) A program that allows interactive analysis of a running program, by allowing the user to pause execution of the running program and examine its variables and path of execution at any point. (2) Program that aids in debugging.

debugging   (1) Locating and correcting errors in a circuit or a computer program. (2) Determining the exact nature and location of a program error, and fixing the error.

decode   The process of isolating the opcode field of a macroinstruction and determining the address in micromemory of the programming corresponding to it.

default The value or status that is assumed unless otherwise specified.

defect The preferred term for an error in requirement, design, or code. *See also* fault, failure.

demand page system Technique where program segments are permitted to be loaded in noncontiguous memory, as they are requested in fixed-size chunks.

density In computer memory, the number of bits per unit area.

dependability System feature that combines such concepts as reliability, safety, maintainability, performance, and testability.

de-referencing The process in which the actual locations of the parameters that are passed using call-by-value are determined.

derivative of f at x Represents the slope of the function f at point x.

deterministic system A system where for each possible state, and each set of inputs, a unique set of outputs and next state of the system can be determined.

digital-to-analog conversion The process of converting discrete (digital) signals into continuous (analog) ones.

direct memory access (DMA) A scheme in which access to the computer's memory is afforded to other devices in the system without the intervention of the CPU.

direct mode instruction Instruction in which the operand is the data contained at the address specified in the address field of the instruction.

disassembler A computer program that can take an executable image and convert it back into assembly code.

discrete signals Logic lines used to control devices.

discriminator An enumerated attribute that indicates which aspect of an object is being abstracted by a particular generalization.

disjunctive normal form A representation of a Boolean expression that involves a logical sum of products (maximum of minima).

dispatcher The part of the kernel that performs the necessary bookkeeping to start a task.

distributed computing An environment in which multiple computers are networked together and the resources from more than one computer are available to a user.

distributed real-time systems A collection of interconnected self-contained processors.

DMA *See* direct memory access.

DMA controller Device that performs bus arbitration.

dormant state In the task-control block model, a state that is best described as a TCB belonging to a task that is unavailable to the operating system.

double buffering A technique using two buffers where one is filled while the data in the other is being used.

double indirect mode A memory addressing scheme similar to indirect mode, but with another level of indirection.

DRAM Dynamic random-access memory.

drive line In core memory, a wire used to induce a magnetic field in a toroid-shaped magnet. The orientation of the field represents either a 1 or a 0.

DSI Delivered source instructions. *See* KLOC.

dynamic memory Memory that uses a capacitor to store logic 1s and 0s, and that must be refreshed periodically to restore the charge lost due to capacitive discharge.

dynamic priority system A system in which the priorities to tasks can change. Contrast with fixed priority system.

Dynamic Systems Development Method (DSDM) A lightweight programming methodology conceived as a methodology for rapid application development, DSDM relies on a set of principles that include empowered teams, frequent deliverables, incremental development, and integrated testing.

effort One of Halstead's metrics (see Chapter 8).

embedded software Software that is part of an embedded system.

embedded system A computing machine contained in a device whose purpose is not to be a computer. For example, the computers in automobiles and household appliances are embedded computers. Embedded computers use embedded software, which integrates an operating system with specific drivers and application software. Their design often requires special software–hardware codesign methods for speed, low power, low cost, high testability, or other special requirements.

emulator (1) The firmware that simulates a given machine architecture. (2) A device, computer program, or system that accepts the same inputs and produces the same outputs as a given system.

enabled state In a data flow architecture when all necessary tokens have arrived and the input lines are full. *Also called* the ready state.

encapsulation Property of a program that describes the complete integration of data with legal process relating to the data.

entity relationship diagram A diagram that describes the important entities in a system and the ways in which they are interrelated.

enumeration A list of permitted values.

environment A set of objects outside the system, a change in whose attributes affects, and is affected by, the behavior of the system.

event Any occurrence that results in a change in the state of a system.

event determinism When the next states and outputs of the system are known for each set of inputs that trigger events.

event flag Synchronization mechanism provided by certain languages.

exception Error or other special condition that arises during program execution.

exception handler Code used to process exceptions.

execute Process of sequencing through the steps in micromemory corresponding to a particular macroinstruction.

executing state In the task-control block model, a task that is currently running.

executive *See* kernel.

external fragmentation   When main memory becomes checkered with unused but available partitions, as in Figure 3.22.

eXtreme Programming (XP)   A lightweight programming methodology based on twelve practices including pair programming (all code developed jointly by two developers), test first coding, having the customer on site, and frequent refactoring. eXtreme programming is, perhaps, the most prescriptive of the lightweight methodologies. Also written "Extreme programming".

failed system   A system that cannot satisfy one or more of the requirements listed in the formal system specification.

failure   Manifestation of an error at system level. It relates to execution of wrong actions, nonexecution of correct actions, performance degradation, and so on as.

failure function   A function describing the probability that a system fails at time t.

fault   The appearance of a defect during the operation of a software system.

fault prevention   Any technique or process that attempts to eliminate the possibility of having a failure occur in a hardware device or software routine.

fault tolerance   Correct execution of a specified function in a system, provided by redundancy, despite faults. The redundancy provides the information needed to negate the effects of faults.

feature-driven development   A lightweight model-driven, short-iteration process built around the feature, a unit of work that has meaning for the client and developer and is small enough to be completed quickly.

fetch   The process of retrieving a macroinstruction from main memory and placing it in the instruction register.

fetch–execute cycle   The process of continuously fetching and executing macroinstructions from main memory.

file fragmentation   Analogous to memory fragmentation, but occurring within files, with the same associated problems.

finite state automaton (FSA)   *See* finite state machine.

finite state machine (FSM)   A mathematical model of a machine consisting of a set of inputs, a set of states, and a transition function that describes the next state given the current state and an input. *Also known as* finite state automaton and state transition diagram.

firing   In Petri nets or in certain multiprocessor architectures, when a process performs its prescribed function.

firm real-time system   A real-time system that can fail to meet one or more deadlines without system failure.

fixed priority system   A system in which the task priorities cannot be changed. Contrast with dynamic priority system.

fixed-rate system   A system in which interrupts occur only at fixed rates.

flip-flop   A bistable logic device.

floating-point number   A term describing the computer's representation of a real number.

flowchart   A traditional graphic representation of an algorithm or a program, in using named functional blocks (rectangles), decision evaluators (diamonds), and I/O symbols (paper, disk) interconnected by directional arrows that indicate the flow of processing. *Syn*: flow diagram.

flush   In pipelined architectures, the act of emptying the pipeline when branching occurs.

foreground   A collection of interrupt driven or real-time processes.

formal parameter   The dummy variable used in the description of a procedure or subroutine.

forward error recovery   A technique (also called roll-forward) of continuing processing by skipping faulty states (applicable to some real-time systems in which occasional missed or wrong responses are tolerable).

framework   A skeletal structure of a program that requires further elaboration.

FSA   Finite state automaton. *See* finite state machine.

FSM   *See* finite state machine.

function points   A widely used metric set in nonembedded environments; they form the basis of many commercial software analysis packages. Function points measure the number of interfaces between modules and subsystems in programs or systems.

function test   A check for correct device operation generally by truth table verification.

functional decomposition   The division of processes into modules.

functional requirements   Those system features that can be directly tested by executing the program.

garbage   An object or a set of objects that can no longer be accessed, typically because all pointers that direct accesses to the object or set have been eliminated.

garbage collector   A software run-time system component that periodically scans dynamically allocated storage and reclaims allocated storage that is no longer in use (garbage).

general polynomial   The modulo-2 divisor of the message polynomial in CRC.

general register   CPU internal memory that is addressable in the address field of certain macroinstructions.

general semaphore   *See* counting semaphore.

generalization   The relationship between a class and one or more variations of that class.

global variable   Any variables that is within the scope of all modules of the software system.

group walkthrough   A kind of white-box testing in which a number of persons inspect the code line-by-line with the unit author.

Hamming code   A coding technique used to detect and correct errors in computer memory.

hard error   Physical damage to memory cell.

hard real-time system   A real-time system in which missing even one deadline results in system failure.

hazard   A momentary output error that occurs in a logic circuit because of input signal propagation along different delay paths in the circuit.

heterogeneous   Having dissimilar components in a system; in the context of computers, having different types or classes of machines in a multiprocessor or multicomputer system.

host   A computer that is the one responsible for performing a certain computation or function.

hybrid system   A system in which interrupts occur both at fixed frequencies and sporadically.

hypercube processors   A processor configuration that is similar to the linear array processor, except that each processor element communicates data along a number of other higher dimensional pathways.

ICE   *See* in-circuit emulator.

immediate mode instruction   An instruction in which the operand is an integer.

implied mode instruction   An instruction involving one or more specific memory locations or registers that are implicitly defined in the operation performed by instruction.

imprecise computation   Techniques involving early termination of a computation in order to meet deadlines. Sometimes called approximate reasoning.

in-circuit emulator (ICE)   A device that replaces the processor and provides the functions of the processor plus testing and debugging functions.

incrementality   A software approach in which progressively larger increments of the desired product are developed.

indirect mode instruction   Instruction where the operand field is a memory location containing the address of the address of the operand.

induction variable   A variable in a loop that is incremented or decremented by some constant.

information hiding   A program design principle that makes available to a function only the data it needs.

inheritance   In object orientation, the possibility for different data types to share the same code.

initialize   (1) To place a hardware system in a known state, for example, at power-up. (2) To store the correct beginning data in a data item, for example, filling an array with zero values before it is used.

in-line patch   A patch that fits into the memory space allocated to the code to the changed.

input space   The set of all possible input combinations to a system.

instance   An occurrence of a class.

instruction issue   The sending of an instruction to functional units for execution.

instruction register CPU internal register that holds the instruction pointed to by the contents of the program counter.

instruction set The instruction set of a processor is the collection of all the machine language instructions available to the programmer. *Also known as* instruction repertoire.

integration The process of uniting modules from different sources to form the overall system.

internal fragmentation Condition that occurs in fixed-partition schemes when, for example, a processor requires 1 kilobyte of memory, while only the 2-kilobyte partitions are available.

interoperability Software quality that refers to the ability of the software system to coexist and cooperate with other systems.

interpreter A computer program that translates and immediately performs intended operations of the source statements of a high-level language program.

interrupt An input to a processor that signals the occurrence of an asynchronous event. The processor's response to an interrupt is to save the current machine state and execute a predefined subprogram. The subprogram restores the machine state on exit and the processor continues in the original program.

interrupt controller A device that provides additional interrupt handling capability to a CPU.

interrupt handler A predefined subprogram that is executed when an interrupt occurs. The handler can perform input or output, save data, update pointers, or notify other processes of the event. The handler must return to the interrupted program with the machine state unchanged.

interrupt handler location Memory location containing the starting address of an interrupt handler routine. The program counter is automatically loaded with its address when an interrupt occurs.

interrupt latency The delay between when an interrupt occurs and when the CPU begins reacting to it.

interrupt register Register containing a big map of all pending (latched) interrupts.

interrupt return location Memory location where the contents of the program counter is saved when the CPU processes an interrupt.

interrupt vector Register that contains the identity of the highest-priority interrupt request.

intrinsic function A macro where the actual function calls is replaced by in-line code.

Jackson Chart A form of structure chart that provides for conditional branching.

Kalman filter A mathematical construct used to combine measurements of the same quantity from different sources.

KDSI *See* KLOC

kernel The smallest portion of the operating system that provides for task scheduling, dispatching, and inertia communication.

kernel preemption A method used in real-time UNIX that provides preemption points in calls to kernel functions to allow them to be interrupts.

key In a mailbox, the data that are passed as a flag used to protect a critical region.

KLOC A software metric measuring thousands of lines of code (not counting comments and nonexecutable statements). *Called* the "clock" metric. *Also known as* thousands of delivered source instructions (KDSI) and noncommented source-code statements (NCSS).

least recently used rule (LRU) The best nonpredictive page-replacement algorithm.

legacy system Applications that are in a maintenance phase but are not ready for retirement.

leveling In data flow diagrams, the process of redrawing a diagram at a finer level of detail.

library A set of precompiled routines that may be linked with a program at compile time or loaded at load time or dynamically at run time.

lightweight programming methodology Any programming methodology that is adaptive rather than predictive and emphasizes people rather than process. Same as agile programming.

linear array processor A processor organized so that multiple instructions of the same type can be executed in parallel.

link The portion of the compilation process in which separate modules are placed together and cross-module references resolved.

linker A computer program that takes one or more object files, assembles them into blocks that are to fit into particular regions in memory, and resolves all external (and possibly internal) references to other segments of a program and to libraries of precompiled program units.

Little's law Rule from queuing theory stating that the average number of customers in a queuing system, N_{av}, is equal to the average arrival rate of the customers to that system, r_{av}, times the average time spent in that system, t_{av}.

live variable A variable that can be used subsequently in the program.

livelock Another term for process starvation.

load module Code that can be readily loaded into the machine.

locality-of-reference The notion that if you examine a list of recently executed program instructions on a logic analyzer, you will see that most of the instructions are localized to within a small number of instructions.

lock-up When a system enters a state in which it is rendered ineffective.

logic analyzer A machine that can be used to send signals to, and read output signals from, individual chips, circuit boards, or systems.

logical operation The machine-level instruction that performs Boolean operations such as AND, OR, and COMPLEMENT.

look-up table An integer arithmetic technique that uses tables and relies on mathematical definition of the derivative to compute functions quickly.

loop invariant optimization The process of placing computations outside a loop that do not need to be performed within the loop.

loop invariant removal An optimization technique that involves removing code that does not change inside a looping sequence.

loop jamming An optimization technique that involves combining two loops within the control of one loop variable.

loop unrolling An optimization technique that involves expanding a loop so that loop overhead is removed.

loosely coupled system A system that can run on other hardware with the rewrite of certain modes.

LRU *See* least recently used rule.

machine code The machine format of a compiled executable, in which individual instructions are represented in binary notation.

machine language The set of legal instructions to a machine's processor, expressed in binary notation.

macro *See* macroinstruction.

macroinstruction A native machine instruction.

macroprogram A sequence of macroinstructions.

mailbox An intertask communication device consisting of a memory location and two operations – post and pend – that can be performed on it.

main memory Memory that is directly addressable by the CPU.

maintainability A software quality that is a measure of how easy the system can be evolved to accommodate new features, or changed to repair errors.

maintenance The changes made on a system to fix errors, to support new requirements, or to make it more efficient.

major cycle The largest sequence of repeating processes in cyclic or periodic systems.

MAR *See* memory address register.

mask register A register that contains a bit map either enabling or disabling specific interrupts.

master processor The on-line processor in a master/slave configuration.

MDR *See* memory data register.

Mealy finite state machine A finite state machine with outputs.

memory address register (MAR) Register that holds the address of the memory location to be acted on.

memory caching A technique in which frequently used segments of main memory are stored in a faster bank of memory that is local to the CPU (called a cache).

memory data register (MDR) Register that holds the data to be written to or that is read from the memory location held in the MAR.

memory-loading The percentage of usable memory that is being used.

memory locking In a real-time system, the process of locking all or certain parts of a process into memory to reduce the overhead involved in paging, and thus make the execution times more predictable.

memory-mapped I/O   An input/output scheme where reading or writing involves executing a load or store instruction on a pseudomemory address mapped to the device. *Contrast with* DMA and programmed I/O.

memory reference instruction   An instruction that communicates with virtual memory, writing to it (store) or reading from it (load).

mesh processor   A processor configuration that is similar to the linear array processor, except that each processor element also communicates data north and south.

message exchange   *See* mailbox.

message-passing system   A multiprocessor system that uses messages passed among the processors to coordinate and synchronize the activities in the processors.

message polynomial   Used in CRC.

metadata   Data that describes other data.

methods   In object-oriented systems, functions that can be performed on objects.

microcode   A collection of low-level operations that are executed as a result of a single macro instruction being executed.

microcontroller   A computer system that is programmable via microcode.

microinstructions   *See* microcode.

microkernel   A nanokernel that also provides for task scheduling.

micromemory   CPU internal memory that holds the binary codes corresponding to macroinstructions.

microprogram   Sequence of microcode stored in micromemory.

MIMD   *See* multiple instruction stream, multiple data stream.

minimal representation   For a positive Boolean function an equivalent representation where no product whose variable set does not contain the variable set of a distinct products can be deleted without changing the function.

minor cycle   A sequence of repeating processes in cyclic or periodic systems.

minterm   In disjunctive normal form, a logical sum of products or conjunctions of Boolean variables is taken. These products are the minterms.

MISD   *See* multiple instruction stream, single-data stream.

mixed listing   A printout that combines the high-order language instruction with the equivalent assembly language code.

mixed system   A system in which interrupts occur both at fixed frequencies and sporadically.

modularity   Design principle that calls for design of small, self-contained code units.

Moore finite state machine   *See* finite state machine.

multiple instruction stream, single data stream (MISD)   A computer that can process two or more instructions concurrently on a single datum.

multiple instruction stream, multiple data stream (MIMD)   A computer characterized by a large number of processing elements, each capable of executing numerous instructions.

multiplexer A device used to route multiple lines onto fewer lines.

multiprocessing operating system An operating system where more than one processor is available to provide for simultaneity. *Contrast with* multitasking operating system.

multiprocessor A computer system that has more than one internal processor capable of operating collectively on a computation. Normally associated with those systems where the processors can access a common main memory.

multitasking operating system An operating system that provides sufficient functionality to allow multiple programs to run on a single processor so that the illusion of simultaneity is created. *Contrast with* multiprocessing operating system.

mutex A common name for a semaphore variable.

MUX *See* multiplexer.

nanokernel Code that provides simple thread-of-execution (same as "flow-of-control") management; essentially provides only one of the three services provided by a kernel, that is, it provides for task dispatching.

NCSS Noncommented source statements. *See* KLOC.

nested subroutine A subroutine called by another subroutine. The programming technique of a subroutine calling another subroutine is called nesting.

nonfunctional requirements System requirements that cannot be tested easily by program execution.

nonvolatile memory Memory whose contents are preserved upon removing power.

non–von Neumann architecture An architecture that does not use the stored-program series fetch–execute cycle.

no-op A macroinstruction that does not change the state of the computer.

NP-complete problem A decision problem that is a seemingly intractable problem for which the only known solutions are exponential functions of the problem size and which can be transformed to all other NP-complete problems; *compare with* NP-hard.

NP-hard A decision problem that is similar to an NP-complete problem (except that for the NP-hard problem it cannot be shown to be transformable to all other NP-complete problems).

N-version programming A technique used to reduce the likelihood of system lock-up by using redundant processors, each running software that has been coded to the same specifications by different teams.

nucleus *See* kernel.

null A special value denoting that an attribute value is unknown or not applicable.

object An instance of a class definition.

object code A file comprising an intermediate description of a program segment.

object-oriented The organization of software into discrete objects that encapsulate both data structure and behavior.

object-oriented analysis   A method of analysis that estimates requirements from the perspective of the classes and objects found in the problem domain.

object-oriented design   A design methodology viewing a system as a collection of objects with messages passed from object to object.

object-oriented language   A language that provides constructs that encourage a high degree of information hiding and data abstraction.

object-oriented methodology   An application development methodology that uses a top-down approach based on the decomposition of a system in a collection of objects communicating via messages.

object-oriented programming   A programming style using languages that support abstract data types, inheritance, function polymorphism, and messaging.

object type   The type of an object determines the set of allowable operations that can be performed on the object. This information can be encoded in a "tag" associated with the object, can be found along an access path reaching to the object, or can be determined by the compiler that inserts "correct" instructions to manipulate the object in a manner consistent with its type.

opcode   Starting address of the microcode program stored in micromemory.

open source code   Source code that is made available to the user community for moderate improvement and correction.

open system   An extensible collection of independently written applications that cooperate to function as an integrated system.

operating system   A set of programs that manages the operations of a computer. It oversees the interaction between the hardware and the software and provides a set of services to system users.

operation   Specification of one or a set of computations on the specified source operands placing the results in the specified destination operands.

organic system   A system that is not embedded.

orthogonal product   In statecharts, a process that depicts concurrent processes that run in isolation.

output dependency   The situation when two sequential instructions in a program write to the same location. To obtain the desired result, the second instruction must write to the location after the first instruction.

output space   The set of all possible output combinations for a system.

overlay   Dependent code and data sections used in overlaying.

overlaying   A technique that allows a single program to be larger than the allowable user space.

overloading   Principle according to which operations bearing the same name apply to arguments of different data type.

oversized patch   A patch that requires more memory than is currently occupied by the code to be replaced.

page   Fixed-size chunk used in demand-paged systems.

page fault   An exception that occurs when a memory reference is made to a location within a page not loaded in main memory.

page-frame *See* page.

page stealing When a page is to be loaded into main memory, and no free pages are found, then a page frame must be written out or swapped to disk to make room.

page table A collection of pointers to pages used to allow noncontiguous allocation of page frames in demand paging.

pair-programming A technique in which two persons write code together.

Parnas Partitioning *See* information hiding.

patching The process of correcting errors in the code directly on the target machine.

pattern A named problem–solution pair that can be applied in new contexts, with advice on how to apply it in novel situations.

PC *See* program counter.

PDL *See* program design language.

peephole optimization An optimization technique where a small window of assembly language or machine code is compared against known patterns that yield optimization opportunities.

pend operation Operation of removing data from a mailbox. If data are not available, the process performing the pend suspends itself until the data become available.

performance A measure of the software's capability of meeting certain functional constraints such as timing or output precision.

Petri net A mathematical/pictorial system description technique.

phase-driven code *See* state-driven code.

Ping-Pong buffering *See* double buffering.

pipeline An intertask communication mechanism provided in UNIX.

pipelining A technique used to speed processor execution that relies on the fact that fetching the instruction is only one part of the fetch–execute cycle, and that is can overlap with different parts of the fetch–execute cycle for other instructions.

polled loop system A real-time system in which a single and repetitive test instruction is used to test a flag that indicates that some event has occurred.

polymorphism In object-oriented programming, polymorphism allows the programmer to create a single function that operates on different objects, depending on the type of object involved.

portability A quality in which the software can easily run in different environments.

positive Boolean function A Boolean function that can be represented as a logical sum of products in which no variables are complemented. *Also called* an increasing Boolean function.

post operation Operation that places data in a mailbox.

power bus The collection of wires used to distribute power to the various components of the computer systems.

power on self-test A series of diagnostic tests performed by a machine (such as the personal computer) when it powers on.

pragma In certain programming languages, a pseudo-op that allows assembly code to be placed in line with the high-order language code.

preempt A condition that occurs when a higher-priority task interrupts a lower-priority task.

preemptive priority system A system that uses preemption schemes instead of round-robin or first-come, first-served scheduling.

primary memory *See* main memory.

priority inversion A condition that occurs because a noncritical task with a high execution rate will have a higher priority than a critical task with a low execution rate.

procedure A self-contained code sequence designed to be reexecuted from different places in a main program or another procedure.

procedure call In program execution, the execution of a machine language routine, after which execution of the program continues at the location following the location of the procedure call.

process The context, consisting of allocated memory, open files, network connections, in which an operating system places a running program.

process control block An area of memory containing information about the context of an executing program. Although the process control block is primarily a software mechanism used by the operating system for the control of system resources, some computers use a fixed set of process control blocks as a mechanism to hold the context of an interrupted process.

processing elements The individual processors in a multiprocessing system such as a systolic or wavefront architecture.

program counter (PC) A CPU register containing the address of the next macroinstruction to be executed.

program design language (PDL) A type of abstract high-order language used in system specification.

programmed I/O Transferring data to or from a peripheral device by running a program that executes individual computer instruction or commands to control the transfer. An alternative is to transfer data using DMA.

propagation delay The contribution to interrupt latency due to limitation in switching speeds of digital devices and in the transit time of electrons across wires.

protection fault An error condition detected by the address mapper when the type of request is not permitted by the object's access code.

prototype A mock-up of a software system often used during the design phase.

prototyping Building an engineering model of all or part of a system to prove that the concept works.

pseudocode A technique for specifying the logic of a program in an English-like language. Pseudocode does not have to follow any syntax rules and can be read by anyone who understands programming logic.

pseudo-exhaustive testing A testing technique that relies on various forms of circuit segmentation and application of exhaustive test patterns to these segments.

pseudo-operation In assembly language, an operation code that is an instruction to the assembler rather than a machine-language instruction. *Also known as* pseudo-op.

pseudorandom testing A testing technique based on pseudorandomly generated test patterns. Test length is adapted to the required level of fault coverage.

pure procedure A procedure that does not modify itself during its own execution. The instructions of a pure procedure can be stored in a read-only portion of the memory and can be accessed by many processes simultaneously.

race condition A situation where multiple processes access and manipulate shared data with the outcome dependent on the relative timing of these processes.

raise Mechanism used to initiate a software interrupt in certain languages, such as C.

RAM scrubbing A technique used in memory configurations that include error detection and correction chips. The technique, which reduces the chance of multiple-bit errors occurring, is needed because in some configurations memory errors are corrected on the bus and not in memory itself. The corrected memory data then need to be written back to memory.

random testing The process of testing using a set of pseudorandomly generated patterns.

random variable A function mapping elements of the sample space into a real number.

rate-monotonic system A fixed-rate, preemptive, prioritized real-time system where the priorities are assigned so that the higher the execution frequency, the higher the priority.

reactive system A system that has some ongoing interaction with its environment.

read/write line Logic line that is set to logic 0 during memory write and to logic 1 during memory read.

ready state In the task-control block model, the state of those tasks that are ready to run, but are not running.

real-time Refers to systems whose correctness depends not only on outputs but the timeliness of those outputs. Failure to meet one or more of the deadlines can result in system failure.

real-time computing Support for environments in which response time to an event must occur within a predetermined amount of time. Real-time systems may be categorized into hard, firm and, soft real time.

reentrant Term describing a program that uses concurrently exactly the same executable code in memory for more than one invocations of the program (each with its own data), rather than separate copies of a program for each invocation. The read and write operations must be timed so that the correct results are always available and the results produced by an invocation are not overwritten by another one.

recovery   Action that restores the state of a process to an earlier configuration after it has been determined that the system has entered a state that does not correspond to functional behavior. For overall functional behavior, the states of all processes should be restored in a manner consistent with each other, and with the conditions within communication links or message channels.

recovery block   Section of code that terminates in checkpoints. If the check fails, processing can resume at the beginning of a recovery block.

recursion   The process whereby a program calls itself.

recursive procedure   A procedure that can be called by itself or by another program that it has called; effectively, a single process can have several executions of the same program alive at the same time. Recursion provides one means of defining functions. The recursive definition of the factorial function is the classic example: for all $n > 0$, factorial$(n) = n^*$ factorial $(n - 1)$.

reduced instruction set computer (RISC)   Architecture usually characterized by a small instruction set with limited addressing modes and hardwired (as opposed to microcoded) instructions.

reduction in strength   Optimization technique that uses the fastest macroinstruction possible to accomplish a given calculation.

redundancy   The use of parallel or series components in a system to reduce the possibility of failure. Similarly, referring to an increase in the number of components that can interchangeably perform the same function in a system. Sometimes it is referred to as hardware redundancy in the literature to differentiate from so-called analytical redundancy in the field of fault detection and isolation/identification. Redundancy can increase the system reliability.

reentrancy   The characteristic of a block of software code that if present, allows the code in the block to be executed by more than one process at a time.

reentrant procedure   A procedure that can be used by several concurrently running tasks in a multitasking system.

refactoring   To perform a behavior-preserving code transformation.

register direct mode   A memory-addressing scheme similar to direct mode except the operand is a CPU register and not an address.

register direct mode instruction   Instruction in which the operand address is kept in a register named in the operand field of the instruction.

register indirect addressing   An instruction-addressing method in which the register field contains a pointer to a memory location that contains the memory address of the data to be accessed or stored.

register indirect mode   A memory-addressing scheme similar to indirect mode, except the operand address is kept in a register rather than in another memory address.

regression testing   A test methodology used to validate updated software against an old set of test cases that have already been passed.

reliability   The probability that a component or system will function without failure over a specified time period, under stated conditions.

reliability matrix   In a multiprocessing system, a matrix that denotes the reliability of the connections between processors.

requirements analysis   A phase of software-development life cycle in which the business requirements for a software product are defined and documented.

response store   In associative memory the tag memory used to mark memory cells.

response time   The time between the presentation of a set of inputs to a software system and the appearance of all the associated outputs.

reusability   The possibility to use or easily adapt the hardware or software developed for a system to build other systems. Reusability is a property of module design that permits reuse.

reuse   Programming modules are reused when they are copied from one application program and used in another.

reverse engineering   The reverse analysis of an old application to conform to a new methodology.

reverse Polish notation   The result of building a binary parse tree with operands at the leaves and operations at the roots, and then traversing it in postorder fashion.

ring buffer   A first-in, first-out list in which simultaneous input and output to the list is achieved by keeping head and tail pointers. Data are loaded at the tail and read from the head.

RISC   *See* reduced instruction set computer.

robustness   A software quality that measures the software's tolerance to exceptional situations, for example, an input out of range.

root   In overlaying memory management, the portion of memory containing the overlay manage and code common to all overlay segments, such as math libraries.

round-robin system   A system in which several processes are executed sequentially to completion, often in conjunction with a cyclic executive.

round-robin system with timeslicing   A system in which each executable task is assigned a fixed time quantum called a time slice in which to execute. A clock is used to initiate an interrupt at a rate corresponding to the time slice.

safety   The probability that a system will either perform its functions correctly or will discontinue its functions in a well-defined, safe manner.

safety-critical system   A system that is intended to handle rare unexpected, dangerous events.

sampling rate   The rate at which an analog signal is converted to digital form.

scale factor   A technique used to simulate floating-point operations by assigning an implicit noninteger value to the least significant big of an integer.

scaled number   An optimization technique where the least significant bit (LSB) of an integer variable is assigned a real number scale factor.

schedualability analysis   The compile-time prediction of execution-time performance.

scheduler   The part of the kernel that determines which task will run.

scratch-pad memory   CPU internal memory used for intermediate results.

screen signature   The CRC of a screen memory.

Scrum   A lightweight programming methodology based on the empirical process control model, the name is a reference to the point in a rugby match where the opposing teams line up in a tight and contentious formation. Scrum programming relies on self-directed teams and dispenses with much advanced planning, task definition, and management reporting.

secondary memory   Memory that is characterized by long-term storage devices such as tapes, disks, and cards.

secondary storage   Computer devices such as hard disks, floppy disks, tapes, and so forth, that are not part of the physical address space of the CPU.

segment   In pipelining a disjoint processing circuit. *Also called* a stage.

self-modifying code   A program using a machine instruction that changes the stored binary pattern of another machine instruction in order to create a different instruction that will be executed subsequently. This is not a recommended practice.

self-test   A test that a module, either hardware or software, runs upon itself.

self-test and repair   A fault-tolerant technique based on functional unit active redundancy, spare switching, and reconfiguration.

semaphore   A special variable type used for protecting critical regions.

semaphore primitives   The two operations that can be performed on a semaphore, namely, wait and signal.

semidetached system   *See* loosely coupled system.

sense line   In core memory a wire that is used to "read" the memory. Expanding on the orientation of the magnetic field in the core, a pulse is or is not generated in the sense line.

sequential fault   A fault that causes a combinational circuit to behave like a sequential one.

serially reusable resource   A resource that can only be used by one task at a time and that must be used to completion.

server   A process used to manage multiple requests to a serially reusable resource.

SEU   *See* single-event upset.

signal operation   Operation on a semaphore that essentially releases the resource protected by the semaphore.

SIMD   *See* single instruction stream, multiple data stream.

single-event upset   Alteration of memory contents due to charged particles present in space, or in the presence of a nuclear event.

single instruction stream, multiple data stream (SIMD)   A computer where each processing element is executing the same (and only) instruction, but on different data.

single instruction stream, single data stream (SISD)   A type of computer where the CPU processes a single instruction at a time and a single datum at a time.

SISD   *See* single instruction stream, single data stream.

slave processor   The off-line processor in a master–slave configuration.

SLOC   *See* source lines of code.

soft computing   An association of computing methodologies centering on fuzzy logic, artificial neural networks, and evolutionary computing. Each of these methodologies provides complementary and synergistic reasoning and searching methods to solve complex, real-word problems.

soft error   Repairable alternation of the contents of memory.

soft real-time system   A real-time system in which failure to meet deadlines results in performance degradation but not necessarily failure.

software   A collection of macroinstructions.

software design   A phase of software development life cycle that maps what the system is supposed to do into how the system will do it in a particular hardware/software configuration.

software development life cycle   A way to divide the work that takes place in the development of an application.

software engineering   Systematic development, operation, maintenance, and retirement of software.

software evolution   The process that adapts the software to changes of the environment where it is used.

software interrupt   A machine instruction that initiates an interrupt function. Software interrupts are often used for system calls because they can be executed from anywhere in memory and the processor provides the necessary return address handling.

software reengineering   The reverse analysis of an old application to conform to a new methodology.

software reliability   The probability that a software system will not fail before some time t.

source code   Software code that is written in a form or language meant to be understood by programmers. Must be translated to object code in order to run on a computer.

source lines of code (SLOC)   A metric that measures the number of executable program instructions – one SLOC may span several lines, for example, as in an if-then-else statement.

spatial fault tolerance   Methods involving redundant hardware or software.

specification   A statement of the design or development requirements to be satisfied by a system or product.

speculative execution   A CPU instruction execution technique in which instructions are executed without regard to data dependencies.

spin lock   Another name for the wait semaphore operation.

sporadic system   A system with all interrupts occurring sporadically.

sporadic task   A task driven by an interrupt that occurs periodically.

spurious interrupt   Extraneous and unwanted interrupt that is not due to time-loading.

SRAM   *See* static random-access memory.

stack   A first-in, last-out data structure.

stack filter   Positive Boolean function used as a filter in conjunction with threshold sets.

stack machine   Computer architecture in which the instructions are centered on an internal memory store called a stack, and an accumulator.

stage   *See* segment.

starvation   A condition that occurs when a task is not being serviced frequently enough.

state diagram   A diagram showing the conditions (states) that can exist in a logic system and what signals are required to go from one state to another state.

state-driven code   Program code based on a finite state automaton.

static memory   Memory that does not rely on a capacitive charge to store binary data.

static random-access memory (SRAM)   Random access memory that does not need to be recharged periodically.

statistically based testing   Technique that uses an underlying probability distribution function for each system input to generate random test case.

status register   A register involved in interrupt processing that contains the value of the lowest interrupt that will currently be honored.

stress testing   A type of testing wherein the system is subjected to a large disturbance in the inputs (for example, a large burst of interrupts), followed by smaller disturbances spread out over a longer period of time.

structure chart   Graphical design tool used to partition system functionality.

subclass   A class that adds specific attributes, behavior, and relationships for a generalization.

subroutine   A group of instructions written to perform a task, independent of a main program and can be accessed by a program or another subroutine to perform the task.

superclass   A class that holds common attributes, behavior, and relationships for generalization.

suspended state   In the task-control block model, those tasks that are waiting on a particular resource, and thus are not ready. *Also called* blocked state.

swapping   The simplest scheme that allows the operating system to allocate main memory to two processes simultaneously.

switch bounce   The physical phenomenon that an electrical signal cannot instantaneously change logic states.

synchronous An operation or operations that are controlled or synchronized by a clocking signal.

synchronous data *See* time-relative data.

synchronous event Event that occurs at predictable times in the flow-of-control.

syndrome bits The extra bits needed to implement a Hamming code.

syntax The part of a formal definition of a language that specifies legal combinations of symbols that make up statements in the language.

system An entity that when presented with a set of inputs produces corresponding outputs.

system implementation A phase of the software development life cycle during which a software product is integrated into its operational environment.

system program Software used to manage the resources of the computer.

system unification A process consisting of linking together the testing software modules in an orderly fashion.

systems engineering An approach to the overall life-cycle evolution of a product or system. Generally, the systems engineering process comprises a number of phases. There are three essential phases in any systems engineering life cycle: formulation of requirements and specifications, design and development of the system or product, and deployment of the system. Each of these three basic phases can be further expanded into a larger number. For example, deployment generally comprises operational test and evaluation, maintenance over an extended operational life of the system, and modification and retrofit (or replacement) to meet new and evolving user needs.

systolic processor Multiprocessing architecture that consists of a large number of uniform processors connected in an array topology.

task-control block (TCB) A collection of data associated with a task including context, process code (or a pointer to it), and other information.

TCB *See* task control block.

template In a data flow architecture a way of organizing data into tokens. *Also called* an activity packet.

temporal determinism A condition that occurs when the response time for each set of outputs is known in a deterministic system.

temporal fault tolerance Techniques that allow for tolerating missed deadlines.

test-and-set instruction A macroinstruction that can atomically test and then set a particular memory address to some value.

test first coding A software engineering technique in which the code unit test cases are written by the programmer before the actual code is written.

test pattern Input vector such that the faulty output is different from the fault-free output.

test probe A checkpoint used only during testing.

test suite A collection of test cases.

testability The measure of the ease with which a system can be tested.

testing   A phase of software development life cycle during which the application is exercised for the purpose of finding errors.

thrashing   Very high paging activity.

throughput   A measure of the number of macroinstructions per second that can be processed based on some predetermined instruction mix.

time-loading   The percentage of "useful" processing the computer is doing. *Also known as* the utilization factor.

time overloaded   A system that is 100% or more time-loaded.

time-relative data   A collection of data that must be timed correlated.

timeslice   A fixed time quantum used to limit execution time in round-robin systems.

timing error   An error in a system due to faulty time relationships between its constituents.

token   In data flow architectures, data items employed to represent the dynamics of a data flow system.

traceability   A software property that is concerned with the relationships between requirements, their sources, and the system design.

tracing   In software engineering, the process of capturing the stream of instructions, referred to as the trace, for later analysis.

transceiver   A transmit/receive hybrid device.

transputer   A fully self-sufficient, multiple instruction set, von Neumann processor, designed to be connected to other transputers.

trap   Internal interrupt caused by the execution of a certain instruction.

tri-state   A high-impedance state that, in effect, disconnects a device from the bus.

UML   *See* Unified Modeling Language.

unconditional branch   An instruction that causes a transfer of control to another address without regard to the state of any condition flags.

Unified Modeling Language (UML)   A collection of modeling tools for object-oriented representation of software and other enterprises.

Unified Process Model (UPM)   Process model that uses an object-oriented approach by modeling a family of related software processes using the Unified Modeling Language (UML) as a notation.

unit   A software module.

unreachable code   Code that can never be reached in the normal flow of control.

UPM   *See* Unified Process Model.

usability   A property of software detailing the ease in which it can be used.

user space   Memory not required by the operating system.

utilization factor   *See* time-loading.

validation   A review to establish the quality of a software product for its operational purpose.

vector processor   *See* linear array processor.

verifiability   Software property in which its other properties (e.g., portability, usability) can be verified easily.

version control software   A system that manages the access to the various components of the system from the software library.

very long instruction word computer (VLIW)   A computer that implements a form of parallelism by combining microinstructions to exploit redundant CPU components.

virtual machine   A process on a multitasking computer that behaves as if it were a stand-alone computer and not part of a larger system.

VLIW   *See* very long instruction word computer.

volatile memory   Memory in which the contents will be lost if power is removed.

von Neumann architecture   A CPU employing a serial fetch–decode–execute process.

von Neumann bottleneck   A situation in which the serial fetch and execution of instructions limits overall execution speed.

WBS   *See* work breakdown structure.

wait-and-hold condition   The situation in which a task acquires a resource and then does not relinquish it until it can acquire another resource.

wait operation   Operation on a semaphore that essentially locks the resource protected by the semaphore, or prevents the requesting task from proceeding if the resource is already locked.

wait state   Clock cycles used to synchronize macroinstruction execution with the access time of memory.

watchdog timer   A device that must be reset periodically or a discrete signal is issued.

wavefront array processor   Similar to a systolic processor, except that there is no external clock.

wavefront processor   A multiprocessing architecture that consists of an array of identical processors, each with its own local memory and connected in a nearest-neighbor topology.

white-box testing   Logic-driven testing designed to exercise all paths in the module. Same as clear-box testing.

work breakdown structure (WBS)   A hierarchically decomposed listing of tasks.

BIBLIOGRAPHY

[Allard91] Allard, J. R., and Hawkinson, L. B. "Real-Time Programming in Common LISP." *Communications of the ACM*, Volume 35, Number 9, September 1991, pp. 64–69.

[Allworth87] Allworth, S. T., and Zobel, R. N. *Introduction to Real-Time Software Design*, 2nd Edition. Springer-Verlag, New York, 1987.

[Amdahl67] Amdahl, G. M. "Velocity of the Single-Processor Approach to Large Scale Computing Capabilities." *Proceedings of AFIPS*, Volume 30. Atlantic City, NJ, April 18–20, AFIPS Press, Reston: VA, 1967, pp. 483–485.

[Avrunin98] Avrunin, G., Corbett, J., and Dillon, L. "Analyzing Partially-Implemented Real-Time Systems." *IEEE Transactions on Software Engineering*, Volume 24, Number 8, August 1998, pp. 602–614.

[Baker90] Baker, T. P. "A Stacked-Based Resource Allocation Policy for Real-Time Processes." *Proceedings of the 11th Real-Time Systems Symposium*. Lake Buena Vista, FL, December 1990, pp. 191–200.

[Ball00] Ball, S. *Embedded Microprocessor Systems*. Newnes, Boston: MA, 2000.

[Barr99] Barr, M. *Programming Embedded Systems in C and C++*. O'Reilly and Associates, Sebastopol, CA, 1999.

[Bartee91] Bartee, T. C. *Computer Architecture and Logic Design*. McGraw-Hill, New York, 1991.

[Baruah90] Baruah, S. K., Mok, A. K., and Rosier, L. E. "Preemptively Scheduling Hard Real-Time Sporadic Tasks on One Processor." *Proceedings of the 11th Real-Time Systems Symposium*. Lake Buena Vista, FL, December 1990, pp. 182–190.

[Beck99] Beck, K. *Extreme Programming Explained: Embrace Change*. Addison-Wesley, New York, 1999.

[Berger02] Berger, A. *Embedded Systems Design: An Introduction to Processes, Tools, & Techniques*. CMP Books, Lawrence: KS, 2002.

[Bergman93] Bergman, G. D. *Electronic Architectures for Digital Processing: Software/Hardware Balance in Real-Time Systems*. Prentice Hall, Englewood Cliffs, NJ, 1993.

[Blackman75] Blackman, M., *The Design of Real-Time Applications*. John Wiley & Sons, New York, 1975.

[Blum92] Blum, B. I. *Software Engineering: A Holistic View*. Oxford University Press, New York, 1992.

[Bodilsen94] Bodilsen, S. "Scheduling Theory and Ada 9X." *Embedded Systems Programming*. December 1994, December pp. 32–52.

[Boehm81] Boehm, B. *Software Engineering Economics*. Prentice Hall, Englewood Cliffs, NJ, 1981.

Real-Time Systems Design and Analysis, By Phillip A. Laplante
ISBN 0-471-22855-9 © 2004 Institute of Electrical and Electronics Engineers

[Bollella00a] Bollella, G., and Gosling, J. "The Real-Time Specification for Java." *IEEE Computer*, June 2000, Volume 33, Number 6, pp. 47–54.

[Bollella00b] Bollella, G., Brosgol, B., Furr, S., Hardin, D., Dibble, P., Gosling, J., and Turnbull, M. *The Real-Time Specification for Java*. Addison-Wesley, Boston, MA, 2000.

[Booch99] Booch, G., Jacobson, I., and Rumbaugh, J. *The Unified Modeling Language User's Guide*. Addison-Wesley, Boston, MA, 1999.

[Booch91] Booch, G. *Object-Oriented Design with Applications*, Benjamin Cummings, New York, 1991.

[Boussinot91] Boussinot, F., and DeSimmi, R. "The ESTEREL Language." *Proceedings of the IEEE*, Volume 79, Number 9, September 1991, pp. 1293–1304.

[Bowen95] Bowen, J. P. and Hinchey, M. G. "Ten Commandments of Formal Methods." *IEEE Computer*, Volume 28, Number 4, April 1995, pp. 56–63.

[Brooks95] Brooks, F. *The Mythical Man-Month*, 2nd Edition. Addison-Wesley, New York, 1995.

[Bucci95] Bucci, G., Campanai, M., and Nesi, P. "Tools for Specifying Real-Time Systems." *Real-Time Systems: The International Journal of Time Critical Systems*, Volume 8, Number 2/3, March/April 1995, pp. 117–172.

[Burns90] Burns, A. and Wellings, A. *Real-time Systems and Their Programming Languages*. Addison-Wesley, New York, 1990.

[Buttazzo00] Buttazzo, G. *Hard Real-Time Computing Systems: Predictable Scheduling Algorithms and Applications*. Kluwer Academic Publishers, Norwell, MA, 2000.

[Calvez93] Calvez, J. P., Wyche, A., and Edmundson, C. (Translators). *Embedded Real-Time Systems/a Specification and Design Methodology*. John Wiley & Sons, New York, 1993.

[Campbell88] Campbell, J. *C Programmer's Guide to Serial Communications*. Howard Sams & Co., Indianapolis, IN, 1988.

[Cardelli96] Cardelli, L. "Bad Engineering Properties of Object-Oriented Languages." *ACM Computing Surveys*, Volume 28A, Number 4, December 1996, pp. 150–158.

[Chiodo94] Chiodo, M., Giusto, P., Jurecska, A., Marelli, M., Hsieh, H., Sangiovanni-Vincentelli, A., and Lavagno, L. "Hardware-Software Codesign of Embedded Systems." *IEEE Micro*, Volume 14, Number 4, August 1994, pp. 26–36.

[Clark91] Clark, E. M. Jr., Long, D. E., and McMillen, K. "A Language for Computational Specification and Verification of Finite State Hardware Controllers." *Proceedings of the IEEE*, Volume 79, Number 9, September 1991, pp. 1283–1292.

[Cnet00] CNet, "10 Great Bugs of History." http://coverage.cnet.com/content/features/DLife/Bugs/SS05i.html, accessed 9/13/2000.

[Cottet02] Cottet, F., Delacroix, J., Kaiser, C., and Mammeri, Z. *Scheduling in Real Time Systems*. John Wiley & Sons, Chichester England, 2002.

[Craigen95] Craigen, D., Gerhart, S., and Ralston, T. "Formal Methods Reality Check: Industrial Usage." *IEEE Transactions on Software Engineering*, Volume 21, Number 2, February 1995, pp. 90–98.

[Crenshaw00] Crenshaw, J. *Math Toolkit for Real-Time Programming*. CMP Books, Lawrence, KS, 2000.

[Daigle92] Daigle, J. N. *Queuing Theory for Telecommunications*. Addison-Wesley, New York, 1992.

[Davari93] Davari, S., Leibfried, T. F. Jr., Natarajan, S., Pruett, D., Sha, L., and Zhao, W. "Real-Time Issues in the Design of the Data Management System for the Space Station Freedom." *Proceedings of the First Real-Time Applications Workshop*, July 1993, IEEE CS Press, New York, pp. 161–165.

[Davis73] Davis, M. *Computability and Unsolvability*. Dover Publishing Co., New York, 1973.

[de la Puente00] de la Puente, J. "Real-Time Object-Oriented Design and Formal Methods." *The International Journal of Time-Critical Computing Systems*, Volume 18, Number 1 2000, pp. 79–83.

[DeMarco78] DeMarco, T. *Structured Analysis and System Specification.* Prentice Hall, Englewood Cliffs, NJ, 1978.

[DeMillo79] DeMillo, R. A., Lipton, R. J., and Perlis, A. "Social Processes and Proofs of Theorems and Programs." *Communications of the ACM*, Volume 22, Number 5, May 1979, pp. 271–280.

[Desmonde64] Desmonde, W. H. *Real-Time Data Processing Systems: Introductory Concepts.* Prentice Hall, Englewood Cliffs, NJ, 1964.

[Dibble02] Dibble, P. *Real-Time JAVA Platform Programming.* Sun Microsystems, Inc., Palo Alto, CA, 2002.

[Dijkstra68a] Dijkstra, E. W. "Goto Statement Considered Harmful." *Communications of the ACM*, Volume 11, Number 3, March 1968, pp. 147–148.

[Dijkstra68b] Dijkstra, E. W. "Solution of a Problem in Concurrent Programming Control." *Communications of the ACM*, Volume 11, Number 3, March 1968, page 569.

[Dijkstra65] Dijkstra, E. W. "Cooperating Sequential Processes." *Technical Report EWD-123*, Technological University, Eindhoven, Netherlands, 1965.

[DOD-STD88] DOD-STD-2167A. *Military Standard Defense System Software Development.* U.S. Department of Defense, Washington: DC, 1988.

[Dougherty95a] Dougherty, E. R., and Laplante, P. A. *Introduction to Real-Time Image Processing*, SPIE Press/IEEE Press, Bellingham, WA, January 1995.

[Douglass03] Douglass, B. P. *Real-Time Design Patterns: Robust Scalable Architecture for Real-Time Systems.* Addison-Wesley, Boston, MA, 2003.

[Douglass99] Douglass, B. P., *Doing Hard Time – Developing Real-Time Systems with UML, Objects, Frameworks, and Patterns.* Addison-Wesley Longman, Reading, MA, 1999.

[Douglass99] Douglass, B. P., *Real-Time UML*, Second Edition, *Developing Efficient Objects for Embedded Systems.* Addison-Wesley Longman, Reading, MA, 1999.

[Edwards93] Edwards, K. *Real-Time Structured Methods: Systems Analysis*, John Wiley & Sons, New York, 1993.

[Electronic Industry98a] Electronic Industry Association, *EIA 709.1-1998, Control Network Protocol Specification.* 1998, Arlington, VA.

[Electronic Industry98b] Electronic Industry Association, *EIA 709.3-1998, Free Topology Twisted Pair Channel Specification.* 1998, Arlington, VA.

[Ellis94] Ellis, J. R. *Objectifying Real-Time Systems.* Prentice Hall, Englewood Cliffs, NJ, 1994.

[Fenton96] Fenton, N. *Software Metrics: A Rigorous Approach.* Chapman & Hall, London, 1996.

[Ferrintino77] Ferrintino, A. B., and Mills, H. D. "State Machines and Their Semantics in Software Engineering." *Proceedings of the IEEE COMPSAC Conference*, 1977, pp. 242–251.

[Fetzer88] Fetzer, J. H. "Program Verification: The Very Idea." *Communications of the ACM*, Volume 31, Number 9, September 1988, pp. 1048–1062.

[Flynn66] Flynn, M. J. "Very High-Speed Computing Systems." *IEEE Transactions on Computing.* Volume 54, Number 12, December 1966, pp. 1901–1909.

[Forestier89] Forestier, J. P., Forarino, C., and Franci-Zannettacci, P. "Ada++: A Class and Inheritance Extension for Ada." *Proceedings of Ada-Europe International Conference*, Madrid. Ada Companion Series. Cambridge University Press, Cambridge, June 1989.

[Foster81] Foster, C. *Real Time Programming- Neglected Topics.* Addison-Wesley, Reading, MA, 1981.

[Fowler00] Fowler, M. *Refactoring.* Addison-Wesley, New York, 2000.

[Frakes96] Frakes, W. B., and Fox, C. J. "Quality Improvement Using a Software Reuse Failure Modes Model." *IEEE Transactions on Software Engineering*, Volume 22, Number 4, April 1996, pp. 274–279.

[Freedman77] Freedman, A. L., and Lees, R. A. *Real-Time Computer Systems.* Crane, Russak & Co., New York, 1977.

[Furht91] Furht, B., Grostick, D., Gluch, D., Rabbat, G., Parker, J., and McRoberts, M. *Real-Time Unix Systems Design and Application Guide.* Kluwer Academic Publishers, Boston, MA, 1991.

[Gamma94] Gamma, E., Helm, R., Johnson, R., and Vlissides, J. *Design Patterns: Elements of Reusable Object-Oriented Software.* Addison-Wesley, New York, 1994.

[Gane79] Gane, C., and Saron, T., *Structured Systems Analysis,* Prentice Hall, Englewood Cliffs, NJ, 1979.

[Garrett94] Garrett, P. H. *Advanced Instrumentation and Computer I/O Design: Real-Time System Computer Interface Engineering.* IEEE Press, Piscataway, NJ, 1994.

[Garver94] Garver, R. "How to Implement ISO 9000." *T & D.* September 1994, pp 36–42.

[Ghezzi91] Ghezzi, C., Jazayeri, M., and Mandriolo, D. *Fundamentals of Software Engineering.* Prentice Hall, Englewood Cliffs, NJ, 1991.

[Gilreath03] Gilreath, W., and Laplante, P. *Computer Architecture: A Minimalist Perspective.* Kluwer Academic Publishers, Norwell, MA, 2003.

[Gomaa00] Gomaa, H. *Designing Concurrent, Distributed and Real-Time Applications with UML.* Addison-Wesley, Boston, MA, 2000.

[Gomaa93] Gomaa, H. *Software Designing Methods for Concurrent and Real-Time Systems.* Addison-Wesley, Reading, MA, 1993.

[Goodenough88] Goodenough, J. B., and Sha, L. "The Priority Ceiling Protocol: A Method for Minimizing the Blocking of High-Priority Ada Tasks." *Technical Report CMU/SEI-88-SR-4.* Carnegie-Mellon University: Software Engineering Institute, Pittsburgh, PA, 1988.

[Gopinath93] Gopinath, P., Bihri, T., and Gupta, R. "Compiler Support for Object-Oriented Real-Time Software." *IEEE Software,* September 1993, pp. 42–49.

[Grehan98] Grehan, R., Moote, R., and Cyliax, I. *Real-Time Programming: A Guide to 32-bit Embedded Development,* Chapter 9. Addison-Wesley, New York, 1998.

[Gustafson88] Gustafson, J. "Reevaluating Amdahl's Law." *Communications of the ACM,* Volume 31, Number 5, May 1988, pp. 532–533.

[Habermann69] Habermann, A. N. "Prevention of System Deadlocks." *Communications of the ACM,* Volume 12, Number 7, July 1969, pp. 171–176.

[Halang94] Halang, W. A., and Stoyenko, A. *Real Time Computing. NATO ASI Series.* Springer-Verlag, Berlin, 1994.

[Halang91] Halang, W. A., and Stoyenko, A. *Constructing Predictable Real-Time Systems.* Kleuwer Academic Publishers, Boston, MA, 1991.

[Halstead77] Halstead, M. H. *Elements of Software Science.* North-Holland, Amsterdam, 1977.

[Harbison91] Harbison, S. P., and Steele, G. L. Jr. *C: A Reference Manual.* Prentice Hall, Englewood Cliffs, NJ, 1991.

[Harel90] Harel, D., Lanchover, H., Naamad, A., Pnueli, A., Politi, M., Sherman, R., and Trauring, A. "STATEMATE: A Working Environment for the Development of Complex Reactive Systems." *IEEE Transactions on Software Engineering,* Volume 16, Number 4, April 1990, pp. 403–414.

[Harel88] Harel, D. "On Visual Formalisms." *Communications of the ACM,* Volume 31, Number 5, May 1988, pp. 514–530.

[Hatley87] Hatley, D., and Pribhai, I. *Strategies for Real-time System Specification.* Dorset House, New York, 1987.

[Henize] Henize, J. *Understanding Real-Time UNIX.* Concurrent Computer Corporation, One Technology Way, Westford, MA 01886 1987.

[Hetzel88] Hetzel, B. *The Complete Guide to Software Testing,* Second Edition. QED Information Sciences, Wellesley, MA, 1988.

[Hill87] Hill, F. J., and Peterson, G. R. *Digital Systems: Hardware Organization and Design,* Third Edition. John Wiley & Sons, New York, 1987.

[Hillis98] Hillis, D. *The Pattern on the Stone.* Basic Books, New York, 1998.

[Holt01] Holt, J. *UML for Systems Engineering*. The Institution of Electrical Engineers, London, 2001.

[Howe03] Howe, D. (Editor). *The Free Online Dictionary of Computing*. (http://foldoc.doc.ic.ac.uk/), Last accessed 4/23/03.

[IEEE01] Proceedings of the IEEE, Object-Oriented Real-Time Dependable Systems (WORDS). Rome, Italy January 2001.

[IEEE98] Institute of Electrical and Electronics Engineers, IEEE Std 830-1998, Recommended Practice for Software Requirements Specifications, IEEE, New York, 1998.

[IEEE87] IEEE Std. 1016. *Recommended Practice for Software Design Description*. IEEE, New York, 1987.

[Jain91] Jain, R. *The Art of Computer Systems Performance Analysis*. John Wiley & Sons, New York, 1991.

[Jensen94] Jensen, D. "Adventures in Embedded Development: Taking the Guesswork out of Tool Selection." *IEEE Software*, Volume 11, Number 6, November 1994, pp. 116–118.

[Joel57] Joel, A. E. "Communication Switching Systems as Real-Time Computers." *Proceedings of the Eastern Joint Computer Conference – 1957* IRE/ACM, 1957.

[Joerg90] Joerg, W. B. "A Subclass of Petri Nets as a Design Abstraction for Parallel Architectures." *ACM Computer Architecture News*, Volume 18, Issue 4, December 1990, pp. 67–75.

[Jones98] Jones, C. *Estimating Software Costs*. McGraw-Hill, New York, 1998.

[Jorgensen02] Jorgensen, P. *Software Testing, A Craftsman's Approach*, Second Edition. CRC Press, Boca Raton, FL, 2002.

[Jovanovic93] Jovanovic, V., and Mrdalj, S. "A Structured Specification Technique for Hyper Media Systems." *Communications of the ACM*, Volume 36, Issue 1, November 1993, pp. 18–20.

[Kernighan90] Kernighan, B. W., and Ritchie, D. M. *The C Programming Language*, Second Edition. Prentice Hall, Englewood Cliffs, NJ, 1990.

[Kernighan81] Kernighan, B. W. "Why Pascal Is Not My Favorite Language." *Computing Science Technical Report No. 100*. Bell Laboratories, Murray Hill, NJ, July 18, 1981.

[Kfoury82] Kfoury, A. J., Moll, R. N., and Arbib, M. A. *A Programming Approach to Computability*. Springer-Verlang, New York, 1982.

[Kim97] Kim, K. "Object Structures for Real-Time Systems and Simulators." *IEEE Computer*, Volume 30, Number 8, August 1997, pp. 62–70.

[Kleinrock75] Kleinrock, L. *Queuing Systems*, Volume 1, *Theory*. John Wiley & Sons, New York, 1975.

[Kopetz98] Kopetz, H. "The Time-Triggered (TT) Model Computation." *Real-Time System Symposium*, IEEE Computer Society Press, Los Alamitos, CA, 1998.

[Kopetz97] Kopetz, H. *Real-Time Systems Design Principles for Distributed Applications*. Kluwer Academic Press, Norwell, MA, 1997.

[Krishna93] Krishna, C., and Singh, A. "Reliability of Checkpointed Real-Time Systems Using Time Redundancy." *IEEE Transaction on Reliability*, Volume 42, Number 3, September 1993, pp. 427–435.

[Krishna91] Krishna, C. M., and Lee, Y. H. "Guest Editor's Introduction: Real-Time Systems." *IEEE Computer*, Volume 24, Number 5, May 1991, pp. 10–11.

[Kung82a] Kung, S.-Y., Arun, K. S., Gal-ezer, R. J., Bhaskar Rao, D. V. "Wavefront Array Processor: Language, Architecture, and Applications." *IEEE Transactions on Computers*, Volume C-31, Number 11, November 1982, pp. 1054–1066.

[Kung82b] Kung, H. T. "Why Systolic Architectures?" *IEEE Computer*, Volume 15, Number 1, January 1982, pp. 37–46.

[Labrosse00] Labrosse, J. *Embedded Systems Building Blocks*. CMP Books, Lawrence, KS, 2000.

[Lala91] Lala, J., Harper, R., and Alger, L. "A Design Approach for Ultrareliable Real-Time Systems." *IEEE Computer*, Volume 24, Number 5, May 1991, pp. 12–22.

[Laplante04] Laplante, Phillip A., "Criteria and an Objective Approach to Selecting Real-Time Operating Systems Based on Published Information," to appear, *Journal of Computers and Applications*, 2004.

[Laplante03a] Laplante, P. A., and Neill, C. J. "Software Specification and Design for Imaging Systems." *Journal of Electronic Imaging*, Volume 12, Number 2, April 2003, pp. 252–262.

[Laplante03b] Laplante, P. and Neill, C. "A Class of Kalman Filters for Real-Time Image Processing." *Proceedings of the Real-Time Imaging Conference*, SPIE, Santa Clara, CA, January 2003, pp. 22–29.

[Laplante03c] Laplante, P. A. *Software Engineering for Imaging Systems*. CRC Press, Boca Raton, FL, 2003.

[Laplante02a] Laplante, P. A., and Neill, C. J. "An Overview of Software Specification Techniques for Real-Time Imaging." *Proceedings of the Real-Time Imaging Conference*, SPIE, San Jose, CA, January 2002, pp. 55–64.

[Laplante02b] Laplante, P. A., Neill, C. J., and Russell, D. W. "Object-Oriented Requirements Specification for Imaging Systems." *Proceedings of the Real-Time Imaging Conference*, SPIE, Seattle, WA, July 2002, pp. 189–199.

[Laplante02c] Laplante, P. A. "A Retrospective on Real-Time Imaging: A New Taxonomy and a Roadmap for the Future." *Real-Time Imaging*, Volume 8, Number 5, October 2002, pp. 413–425.

[Laplante02d] Laplante, P. A., Neill, C. J., and Jacobs, C. "Requirements Specification Practices: Some Real Data." *Proceedings of the 27th NASA/IEEE Software Engineering Workshop*, December 2002, Greenbelt, MD, pp. 121–128.

[Laplante01] Laplante, Phillip A. (editor-in-chief), *Comprehensive Dictionary of Computer Science, Engineering and Technology*, CRC Press, Boca Raton, FL, 2001.

[Laplante00] Laplante, P. A. (Editor) *A Practical Approach to Real-Time Systems*. IEEE Press, Piscataway, NJ, 2000.

[Laplante97] Laplante, P. *Real-Time Systems Design and Analysis: An Engineer's Handbook*, Second Edition. IEEE Press, New York, 1997.

[Laplante96a] Laplante, P., Stoyenko, A., and Sinha, D. Image Processing Methods: Real-Time Imaging, Conference Proceedings, SPIE Press, Bellingham, WA, 1996.

[Laplante96b] Laplante, P. A., and Stoyenko, A. (Editors), *Real-Time Image Processing: Theory, Techniques, and Applications*. IEEE Press, New York, 1996.

[Laplante95] Laplante, P. A. Funck-Rose, E., and Garcia-Watson, M. "An Historical Overview of Early Real-Time System Developments in the U.S." *Real-Time Systems Journal*, Volume 8, Number 3/4, January 1995, pp. 199–214.

[Laplante93] Laplante, P. A. "Fault-Tolerant Control of Real-Time Systems in the Presence of Single Event Upsets." *Control Engineering Practice*, Volume 1, Number 5, October 1993, pp. 9–16.

[Laplante91] Laplante, P. A. "The Heisenberg Uncertainty Principle and the Halting Problem." *ACM SIGACT Newsletter*, Volume 22, Number 3, Summer 1991, pp. 63–65.

[Laplante88] Laplante, P. A. "Some Thoughts on Cleanroom Software Development and Its Impact on System Test." *Proceedings of the Third AT&T Software Quality Symposium*, Holmdel, NJ, December 1988, pp. 86–90.

[Larman02] Larman, C. *Applying UML and Patterns: An Introduction to Object-Oriented Analysis and Design and the Unified Process*, Second Edition. Prentice Hall, Englewood Cliffs, NJ, 2002.

[Lawson92] Lawson, H. W. *Parallel Processing in Industrial Real-Time Applications*. Prentice Hall, Englewood Cliffs, NJ, 1992.

[Lehoczky89] Lehoczky, J., Sha, L., and Ding, Y. "The Rate Monotonic Scheduling Algorithm: Exact Characterization and Average Case Behavior." *Proceedings of the 10th Real-Time Systems Symposium*, Santa Monica, CA, December 1989, pp. 166–171.

[Levenson87] Levenson, N. G., and Stolzy, J. L. "Safety Analysis Using Petri Nets." *IEEE Transactions on Software Engineering*, Volume 13, Number 3, March 1987, pp. 386–397.

[Levi90] Levi, S.-T. and Ashok, K. A. *Real-Time System Design*. McGraw-Hill, New York, 1990.

[Lim95] Lim, S.-S., Bae, Y., Jang, G., Rhee, B.-D., Min, S., Park, C., Shin, H., Park, K., Moon, S.-M., and Kim, C. "An Accurate Worst Case Timing Analysis for RISC Processors." *IEEE Transaction on Software Engineering*, Volume 21, Number 7, July 1995, pp. 593–604.

[Liskov88] Liskov, B. "Data Abstraction and Hierarchy," *SIGPLAN Notices*, Volume 23, Number 5, May 1988, pp. 17–34.

[Liu73] Liu, C. L., and Layland, J. W. "Scheduling Algorithms for Multiprogramming in a Hard Real-Time Environment." *Journal of the ACM*, Volume 20, Number 1, January 1973, pp. 46–61.

[Liu00] Liu, J. *Real-Time Systems*. Prentice Hall. Upper Saddle River, NJ, 2000.

[Lo03] Lo, C.-T. D., Srisa-an, W., and Chang, J. M. "Who is Collecting Your Java Garbage?" *IT Professional*, Volume 5, Number 2, March/April 2003, pp. 45–50.

[Locke88] Locke, C. D., and Goodenough, J. B. "A Practical Application of the Ceiling Protocol in a Real-Time System." *Technical Report CMU/SEI-88-SR-03*, Carnegie-Mellon University: Software Engineering Institute, Pittsburgh, PA, 1988.

[Louden93] Louden, K. C. *Programming Languages Principles and Practice*. PWS-KENT Publishing Company, Boston, MA, 1993.

[Low90] Low, G. C., and Jeffery, D. R. "Function Points in the Estimation and Evaluation of the Software Process." *IEEE Transactions on Software Engineering*, Volume 16, Number 1, January 1990, pp. 64–71.

[Lutz03] Lutz, M. and Laplante, P. A. "An Analysis of the Real-time Performance of C#." *IEEE Software*, Volume 20, Number 1, January 2003, pp. 74–80.

[Lutz01] Lutz, R. R., and Mikulski, I. C. "Evolution of Safety-Critical Requirements Post-Launch." *Proceedings of the Fifth IEEE International Symposium on Requirements Engineering*, Toronto, Canada, August 2001, pp. 222–227.

[Lyu96] Lyu, M. R., (Editor). *Software Reliability Engineering*, IEEE Press, Piscataway, NJ, 1996.

[MacWilliams77] MacWilliams, F. J., and Sloane, N. J. A. *The Theory of Error-Correcting Codes*. North-Holland, Amsterdam, 1977.

[Maher96] Maher, A., Kuusela, J., Ziegler, J. *Object-Oriented Technology for Real-Time Systems: A Practical Approach Using Omt and Fusion*. Prentice Hall, Englewood Cliffs, NJ, 1996.

[Markov84] Markov, J. "RISC Chips." *BYTE*. November 1984, pp. 191–206.

[Marshall96] Marshall, A. D. Programming in C UNIX System Calls and Subroutines using C. Available on-line at http://www.cs.cf.ac.uk/Dave/C/CE.html.

[Martin67] Martin, J. *Design of Real-Time Computer Systems*. Prentice Hall, Englewood Cliffs, NJ, 1967.

[Martin65] Martin, J. *Programming Real-Time Computer Systems*. Prentice Hall, Englewood Cliffs, NJ, 1965.

[Martin96] Martin, R. C. "The Dependency Inversion Principle," *C++ Report*, SIGS publications, Austin, TX May 1996.

[Massa03] Massa, A. J. *Embedded Software Development with eCos*. Prentice Hall Professional Technical Reference, Upper Saddle River, NJ, 2003.

[Mathai96] Mathai, J. (Editor), *Real-Time Systems: Specification, Verification and Analysis*. Prentice Hall, Englewood Cliffs: NJ, 1996.

[McCabe76] McCabe, T. J. "A Complexity Measure." *IEEE Transactions on Software Engineering*, Volume 2, Number 4, December 1976, pp. 308–320.

[Mellor86] Mellor, S. J., and Ward, P. T. *Structured Development for Real-Time Systems*. Volumes I, II, & III. Prentice-Hall/Yourdon, Englewood Cliffs, NJ, 1986.

[Meyer00] Meyer, B. *Object-Oriented Software Construction*, 2nd Edition. Prentice Hall, Englewood Cliffs, NJ, 2000.

[Möeller99] Möller, T., and Haines, E. *Real-Time Rendering*. A. K. Peters, Natick, MA, 1999.

[Möeller93] Möeller, K. H., and Paulish, D. J. *Software Metrics: A Practitioner's Guide to Improved Product Development*. Chapman & Hall, London, 1993.

[Moore95] Moore, D. L. "Object-Oriented Facilities in Ada 95." *Dr. Dobb's Journal*, October 1995, pp. 28–35.

[Moore94] Moore, S. W. "Scalable Temporally Predictable Memory Structures." *Proceedings of the 2nd IEEE Workshop on Real-Time Applications*, Beltsville, MD, July 1994, pp. 99–103.

[Motus94] Motus, L., and Rodd, M. G., *Timing Analysis of Real-Time Software*. Pergamon, New York, 1994.

[Mrva97] Mrva, M. "Reuse Factors in Embedded Systems Design." *IEEE Computer*, Volume 30, Number 8, August 1997, pp. 93–95.

[MTOS89] *MTOS-UX/Ada Product Profile*. Industrial Programming Inc., Jericho: NY, 1989.

[Muppala91] Muppala, J. K., Woolet, S., and Trivedi, K. S. "Real-Time Systems Performance in the Presence of Failures." *IEEE Computer*, Volume 24, Number 5, May 1991, pp. 37–47.

[Musa80] Musa, J. D. "The Measurement and Management of Software Reliability." *Proceedings of the IEEE*, Volume 68, Issue 9. September 1980, pp. 1131–1143.

[Musliner95] Musliner, D., Hendler, J., Agrawala, A., Durfee, E., Strosnider, J., and Paul, C. "The Challenges of Real-Time AI." *IEEE Computer*, Volume 28, Number 1, January 1995, pp. 58–66.

[Myers75] Myers, G. J. *Reliable Software Through Composite Design*. Van Nostrand Reinhold, New York, 1975.

[Naks01] Naks, T., and Motus, L. "Handling Timing in a Time-Critical Reasoning System – A Case Study." *Annual Reviews in Control*, Volume 25, Number 0, January 2001, pp 157–168.

[Nawrocki99] Nawrocki, J., Schwarz, J.-J., and Zalewski, J. (Editors). *Real-Time Systems Education*. IEEE Computer Society, Los Alamitos, CA, 1999.

[Neill03] Neill, C. J., and Laplante, P. A. "Specification of Real-Time Imaging Systems Using UML." *Real-Time Imaging*, Volume 9, Number 2, April 2003, pp. 125–137.

[Neill02] Neill, C. J., and Laplante, P. A. "Modeling Time in Object-Oriented Specifications of Real-Time Imaging Systems." *Proceedings of the Real-Time Imaging Conference*, SPIE, Seattle, WA, July 2002, pp. 200–206.

[Nissanke97] Nissanke, N. *Realtime Systems*, Prentice Hall, New York, 1997.

[Orr77] Orr, K. *Structured System Development*. Yourdon Press, Englewood Cliffs, NJ, 1977.

[özsoyoglu95] özsoyoglu, G., and Snodgrass, R. "Temporal and Real-Time Databases: A Survey." *IEEE Transactions on Knowledge and Data Engineering*, Volume 7, Number 4, August 1995, pp. 513–532.

[Parnas86] Parnas, D. L., and Clements, P. C. "A Rational Design Process – How and Why to Fake It." *IEEE Transactions on Software Engineering*, Volume 12, Number 2, February 1986, pp. 251–257.

[Parnas79] Parnas, D. L., "Designing Software for Ease of Extension and Contraction." *IEEE Transactions on Software Engineering*, Volume SE-5, Number 2, March 1979, pp. 128–138.

[Parnas72] Parnas, D. L. "On the Criteria to be Used in Decomposing Systems into Modules," *Communications of the ACM*, Volume 15, Number 12, December 1972, pp. 1053–1058.

[Patterson96] Patterson, D. A., Hennessy, J. L., and Goldberg, D. *Computer Architecture: A Quantitative Approach*, Second Edition. Morgan Kaufman Publishers, San Francisco, CA, 1996.

[Patterson95] Patterson, J. G. *ISO 9000 Worldwide Quality Standard*. Crisp Publications, Menlo Park, CA, 1995.

[Paulin97] Paulin, P., Liem, C., Cornero, M., Naçabal, F., and Goossens, G. "Embedded Software in Real-Time Signal Processing Systems: Application and Architecture Trends." *Proceedings of the IEEE*, Volume 85, Number 3, March 1997, pp. 419–435.

[Paulish92a] Paulish, D. J. and Moller, K. H. *Best Practices of Software Metrics*. IEEE Press, Piscataway, NJ, 1992.

[Paullish92b] Paulish, D. J., and Moller, K. H. *Software Metrics: A Practitioner's Guide to Improved Product Development*. First published by Chapman & Hall, London, 1992. Exclusive North American distribution rights assigned to IEEE Press, Piscataway, NJ.

[Pham95] Pham, H. *Software Reliability and Testing*, IEEE Press, Piscataway, NJ, 1995.

[POSIX02] IEEE, POSIX Standard, http://standards.ieee.org/reading/ieee/stdpublic/description/posix/, last accessed 8/5/02.

[Pressman00] Pressman, R. S. *Software Engineering: A Practitioner's Approach*, Fifth Edition. McGraw-Hill, New York, 2000.

[Putnam97] Putnam, L. H., and Myers, W. *Industrial Strength Software: Effective Management Using Measurement*. IEEE Computer Society Press, Los Alamitos, CA, 1997.

[Redmond80] Redmond, K. C., and Smith, T. S. *Project Whirlwind – The History of a Pioneer Computer*. Digital Press, Bedford, MA, 1980.

[Rich88] Rich, C., and Waters, R. "Automatic Programming: Myths and Prospets." *IEEE Computer*, Volume 21, Number 8, August 1988, pp. 40–51.

[Ripps90] Ripps, D. L. *An Implementation Guide to Real-Time Programming*. Yourdon Press, Englewood Cliffs, NJ, 1990.

[Ross77] Ross, D. "Structured Analysis (SA): A Language for Communicating Ideas." *IEEE Transactions on Software Engineering*, Volume SE-3, Number 1, January 1977, pp. 16–34.

[Rothstein70] Rothstein, M. F. *Guide to the Design of Real-Time Systems*. Wiley Interscience, New York, 1970.

[Rzucidlo02] Rzucidlo, M., "Fault Tolerance and Reliability in Software." *Proceedings of the Research Institute*, Penn State Great Valley, February 2002, pp. 270–281, Malvern, PA.

[Schildt00] Schildt, H. *C/C++ Programmer's Reference*, Second Edition. Osborne/McGraw Hill, Berkely, CA, 2000.

[Schmidt00] Schmidt, D. C., Stal, M., Robert, H., and Bushmann, F. *Pattern-Oriented Software Architecture: Patterns for Concurrent and Networked Objects*. John Wiley & Sons, New York, 2000.

[Schneider98] Schneider, F., Easterbrook, S. M., Callahan, J. R., and Holzmann, G. J. "Validating Requirements for Fault Tolerant Systems using Model Checking." *Third IEEE Conference on Requirements Engineering*, Colorado Springs, CO, April 6–10, 1998.

[Schoch95] Schoch, D. J., and Laplante, P. A. "A Real-Time Systems Context for the Framework for Information Systems Architecture." *IBM Systems Journal*, Volume 34, Number 1, January 1995. pp. 20–38.

[Selby87] Selby, R. W., Basili, V. R., and Baker, F. T. "Cleanroom Software Development: An Empirical Evaluation." *IEEE Transactions on Software Engineering*, Volume SE-13, Number 9, September 1987, pp. 1027–1037.

[Selic98] Selic, B., and Rumbaugh, J. "Using UML for Modeling Complex Real-Time Systems." ObjecTime Limited/Rational Software Corp. white paper, March 1998. Available at www.rational.com.

[Selic94] Selic, B., Gullekson, G., and Ward, P. T. *Real-Time Object-Oriented Modeling*. John Wiley & Sons, New York, 1994.

[Sha90] Sha, L., Rajkumar, R., and Lehoczky, J. P. "Priority Inheritance Protocols: An Approach to Real-Time Synchronization." *IEEE Transactions on Computers*, Volume 39, Number 9, September 1990, pp. 1175–1185.

[Sha89] Sha, L., and J. B. Goodenough. "Real-Time Scheduling Theory and Ada." *Technical Report CMU/SEI-89-TR-14*, Carnegie-Mellon University: Software Engineering Institute, Pittsburgh, PA, 1989.

[Sha88] Sha, L., and Goodenough, J. B. "Real-Time Scheduling Theory and Ada." *Technical Report CMU/SEI-88-TR-33*, Carnegie-Mellon University: Software Engineering Institute, Pittsburgh, PA, 1988.

[Shaw01] Shaw, A. C. *Real-Time Systems and Software*. John Wiley & Sons, New York, 2001.

[Shaw92] Shaw, A. C. "Communicating Real-Time State Machines." *IEEE Transactions on Software Engineering*, Volume 18, Number 9, September 1992, pp. 805–816.

[Shen90] Shen, C., Ramamritham, K., and Stankovic, J. A. "Resource Reclaiming in Real-Time." *Proceedings of the 11th Real-Time System Symposium*, Lake Buena Vista, FL, December 1990, pp. 41–50.

[Shin89] Shin, K., and Dolter, J. "Alternative Majority-Voting Methods for Real-Time Computing Systems." *IEEE Transactions on Reliability*, Volume 38, Number 1, April 1989, pp. 58–64.

[Simon99] Simon, D. *An Embedded Software Primer*. Addison-Wesley, Boston, MA, 1999.

[Sinha96] Sinha, P., Gorinsky, S., Laplante, P. A., and Stoyenko, A. D. "A Survey of Real-Time Imaging." *The Journal of Electronic Imaging*, Volume 5, Number 4, October 1996, pp. 466–478.

[Som90] Som, S., Mielke, R. R., and Stoughton, J. W. "Strategies for Predictability in Real-Time Data-Flow Architectures." *Proceedings of the 11th Real-Time Systems Symposium*, Lake Buena Vista, FL, December 1990, pp. 226–235.

[Somerville00] Somerville, I. *Software Engineering*, Fourth Edition. Addison-Wesley, New York, 2000.

[Sparks03] Sparks, G., and Mathers, P. *Enterprise Architect 3.51 User Guide*. Sparx Systems, Creswick, Victoria, Australia, 2003.

[Sperry95] Sperry, T. "Real-Time Operating Systems: Let the Buyer be Aware." *Embedded Systems Programming Product News*, Summer 1995, pp. 12–21.

[Spivey89] Spivey, J., *The Z Notation: A Reference Manual. Series in Computer Science*. Prentice Hall, Englewood Cliffs, NJ, 1989.

[Stallings01] Stallings, W. *Operating Systems – Internals and Design Principles*, Fourth Edition. Prentice Hall, Upper Saddle River, NJ, 2001.

[Stankovic95] Stankovic, J. A., Spuri, M., DiNatale, M., and Buttazzo, G. "Implications of Classical Scheduling Results for Real-Time Systems." *IEEE Computer*, Volume 28, Number 6, June 1995, pp. 16–25.

[Stankovic93] Stankovic, J. A., and Ramamritham, K. *Advances in Real-Time Systems*, IEEE Computer Society Press, Los Alamitos, CA, 1993.

[Stankovic88] Stankovic, J., and Ramamritham, K. *Hard Real-Time systems – A Tutorial*. Computer Science Press (IEEE), Washington, DC, 1988.

[Steele94] Steele, R., and Backes, P. "Ada and Real-Time Robotics: Lessons Learned." *IEEE Computer*, Volume 27, Number 4, April 1994, pp. 49–54.

[Steininger91] Steininger, A., and H. Schweinzer. "Can the Advantages of RISC be Utlized in Real-Time Systems?" *Proceedings of the Euromicro '91 Workshop on Real-Time Systems*, Paris, September 1991, pp. 30–35.

[Stewart99] Stewart, D. B. "Twenty-Five Most Common Mistakes with Real-Time Software Development." Class #304, *Proceedings of the 1999 Embedded Systems Conference*, San Jose, CA, September 1999.

[Stimler69] Stimler, S. *Real-Time Data-Processing Systems*. McGraw-Hill, New York, 1969.

[Stoyenko96] Stoyenko, A., Marlowe, T., and Laplante, P. A. "A Description Language for Engineering of Complex Real-Time Systems." *Real-Time Systems Journal*, Volume 11, Number 3, November 1996, pp. 223–244.

[Stoyenko92] Stoyenko, A. D. "Evolution and State-of-the-Art of Real-Time Languages." *Journal of Systems and Software*, Volume 18, Number 4, April 1992, pp. 61–84.

[Stoyenko86] Stoyenko, A. D., and Kligeman, E. "Real-Time Euclid: A Language for Reliable Real-Time Systems." *IEEE Transactions on Software Engineering*, Volume SE-12, Number 9, September 1986, pp. 940–949.

[Stroustrup01] Stroustrup, B., and Ellis, M. A. *The Annotated C++ Reference Manual*. Addison-Wesley, Boston, MA, 2001.

[Tsai97] Tsai, J., Bi, Y., Yang, S., Smith, J. H., and Ross, A. W. *Distributed Real-Time Systems : Monitoring, Visualization, Debugging, and Analysis*, John Wiley & Sons, New York, 1997.

[Tsourkarellas95] Tsourkarellas, M., Gerogiannis, V., and Economides, K. "Systematically Testing a Real-Time Operating System." *IEEE Micro*, Volume 15, Number 5, October 1995, pp. 50–60.

[Vahid02] Vahid, F., and Givargis, T. *Embedded System Design: A Unified Hardware/Software Introduction*. John Wiley & Sons, Hoboken, NJ, 2002.

[Ward85] Ward, P. T., and Mellor, S. J., *Structured Development for Real-Time Systems*, Volume 1: *Introduction & Tools*; Volume 2: *Essential Modeling Techniques*; Volume 3: *Implementation Modeling Techniques*, Yourdon Press, New York, 1985.

[Washabaugh90] Washabaugh, D., and Kafura, D. "Incremental garbage collection of concurrent objects for real-time applications." *Proceedings of the 11th Real-Time Systems Symposium*, Lake Buena Vista, FL, December 1990, pp. 21–30.

[Westwater96] Westwater, R., and Furht, B. *Real-Time Video Compression: Techniques and Algorithms*, Kluwer Academic Publishers, Dordrecht, The Netherlands, 1996.

[Wilson97] William Wilson, "Writing Effective Requirements Specifications," *Software Technology Conference*, Utah, April 1997.

[Wirth83] Wirth, N. *Programming in Modula-2*, Second Edition. Springer-Verlag, New York, 1983.

[Wolf94] Wolf, W. "Hardware-Software Codesign of Embedded Systems." *Proceedings of the IEEE*, Volume 82, Number 7, July 1994, pp. 967–989.

[Wulf73] Wulf, W., and Shaw, M. "Global Variables Considered Harmful." *SIGPLAN Notices*, Volume 8, Number 2, February 1973, pp. 28–34.

[WWW Consortium01a] World Wide Web Consortium, *XML Schema Part 0: Primer*. http://www.w3.org/TR/xmlschema-0/. 2001.

[Wyliue03] Wyliue, M., Stoecklin, S., and Allen, C. L. "Extending UML for a Real-Time System Specification," www.cis.famu.edu/, last accessed 5/1/03.

[Xilinx98] Xlinx, Inc. *Core Solutions Databook*, San Jose, CA, 1998.

[Yourdon91] Yourdon, E. *Modern Structured Analysis*. Prentice Hall, Englewood Cliffs, NJ, 1991.

INDEX

Real-Time Systems Design and Analysis, By Phillip A. Laplante
ISBN 0-471-22855-9 © 2004 Institute of Electrical and Electronics Engineers

ABOUT THE AUTHOR

Dr. Phillip A. Laplante is Associate Professor of Software Engineering and a member of the Graduate Faculty at The Pennsylvania State University. He is also the Chief Technology Officer of the Eastern Technology Council, a nonprofit business advocacy group serving the Greater Philadelphia Metropolitan Area. Before joining Penn State he was a professor and senior academic administrator at several other colleges and universities.

Prior to his academic career, Dr. Laplante spent nearly eight years as a software engineer and project manager working on avionics (including the Space Shuttle), CAD, and software test systems. He has authored or edited 19 books and more than 130 papers, articles, and editorials. He co-founded the journal *Real-Time Imaging*, which he edited for five years, and he created and edits the CRC Press book series on Image Processing.

Dr. Laplante received his B.S., M.Eng., and Ph.D. in Computer Science, Electrical Engineering, and Computer Science, respectively, from Stevens Institute of Technology and an MBA from the University of Colorado. He is a Senior Member of the IEEE and a member of numerous professional societies, program committees, and boards. He is a Licensed Professional Engineer in the state of Pennsylvania, and has provided consulting services to Fortune 500 companies, the U.S. DOD, and NASA on real-time systems, image processing, and software engineering.